95.00 (DR)
v 2 only

AF774528

BEIHEFTE ZUM TÜBINGER ATLAS
DES VORDEREN ORIENTS

herausgegeben im Auftrag des Sonderforschungsbereichs 19
von Wolfgang Frey

Reihe A
(Naturwissenschaften)
19/2

Richard H. Meadow
and Hans-Peter Uerpmann (Eds.)

EQUIDS IN THE ANCIENT WORLD

Volume II

1991

DR. LUDWIG REICHERT VERLAG · WIESBADEN

EQUIDS
IN THE ANCIENT WORLD
Volume II

Richard H. Meadow
and Hans-Peter Uerpmann (Eds.)

1991

DR. LUDWIG REICHERT VERLAG · WIESBADEN

Gedruckt auf Veranlassung des SFB 19, Tübingen,
mit Mitteln der Deutschen Forschungsgemeinschaft

Gedruckt auf säurefreiem Papier
(alterungsbeständig - pH 7, neutral)

CIP-Titelaufnahme der Deutschen Bibliothek

Equids in the ancient world / [im Sonderforschungsbereich 19, Tübingen, entstanden]. Richard H. Meadow and Hans-Peter Uerpmann (eds.). - Wiesbaden : Reichert.

(Beihefte zum Tübinger Atlas des Vorderen Orients : Reihe A, Naturwissenschaften ; 19)

NE: Meadow, Richard H. [Hrsg.]; Sonderforschungsbereich Orientalistik mit Besonderer Berücksichtigung des Vorderen Orients ⟨Tübingen⟩; Tübinger Atlas des Vorderen Orients / Beihefte / A

Vol. 2 (1991)

ISBN 3-88226-457-8

© 1991 Dr. Ludwig Reichert Verlag, Wiesbaden

Das Werk einschließlich aller seiner Teile ist urheberrechtlich geschützt. Jede Verwertung außerhalb der engen Grenzen des Urheberrechtsgesetzes ist ohne Zustimmung des Verlages unzulässig und strafbar. Das gilt insbesondere für Vervielfältigungen, Übersetzungen, Mikroverfilmungen und die Einspeicherung und Verarbeitung in elektronischen Systemen.

Gesamtherstellung: Hubert & Co., Göttingen. Printed in Germany

Preface

The present volume comprises the second part of the expanded proceedings of what has come to be known as "the equid conference" which was held in Tübingen from the 28th of April to the 2nd of May, 1982. The full, rather cumbersome, title of that gathering expresses clearly the focus of the conveners who are also the editors of this compendium: "Symposium on the Upper Pleistocene and Holocene Distribution and Discrimination of Equids in the Palearctic Region with Special Emphasis on the Middle East."

All the papers included here were stimulated by that symposium, although only three – those of Buitenhuis, Gilbert, and Payne – directly reflect presentations made in Tübingen. As with the first volume, we have intentionally provided the reader with differing points of view and varying conclusions as a comparison of, for example, the papers of Uerpmann and Bökönyi on the ass or of Gilbert, Bökönyi, and Anthony on horse domestication will make clear. We have insisted, however, that the authors include clear documentation of the data on which their discussions are based and thus provide the reader with the means to reach an independent conclusion.

The first paper in this volume (by Burleigh, Clutton-Brock, and Gowlett) is a direct continuation of the one by Burleigh in the first volume that was titled "Chronology of Some Early Domestic Equids in Egypt and Western Asia." In the short time between the composing of the two contributions, accelerator (AMS) dating became available, and it was possible to obtain results on previously undatable bone material. The potential of this technique for ascertaining the age of specimens that hitherto have depended for temporal placement upon apparent stratigraphic association is great.

Beginning with the Uerpmann paper, we have arranged the contributions approximately in geographical order from south to north and more or less from asses through hemiones and hydruntines to horses. Thus we pass from asses in Arabia through the northern fringes of the Arabian desert into highland Iran and Anatolia where the presence of all four species of equids must be considered. From there we leave the Middle East and proceed to hydruntines in France, asses in Italy, and horses in the Netherlands, Britain, and the North Pontic region of the Soviet Union. We end with another in the continuing series of invaluable contributions by Véra Eisenmann on the comparative osteology of equids, both ancient and modern. This work on the first phalanges follows directly on that of the metapodials that was included as the third paper of the first equid volume.

Conventional wisdom has the horse being first domesticated in the North Pontic region of the Soviet Union and the ass in Egypt, partly because those areas are known to have been the home of wild relatives of the domestic forms. Yet, as more and more faunal remains become available for study from archaeological sites, it becomes increasingly clear that premodern ranges were considerably different from those of the second two-thirds of the twentieth century. Furthermore, as Groves pointed out in his contribution to the first equid volume, there is no reason why there should not have been overlap in the ranges of such species as the wild ass and the hemione. Thus one important contribution that can be made by the study of zooarchaeological (archaeozoological) materials is a better knowledge of Holocene zoogeography. In order to achieve this, however, it is necessary to identify osteological characters, both metric and non-metric, that permit the analyst to distinguish between closely similar forms. This attempt is clearly reflected in many of the papers included in this and the preceding volume. Many vexing problems remain, however, not the least of which is successfully characterizing *Equus hydruntinus.* In this regard, note should be taken of the papers by Bonifay and Payne.

A traditional contribution of zooarchaeological research has been to the documentation of equid domestication. A great deal has been written on this matter, but we believe that the time is ripe to lay aside all preconceived notions and to re-evaluate what evidence is available in an objective fashion. A beginning has been made in these volumes, and we can only hope that it will continue. In this regard the paper by Anthony is deserving of particular attention.

In concluding, we note that the division of labor in preparing this volume for the publisher was the same as for the previous one. Thus, RHM had responsibility for editing all texts and preparing the bibliographies, while HPU saw to the tabulation of the many pages of measurements and shepherded the book through the press. Again, we wish to express our thanks to the Tübinger Atlas project and to its director Wolfgang Röllig for supporting the publication of these volumes and to Mary Littauer for her support and encouragement. Finally, it was with sadness and a great sense of loss that we learned of the death of Priscilla F. Turnbull in December of 1985. Her contributions to both paleontology and zooarchaeology have been invaluable, and we particularly miss her integrity in dealing with all matters personal and professional. In small token of our appreciation, we dedicate this volume to her.

Richard H. Meadow
Hans-Peter Uerpmann

Cambridge, Mass.
Tübingen
September, 1988

RICHARD BURLEIGH, JULIET CLUTTON-BROCK,
AND JOHN GOWLETT

Early Domestic Equids in Egypt and Western Asia: An Additional Note

Contents

1. Introduction

In a previous contribution to this series (BURLEIGH 1986), attempts at dating the remains of three equids were discussed. In two instances, the bones could not be dated using conventional radiocarbon techniques due to poor preservation of collagen. In the third case, the specimen provided what was evidently an erroneously old date. The equids involved were, respectively, the remains of an ass from Badari, Egypt, that could possibly have been Predynastic, an ass from a tomb (2052) of the First Dynasty at Tarkhan, Egypt, and the remains of a horse dating to the Hyksos period from Tell el Ajjul, Gaza, Palestine. The first two proved to contain insufficient collagen for conventional radiocarbon dating and the third had been impregnated with paraffin wax as a preservative when excavated in the mid-1930s. In sample preparation, this wax evidently had not been fully removed, resulting in an apparent age for these remains of around 8000 bp compared with the date of *circa* 1650–1550 BC expected.
As it was of considerable interest and importance to ascertain or confirm the dating of these three specimens, they were resubmitted for dating by the accelerator (AMS) technique as part of a program for investigating the chronology of early domesticates in North Africa and Western Asia by this means.

2. Measurements and results

Details of the remains are given more fully in the previous paper (BURLEIGH 1986, Table 1 a, 1 b). The bones are held in the collection of the Department of Zoology in the Osteology Room at the British Museum (Natural History). From among these, an axis vertebra of the ass from Badari, part of the incisor region of the mandible of the ass from Tarkhan, and an epiphysis of the left metacarpal of the horse from Gaza (the same bone as that used for BM-

2114, the erroneously old date) were selected for accelerator dating. Amino acids were extracted from samples drilled from these bones, and from these in turn the graphite samples (targets) for the accelerator measurements were prepared (Gillespie/Hedges/Wand 1984; Wand/Gillespie/Hedges 1984). The amount of carbon needed is only about one one-thousandth of that required for so-called conventional radiocarbon measurements, hence the remarkable success of the accelerator method when applied to this kind of hitherto undatable material. The results obtained were as follows (dates in radiocarbon years based on the 5570 year half-life):

Badari ass	700 ± 120 bp	(OxA-564)
Tarkhan ass	4390 ± 130 bp	(OxA-566)
Gaza horse	3400 ± 120 bp	(OxA-565)

3. Discussion

These results are of great interest for several reasons and exemplify the positive and now essentially non-destructive role of radiocarbon determinations in substantiating the dating of the remains of genuine early domesticates, as well as in eliminating other, potentially misleading, specimens from consideration by showing them to be spurious. In the present instance one of the results obtained, the date of 700 ± 120 bp for the ass remains from Badari, also raises another important doubt for these remains were associated with other faunal material (for example, the remains of bovids) considered by the excavators to be definitely Predynastic (Brunton/Caton—Thompson 1928). This attribution is perhaps now less certain, and it would be worth dating some of these other remains as well to help resolve this question.

The date of 4390 ± 130 for the ass from Tarkhan is consistent with its attribution to the First Dynasty based on archaeological evidence, when due allowance is made for natural ^{14}C variations, and the same is true of the date of 3400 ± 120 for the Gaza horse which falls within the Hyksos period when calibrated. This is the first positive evidence from directly dated and unequivocally identified skeletal remains, as opposed to trappings and other less direct evidence, for the use of horses by the Hyksos. Both these results emphasize the enhanced value of dates obtained directly from positively identified animal remains, as opposed to archaeologically associated material such as charcoal. Recognition of the importance of direct dates forms the foundation for our research into the chronology of both domesticates and wild mammalian species (Clutton-Brock/Burleigh 1983).

4. Acknowledgements

The Oxford Radiocarbon Accelerator is supported by the Science and Engineering Research Council (SERC).

5. Literature

BRUNTON, G./CATON-THOMPSON, G., 1928, The Badarian Civilization and Predynastic Remains near Badari; London: British School of Archaeology in Egypt/Bernard Quaritch.

BURLEIGH, R., 1986, Chronology of some early domestic equids in Egypt and Western Asia. In: MEADOW, R. H./UERPMANN, H.-P., eds., Equids in the Ancient World; (Beihefte zum Tübinger Atlas des Vorderen Orients, Reihe A **19/1**); Wiesbaden: Dr. Ludwig Reichert Verlag, 230–236.

CLUTTON-BROCK, J./BURLEIGH, R., 1983, Some archaeological applications of the dating of animal bone by radiocarbon with particular reference to post-Pleistocene extinctions. In: MOOK, W. G./WATERBOLK, H. T., eds., Proceedings of the First International Symposium on C-14 and Archaeology; (PACT Journal **8**); Strasbourg: Council of Europe, 409–419.

GILLESPIE, R./HEDGES, R. E. M./WAND, J. O., 1984, Radiocarbon dating of bone by accelerator mass spectrometry. Journal of Archaeological Science **10**, 165–170.

WAND, J. O./GILLESPIE, R./HEDGES, R. E. M., 1984, Sample preparation for accelerator-based radiocarbon dating. Journal of Archaeological Science **10**, 159–163.

RICHARD BURLEIGH, (Research Laboratory, The British Museum, London WC1B 3DG, England); current address: The Old George, Broadwindsor, Beaminster, Dorset DT8 3QD, England.

JULIET CLUTTON-BROCK, British Museum (Natural History), Cromwell Road, London SW7 5BD, England.

JOHN GOWLETT, (Radiocarbon Accelerator Unit, Research Laboratory for Archaeology and the History of Art, Oxford University, 6 Keble Road, Oxford OX1 3QJ, England); current address: Institute of Prehistoric Sciences & Archaeology, P. O. Box 147, The University, Liverpool L69 3BX, England.

HANS-PETER UERPMANN

Equus africanus in Arabia

Contents

1. Introduction

The former existence of the wild ass, *Equus africanus,* in Arabia has been discussed among archaeozoologists for several years (*q. v.,* MEADOW/UERPMANN 1986). This idea, first put forward by DUCOS (1968), was not readily accepted (e.g., UERPMANN 1981). However, the reluctance to accept this view faded with the discovery of convincing finds of wild ass remains in areas of Arabia likely to have had suitable habitats, similar to those in northeast Africa where the species has survived into recent times (UERPMANN 1987). These finds need to be documented, however, in order to establish a more complete view of the ancient distribution of the ancestor of domestic donkeys.

There are considerable difficulties in distinguishing the bones of *Equus africanus* from those of other small equids, particularly *Equus hemionus* and *Equus hydruntinus.* Both these other species existed in the Levant during the Late Pleistocene (UERPMANN 1987), and it is not known how far their ranges extended into Arabia. One must not, in any case, identify equid remains from that area based on the mere assumption of an exclusive distribution of the different equid species. Thus, among the remains of small equids found at several archaeological sites in Arabia (Figure 1), only some well-preserved bones could be securely identified as *Equus africanus.* In some cases it is obvious from both the dating of the finds and the morphology of the bones that they must have come from wild animals. In other cases an early state of domestication cannot be excluded. The sites in Arabia from which equid materials were studied by this author are discussed in the first part of this paper.

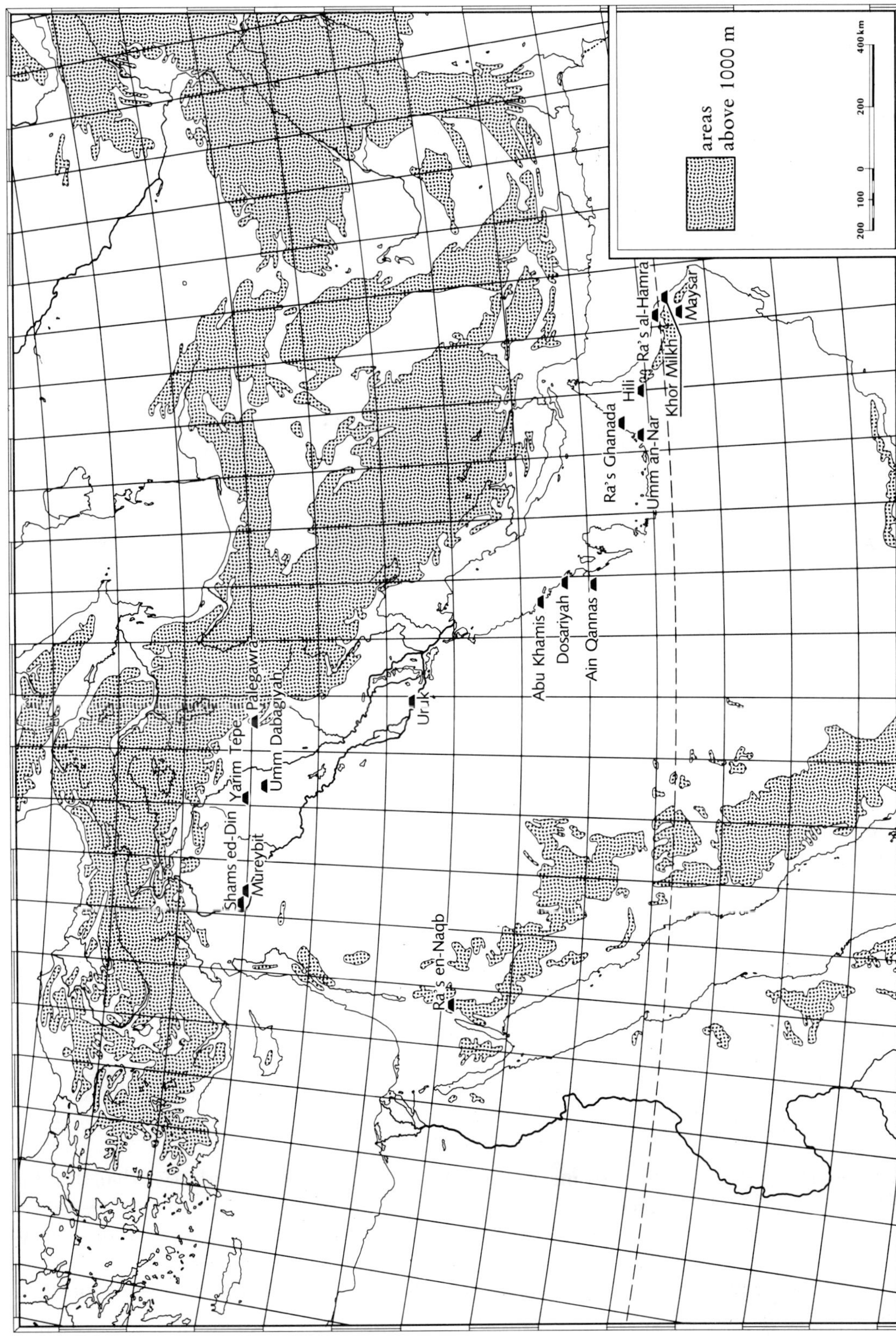

Figure 1: Sites mentioned in the text

2. Sites and Materials

2.1. *Ra's al-Hamra,* Sultanate of Oman

The shell middens of Ra's al-Hamra near Muscat, the capital of Oman, are dated by radiocarbon to the time span between about 5500 and 2000 BC (calibrated). Archaeological investigations at these sites were carried out by an Italian expedition directed by Maurizio Tosi, IsMEO, Rome. Reports were published by DURANTE/TOSI (1977), BIAGI *et al.* (1984 a + b), and COPPA *et al.* (1985). The study of the faunal remains is not yet complete, but equid bones have been isolated among the finds of the middens RH-5, -6, and -10.
Equid bones from Ra's al-Hamra-5 include 2 thoracic vertebrae, 1 rib, 1 tibia, 1 first phalanx, and 1 second phalanx.
From RH-10 there is only a lower premolar, and from RH-6 a fragment of a metapodial of a very young individual which is not identified with certainty.
The preservation of most of the specimens from Ra's al-Hamra is not good enough for specific identification. The second phalanx from RH-5, however, could be definitely identified as *Equus africanus* (see below).

2.2. *Khor Milkh,* Sultanate of Oman

A large shell midden (KM-1) and a smaller concentration of archaeological finds (KM-2) were reported by PHILLIPS/WILKINSON (1979) from the area of Khor Milkh near Quriyat, about 80 km southeast of Muscat on the coast of northern Oman. A radiocarbon date of 4715 ± 160 bp for shells from the surface of KM-1 was published by VITA FINZI (1982). Two small sondages at KM-1 and systematic surface collections at both sites were made by the author in 1986. Among the faunal remains from the sondages at KM-1 are some small fragments of equid bones which could not be identified to species. According to the archaeological finds from KM-1, this site is of an age similar to Ras al-Hamra-5, i.e., dating from about 4500 to 3400 calBC. At this early time, domestic equids cannot yet be expected to have occurred in Eastern Arabia. Thus, the finds from KM-1, although unidentified in terms of species, are another hint as to the existence of a wild equid in the area.
From the surface of KM-2, there are several ass remains. Although not found in a secure context, they will be discussed below. Other finds in the same state of preservation as the ass bones include the mandible of a fairly big wolf, an animal that no longer exists in the area. This is an indication of the prehistoric nature of the faunal remains, which are likely to belong to the same Stone Age context as the artifacts found at the site. The assumed age of KM-2 is thought to be slightly younger than KM-1.*

2.3. *Maysar,* Sultanate of Oman

The Bronze Age mining and smelting site of Maysar near Samad, a little less than 100 km south of Muscat in the interior of Oman, was excavated by an expedition of the "Deutsches Bergbau-Museum", Bochum, directed by G. WEISGERBER (1981). The settlement area of May-

*) Note added in print: The average radiocarbon age is 5250 ± 50 for KM-1 and 4925 ± 50 for KM-2 (uncalibrated years BP).

sar-1 (M-1) yielded only a few, badly preserved animal remains including the fragments of an upper cheektooth of an equid. From the settlement of Maysar-6 (M-6), about one kilometer south of Maysar-1, there are fragments of an equid humerus, metacarpus, pelvis, and femur, as well as three articulating tarsal bones (central and tarsals III and IV) joined with the proximal ends of metatarsals III and IV. The third metatarsal and the *caput femoris* can be identified as *E.africanus.*

Most of the equid remais from Maysar, however, come from inside a well that was excavated inside a Bronze Age fortification called Maysar-25 (M-25). This well contained a large number of animal bones (mostly ruminants including the fairly complete skeleton of an ox). Equid remains comprise articulating elements of a hind leg (tibia to phalanx 1) and a number of isolated fragments (2 ribs, 1 scapula, 1 metacarpal, 2 femora, 1 astragalus, 1 calcaneum, 1 tarsal, 1 third phalanx) as well as a set of ass teeth.

Whereas the architectural context of the well is obviously from the late 3rd or early 2nd millennium BC, a radiocarbon date obtained from the ribs of the ox skeleton falls in the middle of the first millennium AD. Since the bone collagen was completely decayed, contamination with carbonic matter from the ground water seems possible. Archaeologically, a Bronze Age date in the later part of the third millennium BC for the ass bones from the well is more likely than a date in the first millennium AD. The equid bones from the other areas of Maysar also belong to the Umm an-Nar phase of the early Bronze Age.

Judging from their date, there is no doubt that the ass bones from Maysar could have come from domestic donkeys. As shown by the measurements, the animals were not particularly large. Compared to recent donkeys from Oman, however, they were quite big. Therefore, the extremity bones from Maysar will be discussed in the following chapter as possibly belonging to wild animals. The mainly unerupted set of equid teeth from Maysar-25 reveals a strange pathologic condition. The upper cheekteeth in particular are twisted and jolted longitudinally. Apparently they did not have enough space for their development, and obviously they could not have been functional. The animal died before its pathologic permanent teeth broke through. According to the characters of the less distorted lower cheekteeth, this animal also was an ass. Such pathologic condition of the teeth is more likely to be found in a domestic donkey than in a wild ass. For this reason the equid teeth from Maysar-25 will not be discussed any further in this paper.

2.4. *Hili 8,* Abu Dhabi, United Arab Emirates

The site of Hili 8 is part of the prehistoric settlement area in the north of the Buraimi oasis close to the border between the United Arab Emirates and the Sultanate of Oman. It is situated in the narrow belt between the barren mountains to the east and the desert sand dunes to the west. Site no. 8 of prehistoric Hili is one of the typical round fortification structures found at most Bronze Age sites in Eastern Arabia. It was excavated by a French archaeological mission under the direction of Serge Cleuziou who has published two preliminary reports on the site (Cleuziou 1977, 1979). Almost 3000 animal bones were excavated mainly from the early phases of the complex building. With two radiocarbon dates of 4400 and 4440 ± 100 BP these early layers must have formed c. 3000 calBC. The building, however, was used during the whole of the third millennium. Study of the animal remains from the site by the author of this paper is nearing completion. Cattle, sheep, and goat bones dominate the faunal material.

Gazella gazella is the most important wild mammal. A few camel remains and the equid bones listed below are likely to have come from hunted animals as well, but an early domestic state for both animals cannot be completely excluded.

Equid bones from Hili 8

Cheekteeth	6
Scapula	2
Humerus	1
Ulna	1
Metacarpus III	1
Phalanx 3 ant.	1
Pelvis	2
Femur	4
Patella	1
Tibia	1
Astragalus	3
Calcaneus	1
Phalanx 1 anterior or posterior	2
Atlas	1
Sacrum	1
Rib	4
total	32

Several of these bones are well preserved and thus useful for secure identification of the equid species involved. The morphology of the finds from Hili 8 provides substantial evidence for the former presence of wild asses in eastern Arabia, although their dating is too late to exclude domestication on chronological grounds alone.

2.5. *Dosariyah, Ain Qannas, and Abu Khamis,* Kingdom of Saudi Arabia

Faunal remains from these sites in the Eastern Province of Saudi Arabia were studied by M. ZEDER (in MASRY 1974). The equid remains were not identified to species at that time. The present author had the opportunity to study and record the equid teeth in 1983. Some of them were obviously ass teeth, others could not be securely identified to species.

The major occupation of these sites can be paralleled to the 'Ubaid Period of Mesopotamia. Ain Qannas, from where most of the equid teeth come, has an additional earlier phase belonging to the 6th or even 7th millennium BC. Thus, there is little doubt that the equid remains from these sites did not belong to domesticates.

2.6. *Ra's en-Naqb,* Hashemite Kingdom of Jordan

The faunal remains from excavations made by D. O. HENRY at a Chalcolithic and a Natufian site at Ra's en-Naqb in the extreme south of Jordan were studied by Priscilla TURNBULL. Her manuscript on these finds was received by the author in 1981 and is in press.

The equid remains from the Natufian site were studied again by the author in 1987. Contrary to the opinion of TURNBULL, who has ascribed all the equid remains to *Equus hemionus,* there is evidence that most of them belong to *E.africanus,* although the onager seems to be present as well. The specimens well-enough preserved for identification will be discussed in the following section. A complete listing of the equid bones from this site can be found in the forthcoming paper by TURNBULL.

2.7. *Shams ed-Din,* Arab Republic of Syria

The animal remains from the Halafian site of Shams ed-Din in the Tabqa dam area were published earlier, and a separate paper dealt with the equid bones from this site (UERPMANN 1982, 1986). Some few remains of asses among with a large majority of hemione bones were then identified only tentatively. In the context of new finds from Arabia these identifications become more secure now and should be discussed again. These remains might indicate the former northern limits of the distribution of wild asses.

2.8. Comparative material

Two skeletons of African wild asses which died in the Hellabrunn Zoo, Munich, were made available by the Zoologische Staatssammlung München (= ZSM). The larger male "Max" (ZSM 1963/133) was caught in northern Ethiopia in 1930, and the male (ZSM 1963/134) is his offspring born in Munich. They belong to the nominal subspecies of the wild ass, *Equus africanus africanus.* While studying the finds from Ra's en-Naqb in the Field Museum of Natural History (= FMNH), Chicago, a skeleton of the Somali subspecies, *Equus africanus somaliensis,* was used for comparison (FMNH 18851).

In addition, the skeleton of a feral ass from Ethiopia (ZSM 1952/9; also classified as *E.a.africanus,* but obviously too small to be a real wild ass) was used for some comparisons together with the author's own collection of donkey skeletons from Oman and Jordan.

For a comparison with hemiones, three skeletons of the Munich collection (ZSM 1962/203, 1974/157, 1978/184) and a partial skeleton from the Naturkundemuseum Stuttgart (NKMS 32027) were used together with the archaeological collection from Shams ed-Din. Additionally, the recent and archaeological hemione materials of the Museum of Comparative Zoology and the Peabody Museum of Harvard University were studied and the information used for this publication.

3. Identification of the Ass Remains

Among the equid remains from eastern Arabia those from the shell middens on the Omani coast are most important because of an early date that most probably excludes a domestic origin for these bones. A fairly well-preserved second phalanx from the site RH-5 is the most diagnostic specimen (Figure 2). Therefore, discussion of specific identification will start with this part of the skeleton and then follow the anatomical order of elements up the extremities.

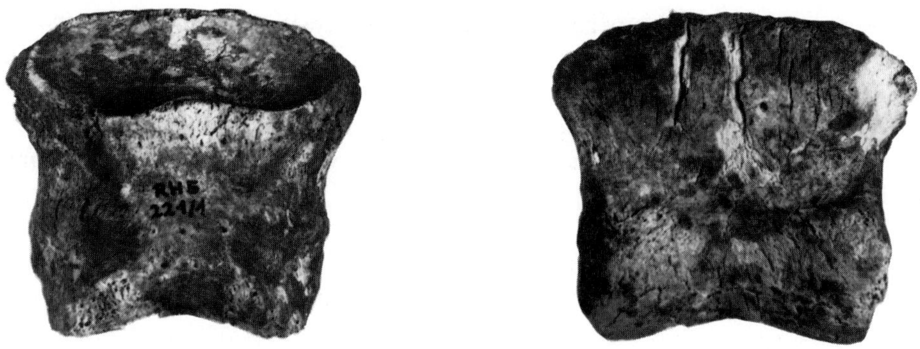

Figure 2: Phalanx 2 anterior from Ra's al-Hamra 5

3.1. *Phalanx 2*

The bone from RH-5 appears to be from the anterior leg, although its shape is not absolutely diagnostic. Another second phalanx from the site KM-2 may be intrusive, but will be dealt with here. From the western margin of Arabia, the finds from Ra's en-Naqb are also of great importance. Dating from the Natufian, they must belong to wild equids. There are two well-preserved second phalanges, one anterior (RN-a) and the other posterior (RN-p). The measurements of all four of these specimens are listed in Table 1 together with comparative measurements from some wild African asses:

TABLE 1

Phalanx 2 – measurements

	RH-5	KM-2	RN-a	RN-p	*E.a.som.*[1]		*E.a.africanus*[2]			
							133		134	
					a	p	a	p	a	p
Bp	45.5	(41.0)	45.3	43.5	44.0	40.5	42.8	41.0	41.2	37.9
Dp	–	–	(27.5)	28.0	27.5	26.0	26.9	26.6	25.3	25.0
SD	37.0	(36.0)	38.6	36.0	38.0	33.0	36.5	33.7	35.7	32.7
DS	–	(16.5)	20.0	(20.5)	19.7	18.5	19.6	19.4	19.0	17.6
Bd	(40.0)	–	(40.0)	35.5	(41.5)	35.5	38.5	34.6	35.8	31.6
Dd	–	–	24.0	23.5	25.0	24.0	24.0	22.8	23.1	20.4
GL	40.5	(37.5)	43.0	41.0	42.5	41.0	38.3	38.8	35.7	35.4

[1] *Equus africanus somaliensis* FMNH 18851
[2] *Equus africanus africanus* ZSM 1963/133 + 134

The specimen from KM-2 is highly corroded and its original measurements can only be estimated. Although the estimates listed above may be a little bit low, the specimen is still in the size range of recent wild asses from northeast Africa. General agreement in size, however, is not sufficient for a specific identification because the hemiones of Southwest Asia are quite close in size to the wild African asses. Their lower legs are more slenderly built, however, and the length/breadth proportions of the leg bones are potentially useful for discriminating between the two forms.

There are no wild hemiones today surviving in the Middle East except for the small populations of onager, khur, and kulan on its eastern margins. The extinct Syrian onager, *E.hemionus hemippus,* is not well represented in zoological collections. However, there are extensive archaeological collections from the region where its ancestors are supposed to have had their range. These collections are themselves questionable with regard to identification of the represented equid species. Nevertheless, comparison of the Arabian specimens with the collections of second phalanges of the equids from Palegawra in Iraq (TURNBULL 1986) and Mureybit in Syria (DUCOS 1978) sheds some light on the nature of all the finds discussed here.

In Figure 3 the proximal breadth (Bp) of second phalanges is plotted against their greatest length (GL). Both anterior and posterior specimens are included, the latter being in general more narrow but having a broad area of overlap with the anterior second phalanges. From the figure it is clear that the Arabian specimens fall outside the cloud formed by the bulk of the Mureybit and Palegawra finds. Of those scatters, the one for Mureybit is completely encompassed by that for Palegawra. There are no differences in length/breadth proportions or general size between the second phalanges of the two populations. Most of the plotted specimens

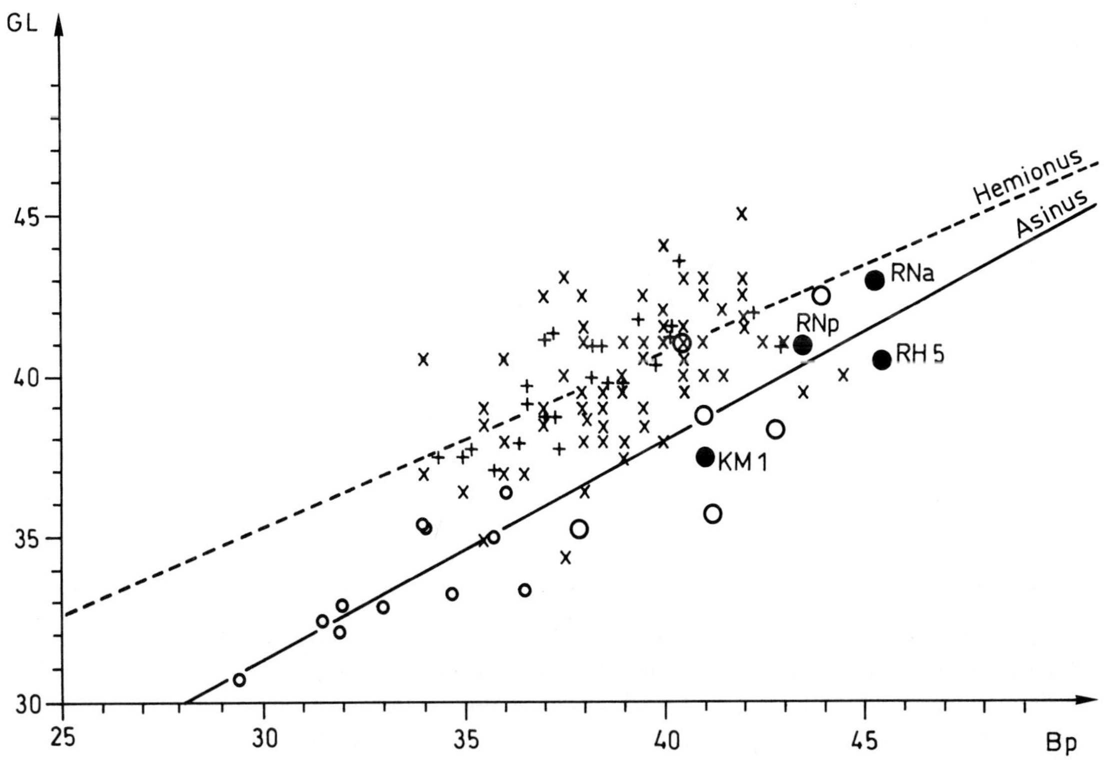

Figure 3: Phalanx 2 anterior and posterior – ● finds from Arabian sites – O recent *E.africanus* – o recent donkeys – × finds from Palegawra – + finds from Mureybit

appear to be from *Equus hemionus.* The two anterior specimens from Ra's al-Hamra and Ra's en-Naqb are broader than any of the finds from farther north. The point for the specimen from Khor Milkh also lies well above the main scatter. There is no doubt that the Arabian finds come from an animal other than that from which most of the Syrian and Iraqi collections derived. The position of the points for the second phalanges of the comparative African wild asses in Figure 3 agrees with the well-known relative stoutness of the foot-bones of asses as compared to hemiones and corroborates the identification of the Arabian specimens as *Equus africanus.*

3.2. *Phalanx 1*

From Ra's en-Naqb there is an almost complete anterior first phalanx. Unfortunately the specimens from Ra's al-Hamra and Hili 8 are fragmented. The only well-preserved find from eastern Arabia is the posterior first phalanx from Maysar-25. There is, however, the specimen from Shams ed-Din in Syria which was found to differ from the more common hemiones of that area (UERPMANN 1986).

TABLE 2

Phalanx 1 – measurements

	RN	M-25	SeD	Hili	*E.a.som.*[1]	*E.a.africanus*[2] 133		134	
	a	p	p	p ?	a	a	p	a	p
Bp	44.0	(41.5)	–	–	43.5	43.3	44.8	41.5	42.2
Dp	30.5	30.2	(30.0)	–	30.5	31.8	32.5	32.2	31.6
SD	27.5	25.2	25.5	(26.0)	27.0	26.1	25.9	23.6	23.4
DS	23.0	18.3	21.0	–	20.5	22.2	22.5	21.0	20.3
BFd	–	–	32.5	37.0	36.0	38.9	35.1	36.6	33.5
Dd	–	–	19.0	20.0	22.0	20.7	19.7	19.3	18.6
GL	78.0	71.0	66.5	–	80.0	77.8	75.6	72.7	69.1

[1] *Equus africanus somaliensis* FMNH 18851
[2] *Equus africanus africanus* ZSM 1963/133 + 134

Plotted onto a diagram (Figure 4) drafted similarly to that for the finds from Shams ed-Din (UERPMANN 1986), the Arabian specimens fall well above the main scatter of the Mureybit bones and together with the tentatively identified ass phalanx from Shams ed-Din. Compared to the majority of the finds from Mureybit (and Shams ed-Din), it is obvious that the first phalanges of asses are shorter and broader than those of the equids dominating the faunal remains of those Neolithic sites in the Euphrates valley.

As in the case of the second phalanges, there are individual specimens in the compared populations that seem to belong to the ass group. For Mureybit the presence of *E.africanus* has

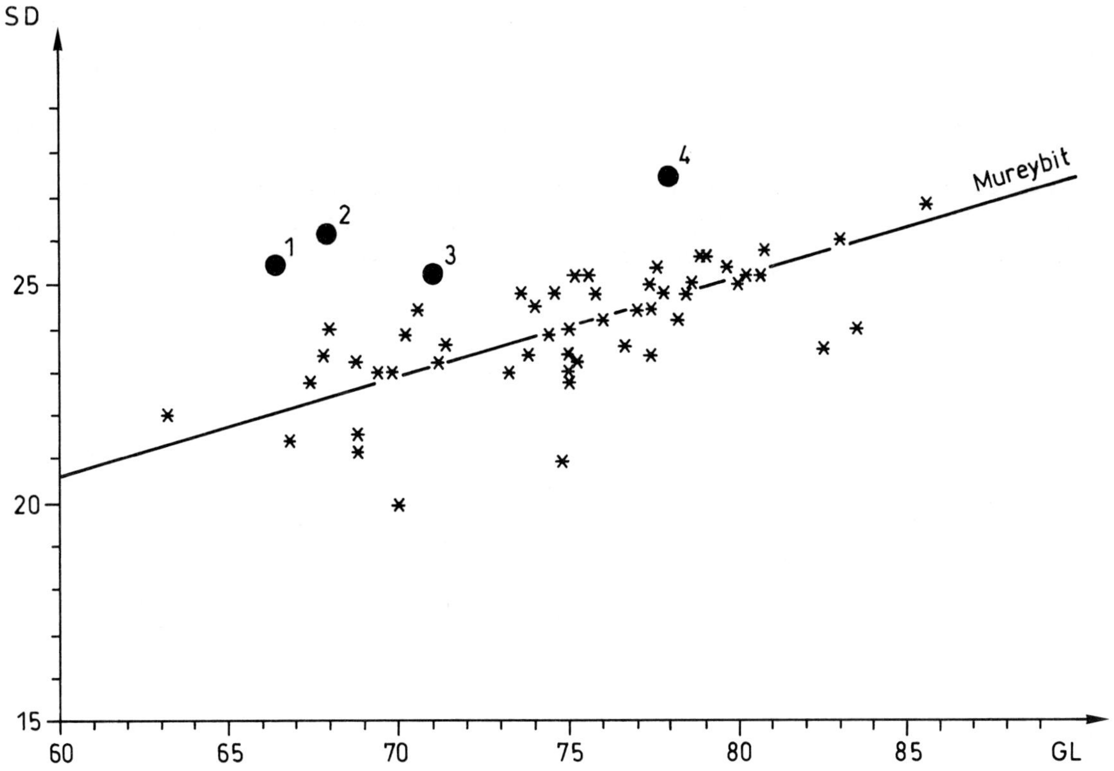

Figure 4: Phalanx 1 anterior and posterior – ● *E.africanus* from 1 – Shams ed-Din, 2 – Mureybit, 3 – Maysar 25, 4 – Ra's en-Naqb * *E.hemionus* from Mureybit

already been postulated by Ducos (1975, 1986). The proximity of that site to Shams ed-Din calls for the same identification for the short and broad first phalanges from both sites. Thus, the presence of a true wild ass in the Syrian Euphrates valley during the Early Holocene must be considered as highly probable.

As for the second phalanges from Palegawra, however, it is necessary to check the original finds (in particular PM 11173 and PM 11700 in the Field Museum, Chicago) before postulating an extension of the ancient range of *E.africanus* to the western foothills of the Zagros mountains. Given the bad preservation of most of the Palegawra bones, longitudinal cracking may have increased their breadth and corrosion may have shortened them slightly, thus shifting their proportions in the ass direction. However, as will be discussed later, the possibility of asses occuring in that area cannot be ruled out *a priori.*

3.3. *Metapodials*

The only well-preserved metapodial to be dealt with here is the metatarsal from Maysar-25 which belonged to the same animal as the first phalanx mentioned above. Although damaged at its proximal end, its length and most breadth and depth measurements could be taken.

There is no need to repeat the discussions about the discrimination of equid metapodials already present in the literature (e.g. Eisenmann/Bekouche 1986; Ducos 1975, 1986). Without any doubt the almost complete metatarsal from Maysar-25 belonged to an ass. Its size is within the limits of the variation for African wild asses. The proximal fragment from Maysar-6 is ass-like in the shape of its articular surfaces. The distal metacarpus fragments from Maysar-25 and Ra's en-Naqb were identified as ass bones because of their straightness in the distal portion of the shaft. In hemiones the lower end of the metacarpal is bent slightly backwards (as it is also in metatarsals of both species, where, however, the bending is more pronounced in hemiones than in asses).

TABLE 3

Metapodials – measurements

	M-25	M-6	M-25	RN	*Equus afr. afr.*[1] 133		134	
	p	p	a	a	a	p	a	p
Bp	–	42.5	–	–	45.2	43.3	44.0	40.5
Dp	–	37.5	–	–	31.0	39.0	29.5	36.0
SD	26.2	(29.0)	–	–	28.5	27.5	24.8	24.0
DS	25.3	–	–	–	22.0	25.5	20.5	23.5
Bd	38.1	–	39.0	42.0	40.0	40.5	38.0	38.0
BFd	38.4	–	38.0	(40.0±.5)	41.0	39.5	37.0	36.5
Dd	29.5	–	(29 .0±1)	32.5	29.4	30.6	28.4	29.5
dTm[2]	23.3	–	(24.0)	23.7	23.5	23.5	23.1	23.1
DTm[3]	26.1	–	25.0	27.2	25.5	26.1	24.4	25.2
GL	(225.0)	–	–	–	198.0	233.0	180.0	212.0

[1] *Equus africanus africanus* ZSM 1963/133 + 134
[2] dTm = smallest diameter of the medial trochlea (Eisenmann 13)
[3] DTm = largest diameter of the medial trochlea (Eisenmann 14)

3.4. *Astragalus*

Among the bones of the ankle joints, the astragalus has the best characters for discrimination between hemiones and asses. As shown in Figure 5, the medial tuberosity for the collateral ligament is generally more protruding in asses than in hemiones. The distal articular surface is deeper and less broad in hemiones than in asses. This difference, although better seen than measured, can also be illustrated in a scatter plot. In Figure 6, the breadth and depth dimensions for the distal articular surface of the astragali of several asses and donkeys are plotted together with the same measurements for some hemiones. Included into this diagram are the respective measurements of Arabian finds from Khor Milkh 2, Maysar-25, and Hili 8 in the east and Shams ed-Din** in the north. All the Arabian points fall close to the regression line for the asses. Of the two Shams ed-Din finds, one is from an onager and the other obviously came from an ass.

**) In the original publication no species identification of this specimen was made (Uerpmann 1986:259). With more material at hand it must now be identified as *E.africanus.*

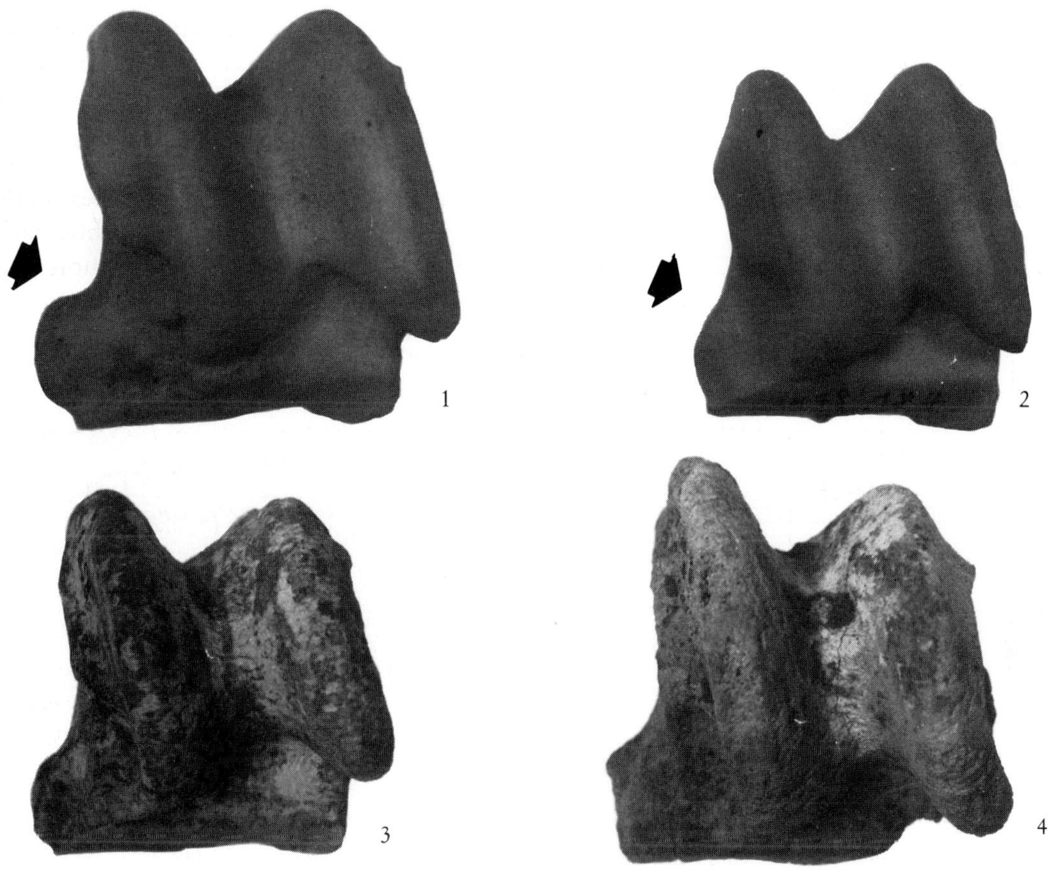

Figure 5: Astragalus 1 – recent *E.africanus* 2 – recent *E.hemionus* 3–4 – finds from Hili 8

From Ra's en-Naqb there are no complete astragali. To judge from the shape of the medial tubercle there is one fragment each of an ass (PM41335) and a hemione (PM41334). The greatest height of the hemione talus is 49.0. The ass bone, although not measurable, is larger than the hemione specimen.

TABLE 4

Astragalus – measurements

	KM 2[1]	M–25	Hili	Hili	SeD	*E.afric.*[2]	
unit	i		890	1358	bu	133	134
GH	(47.0)	52.0	54.0	48.5	48.5	54.7	49.5
GB	(47.0)	49.1	(56.0)	48.8	46.0	55.6	51.3
LmT	(44.0)	49.9	53.4	48.8	48.5	54.1	47.8
BFd	(39.3)	40.9	–	40.3	39.1	44.3	41.3
DFd	(26.0)	39.1	–	27.6	(25.8±0.2)	31.0	28.1

[1] surface highly corroded; measurements rather larger than given here.

[2] *Equus africanus africanus* ZSM 1963/133 + 134

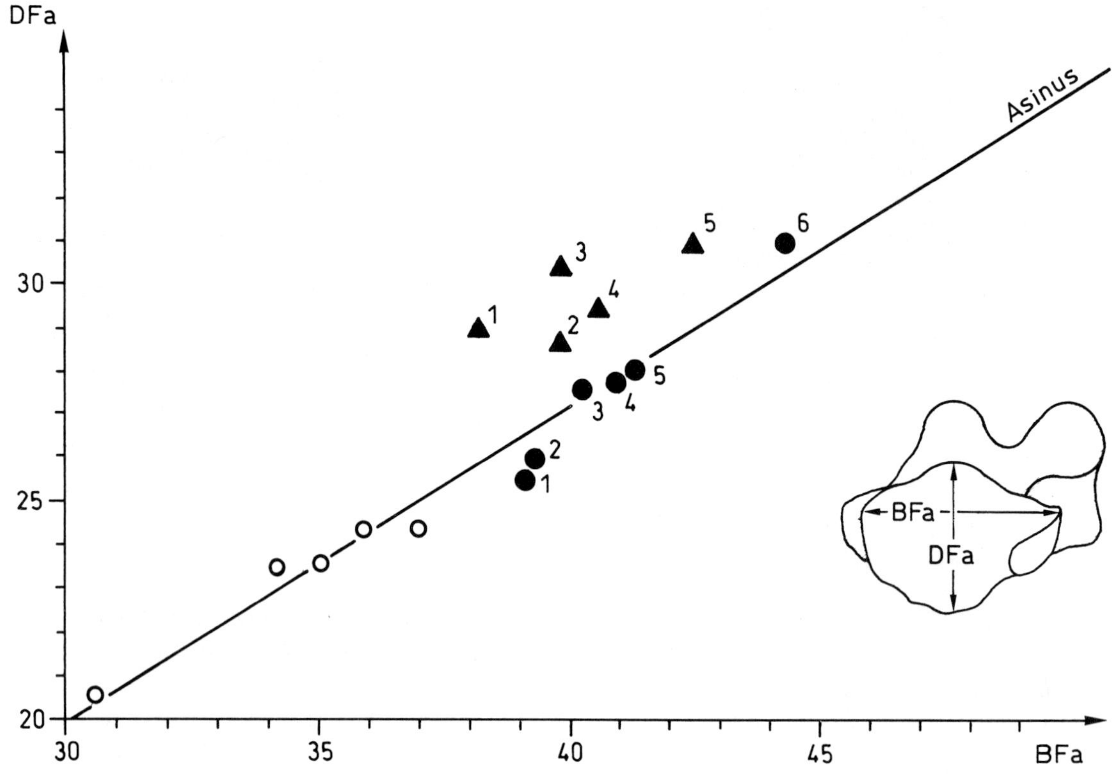

Figure 6: Astragalus – ● *E.africanus* from 1 – Shams ed-Din, 2 – Khor Milkh 2, 3 – Hili 8, 4 – Maysar 25, 5–6 recent. **O** recent donkeys. ▲ *E.hemionus* from 1 – Shams ed-Din, 2–5 recent.

3.5. *Tibia*

Both the smashed but generally complete tibia from the well of Maysar-25 and a loose distal epiphysis from Hili-8 can be identified as *E.africanus* using the characters described in an earlier publication (Uerpmann 1986:259). There is a distinct notch in the medial contour of the articular groove for the astragalus, and the rim of this articular surface flattens out on the medial malleolus as is typical for asses. There is also a distal fragment of a tibia from Ra's al-Hamra-5, but it cannot be identified with certainty. It was from a slightly larger animal than the specimen from Maysar.

TABLE 5

Tibia – measurements

	M-25	Hili	*E.afr.afr.*[1] 133	*E.afr.afr.*[1] 134
Bp	(77.5)	–	82.5	74.0
SD	34.0	–	37.5	33.5
Bd	61.5	(57.5)	65.0	61.0
Dd	42.0	39.5	45.5	39.0
GL	296.0	–	309.0	289.0

[1] *Equus africanus africanus* ZSM 1963/133 + 134

3.6. *Radius*

A radius fragment from Ra's en-Naqb has been identified as coming from *E.africanus* on the basis of the breadth/depth proportion of the medial facet of the distal articular surface. In general this surface is broader and less deep in asses than in hemiones, but there is an overlap between the two species if the total breadth of the articular surface is used for the calculation of an index. The breadth/depth proportion of the prominent medial facet gives a better separation, although particularly the breadth of this facet is difficult to measure in a consistent fashion. However, with the callipers pressed into the volar notch in the middle of the articulation and then closed against the medial rim of the articular surface, replicable results can be obtained. Figure 7 indicates the position of the measuring points.
An index which expresses the breadth as a percentage of the depth (Bx100/D) gives a mean of 73.2% (st.dev. = 3.38; min. = 69.6; max. = 79.3) for seven hemione radii from Iraq collected from sinkhole-deposits by H. FIELD (now in the Peabody Museum of Harvard University). For the three onager and kulan radii of the Munich collection this index is 69.8%, 74.7%, and 75.0%. An archaeological specimen from Shams ed-Din gives a value of 75.7%. In asses on the

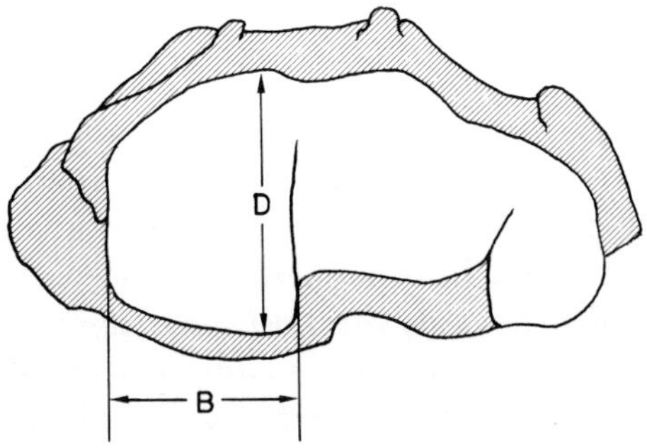

Figure 7: Radius, distal view (adapted from VON DEN DRIESCH 1976, Fig. 33 g) with breadth (B) and depth (D) of the medial condyle as used in calculation of the index Bx100/D

other hand, this index is greater than 75%. The two wild asses of the Munich collection (ZSM 1963/133 + 134) have values of 80.6% and 88.8%. Eight domestic donkeys of the author's collection vary from 76.5% to 84.5% with an average of 79.6% (st. dev. = 2.9).
The specimen from Ra's en-Naqb was studied before the importance of this index was recognized. Thus the breadth measurement of the medial facet was only roughly taken from the separating central ridge to the medial edge, and the depth was measured as the total depth of the distal articular surface. Although this depth is greater and the breadth less than the measurements defined above, the index value as calculated from the existing figures is just over 80% and thus within the range of variation for asses. In addition, the regular spacing of the ridges on the dorsal surface of the distal end of the shaft, which is also diagnostic for the ass (UERPMANN 1986:258), makes its identification fairly certain.
A loose distal epiphysis from Shams ed-Din was also tentatively identified as an ass bone (Uerpmann 1982; 1986:259). The index for the medial articular facet of this specimen is 86.5% which is definitely in the ass range. Thus, this find can be taken as further evidence for the existence of *E.africanus* in northern Syria during the Early Holocene.

TABLE 6

Radius – measurements

	RN	SeD	*E.afr.afr.*[1]	
			133	134
Bd	65.0	63.0	68.5	64.0
BFd	53.5	55.0	56.5	53.0
DFd	30.5	32.0	30.0	30.0
GL	–	–	296.5	282.5

[1] *Equus africanus africanus* ZSM 1963/133 + 134

3.7. *Femur*

The head of a femur from Maysar-6 was identified as an ass bone because of its well-rounded shape. There is a difference between asses and hemiones in this area of the femur which is similar to the difference between goat and sheep. In asses, as in goats, the *caput femoris* is semispherical and not laterally elongated. In hemiones, as in sheep, the articular surface tends to become cylindrical and elongated towards the *trochanter major.* Although quite variable, the shape of the incision for the central ligament gives additional evidence for the identification of the Maysar specimen. It agrees well with comparative ass femora. The diameter of the caput is 45.0 mm. In the Nubian wild asses of the Munich collection (ZSM 1963/133 + 134) this measurement is 45.5 and 43.5 mm respectivly. Thus, the find from Maysar-6, which is from a good Bronze Age context, again falls within the range of wild asses.

3.8. *Scapula*

The articular process of a right ass scapula was excavated from the well of Maysar-25 in addition to the hind leg bones already discussed. There is a slight acromion developed on the distal end of the spine which, itself, overhangs cranially. Both are ass characters (UERPMANN 1986:257). The shape of the *tuber scapulae* is also ass-like, as is the form of the articular surface.

TABLE 7

Scapula – measurements

	M–25	*E.afr.afr.*[1]	
		133	134
SLC	48.5	53.0	49.0
GLP	73.5	73.5	71.0
LG	46.0	46.5	45.5
BG	37.0	41.5	40.0

[1] *Equus africanus africanus* ZSM 1963/133 + 134

In relation to the Munich wild ass skeletons, the scapula from Maysar is of a size similar to the hind leg bones from the same location. Most likely it belongs to the same individual.

3.9. *Teeth*

Teeth are not very useful for the discrimination of ass and hemione remains from Southwest Asia (DAVIS 1980; TURNBULL 1986:326). There are only slight differences between the two subgenera and individual variation causes a great deal of overlap. In general, hemione teeth are intermediate between horses and asses in their enamel patterns (apparently tending more towards the ass in Southwest Asia and more towards the horse in areas farther east).

Typical characters of lower cheekteeth of asses are well-rounded and fairly symmetrical loops of the metaconid and metastylid with a sharp V-shaped sulcus between them. The external valley opposite to this sulcus usually does not penetrate the isthmus (except in milk teeth). In the upper cheekteeth it is mainly the shape of the protocone which characterizes ass dentition. Except for P^2 and M^3 the protocone is short with almost symmetrical anterior and posterior lobes joined by a flat lingual sulcus.

These characters can be observed in some of the equid teeth from the sites discussed so far. The fact that there are no indications of the presence of hemiones at these sites together with the fairly typically ass enamel patterns of most of the teeth found may be reason enough to include the less typical teeth also as ass remains.

More problematic are the equid teeth from the sites in Saudi Arabia. There are no other bones to confirm their identification as *E.africanus.* However, the ass characters of several of these

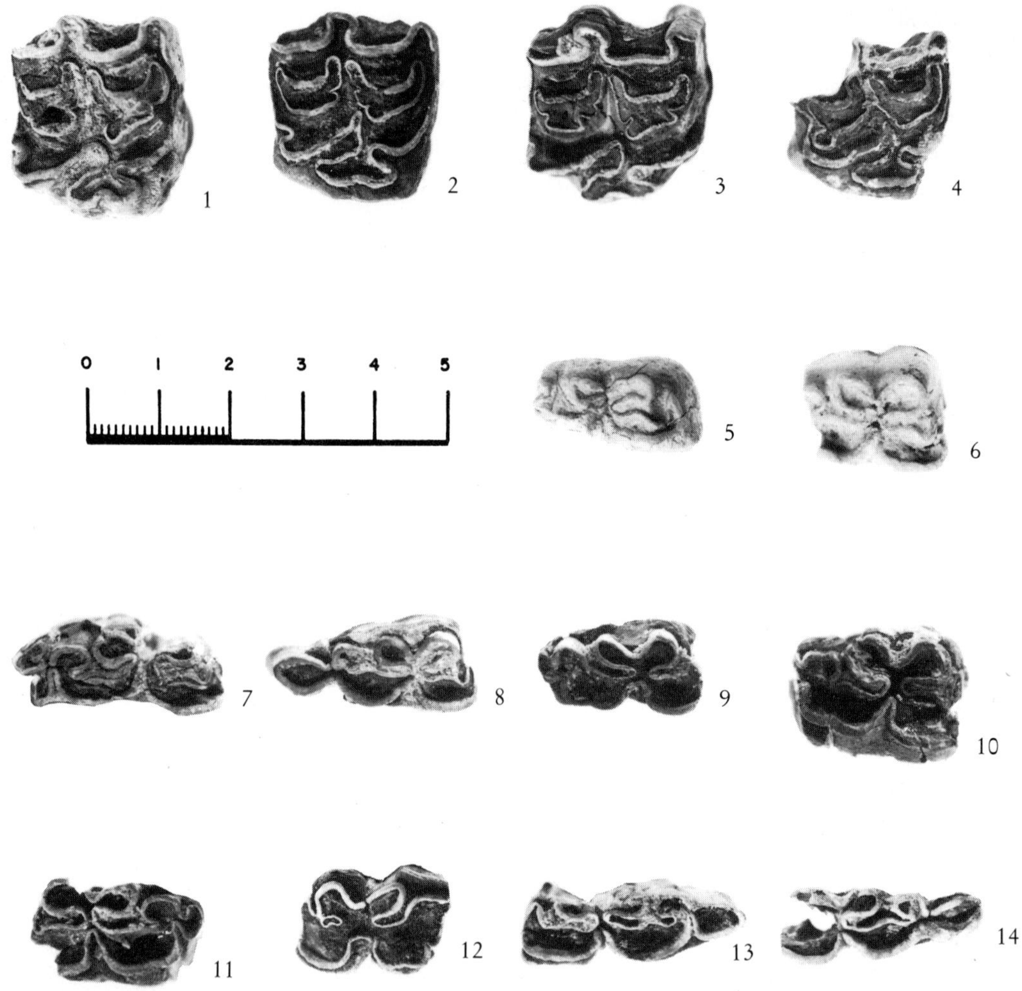

Figure 8: Selection of equid teeth from sites in Saudi Arabia 1–4 Ain Qannas, 5–6 Dosariyah, 7–14 Ain Qannas

finds, particularly Figure 8: 7–11, are quite clear. There is no doubt that they belong to this species.

Whereas asses and hemiones have similar enamel patterns, the third prehistoric species of small equid in the Middle East, *Equus hydruntinus,* is recognizable mainly because of its characteristic cheekteeth (comp. Payne, this vol.). None of the sites discussed in this paper has yielded any finds which could be used as evidence for the occurrence of this species in Arabia during the Holocene.

4. Conclusions

Small as they are, osteological differences between the equid species of the Middle East are still definable. The presence of true wild asses in Southwest Asia during the Early Holocene cannot be questioned any longer. There are a number of sites now from which bone remains of *Equus africanus* are well-documented. They date from a time when domestication did not yet interfere with the natural distribution of this species. The shell middens in Oman, Ra's en-Naqb in southern Jordan, and Shams ed-Din and Mureybit in northern Syria mark the corners of a huge triangle which has to be added to the African area of distribution of true wild asses. Most probably, this triangle will eventually have to be enlarged into a quadrangle once fauna from sites in the Yemen becomes available. The mountainous margins of the Arabian Peninsula provided suitable habitats for *Equus africanus,* which is adapted to the same sort of stony and arid environment on the other side of the Red Sea. The interior of the peninsula – with exception of the sandy areas – will also have had a wild ass population (comp. ZARINS 1986). The same is true for the Gulf area as shown by the finds from Ain Qannas. It remains open for future research to determine whether the "feral" asses of Arabia still contain remnants of their original wild progenitors.

Another important question to be solved in the future is the extent of the zone of overlap between the ranges of hemiones and asses. Apparently onagers ranged as far south as Ra's en-Naqb, but their remains there are fewer than those of the wild ass. At Shams ed-Din, the hemione remains outnumber the ass remains by far, and the same seems to be the case at Mureybit to judge from the distribution of the phalanx measurements. As discussed above, there is even a possibility that ass remains are present among the Final Pleistocene finds from Palegawra. Looking at other early sites north of the Euphrates, BIBIKOVA (1981) lists "*Asinus domesticus*" for the Neolithic site of Yarim Tepe together with "*Equus* (*Asinus*) *onager*". While the presence of domestic donkeys at this early site does not seem to be very likely, the identification of two different equid species by this experienced investigator may well have a base in reality. For the nearby site of Umm Dabagiyah, BÖKÖNYI identified only *Equus hemionus.* However, his graph of the metatarsals of the Umm Dabagiyah equids (BÖKÖNYI 1986:310) contains a specimen which by no means can be ascribed to a hemione. If there is no mistake in this graph, that metatarsal is 240 mm long and has a smallest breadth of 29 mm. These dimensions definitely fall in the range of wild asses. Geographically, wild asses may well be expected to have occurred there. Found on the left bank of the Euphrates in northern Syria they must also have existed in the Djezirah area of northern Mesopotamia and probably throughout all of Mesopotamia. The most likely natural limits for the ancient distribution of the wild ass now seem to be the highlands of the Fertile Crescent. This is a suggested range which even extends that recently postulated by the present author (UERPMANN 1987), where the Euphrates valley was seen as the northern border of the natural distribution of *Equus africanus.*

The southern border of the range of *Equus hemionus* can still be drawn from the head of the Gulf up along the Euphrates to about 35° N, and from there to the Palmyra area and south along the eastern margins of the Levantine highlands to southern Jordan where the species is now definitely identified from Ra's en-Naqb. No clear evidence was found for an extension of the former range of hemiones along the south coast of the Gulf, as proposed by ZARINS (1986: fig.7). The equid teeth from Dosariyah, Abu Khamis and Ain Qannas in the Gulf area of Saudi Arabia (ZEDER in MASRY 1974) were examined by the author in 1983 (see above). Some of them are obvious ass teeth, others might belong to hemiones. It is necessary to study postcranial remains from these sites in order to obtain satisfactory species identifications. An occurrence of hemiones in the Gulf area of Saudi Arabia seems possible if one considers the natural habitats available there. Farther east on the southern Gulf coast, no equid remains at all were found at Umm an-Nar (Hoch 1979) and Ra's Ghanada (Uerpmann in prep.).

The recognition of a broad area of overlap between the two equids has several implications. In a zoogeographical respect, the two species must now be seen as partially sympatric. The view sometimes found in the literature that the so-called Asiatic and African wild asses are merely the two extremes of a cline (e.g. Davis 1980) cannot be supported any longer. As another implication, archaezoologists working in the Middle East (and other parts of the world) will have to abandon the 'principle of exclusive areas of distribution of closely related species'. This principle is unfounded in any event since reproductive boundaries are accepted as the principal criteria for species definitions. As with bovines and caprines, to use only two well known examples, different equids may well occur within the same macro-habitat, and as with different gazelles they may well be very similar to each other in their osteological characters. The fact that equid bone assemblages, even those as large as the ones from Mureybit or Umm Dabagiyah, appear to be uniform metrically does not guarantee their taxonomic uniformity.

With regard to the history of domestication the likelihood that wild asses occurred in Mesopotamia and most of the Levant increases the probability that Southwest Asia—and not Northeast Africa—was the region where the donkey first came under human control. The earliest osteological evidence for a domestic donkey available at present is from Uruk in southern Mesopotamia (Boessneck *et al.* 1984). Numerous skeletal remains of Early Dynastic (or Early Bronze Age) donkeys, which are still close to the wild ancestor in their size, from an area including Mesopotamia (von den Driesch/Amberger 1981; Clutton-Brock 1986; see also Zarins 1986), northern Syria (Boessneck/Kokabi 1981) and the the southern Levant (Ducos 1968, Davis 1976) indicate that domestication of this species happened locally in this region. As indicated by the finds from Maysar, Eastern Arabia may have to be added to this potential area of origin of the domestic donkey.

The observed overlap between wild asses and hemiones has an implication for the process of domestication as well. If both wild species were locally available, one has to assume that both were tested for their manageability. Like some other ungulates, the hemione excluded itself from the process of domestication by its more difficult behavior. However, already in Sumerian times, hemiones seem to have been used for the production of "mules", i.e. hybrids between asses and onagers (Zarins 1986, Postgate 1986). Undoubtedly, the discovery of hybridization between different species as a means for improvement of animal performance is a landmark in the development of man/animal relationships. This dicovery is much easier to understand now. If asses and hemiones were reared together in order to tame them, chance hybridization must have happened every now and then, eventually leading to the discovery of the advantages of this process. On one hand, this result of the common occurrence of the two equid species causes insuperable problems for the osteological treatment of equid remains from early historic sites in the Middle East, because hybridization eliminates any chance for clear-cut species discriminations. On the other hand, general problems surrounding the interpretation of equid remains from that particular area decrease considerably if the idea is accepted that not only the hemione or only the ass but both these species occurred there naturally.

5. Acknowledgements

Thanks are due to the excavators of the materials described for the first time in this paper, in particular to Paolo Biagi, Serge Cleuziou, Maurizio Tosi, and Gerd Weisgerber. William Turnbull and Charles A. Reed provided access to the finds from Ra's en-Naqb and the comparative skeletons at the Field Museum of Natural History in Chicago. Melinda A. Zeder made available her drawings of equid teeth from sites in Saudi Arabia and enabled me to inspect those finds in 1983. For their generous loans of recent equid skeletal materials I want to thank the curators R. Kraft of the Zoologische Staatssammlung, München, and F. Dieterlen of the Naturkundemuseum, Stuttgart. At the Museum of Comparative Zoology of Harvard University, M. Rutzmoser provided facilities to study the comparative collections housed there. The most valuable contributions to this paper were made by Richard H. Meadow who provided not only access to his unpublished collections of equid materials at the Peabody Museum of Harvard University, but also took an active part in the development of the ideas published here.

6. Literature

Biagi, P./Nisbet, R./Maggi, R., 1984a, Excavation at the RH5 Settlement, Qurm, Winter 1984–1985. East and West (n.s.) **34**(4), 455–465.

Biagi, P./Torke, W./Tosi, M./Uerpmann, H.-P., 1984b, Qurum: a case study of coastal archaeology in Northern Oman. World Archaeology **16**, 43–61.

Bibikova, V. I., 1981, Animal husbandary in northern Mesopotamia in the 5th millennium B.C. (Materials from Halafian settlements at Yarim Tepe). In: Munchaev, R. M./Merpert, N. J., comp., Earliest Agricultural Settlements of Northern Mesopotamia; Moscow, 299–307. (In Russian).

Boessneck, J./Driesch, A. v. d./Steger, U., 1984, Tierknochenfunde der Ausgrabungen des Deutschen Archäologischen Instituts Baghdad in Uruk-Warka, Iraq. Baghdader Mitteilungen **15**, 149–190.

Boessneck, J./Kokabi, M., 1981, Tierknochenfunde. Saarbrücker Beiträge zur Altertumskunde **31** (Halawa 1977–1979), 89–104.

Bökönyi, S., 1986, The Equids of Umm-Dabagiyah, Iraq. In: Meadow, R. H./Uerpmann, H.-P., eds., Equids in the Ancient World; (Beihefte zum Tübinger Atlas des Vorderen Orients, Reihe A **19/1**); Wiesbaden: Dr. Ludwig Reichert Verlag, 302–317.

Cleuziou, S., 1977, French Archaeological Mission, 1st Campaign, December 1976-February 1977. Archaeology in the United Arab Emirates **1**.

Cleuziou, S., 1979, The second and third seasons of excavations at Hili 8. Archaeology in the United Arab Emirates **2/3**, 30–69.

Clutton-Brock, J., 1986, Osteology of the Equids from Sumer. In: Meadow, R. H./Uerpmann, H.-P., eds., Equids in the Ancient World; (Beihefte zum Tübinger Atlas des Vorderen Orients, Reihe A **19/1**); Wiesbaden: Dr. Ludwig Reichert Verlag, 207–229.

Coppa, A./Macchiarelli, S./Salvatori, S./Santini, G., 1985, The prehistoric graveyard of Ra's al-Hamra (RH5). Journal of Oman Studies **8**, 97–102.

Davis, S. J., 1976, Mammal bones from the Early Bronze Age city of Arad, northern Negev, Israel: some implications concerning human exploitation. Journal of Archaeological Science **3**, 153–164.

Davis, S. J., 1980, Late Pleistocene and Holocene equid remains from Israel. Zoological Journal of the Linnean-Society **70**, 289–312.

Driesch, A. v. d., 1976, A Guide to the Measurement of Animal Bones from Archaeological Sites; (Peabody Museum Bulletin **1**); Cambridge (MA): Peabody Museum, Harvard University.

Driesch, A. v. d./Amberger, G., 1981, Ein altbabylonisches Eselskelett vom Tell Ababra, Iraq. Bonner Zoologische Beiträge **32**, 67–74.

Ducos, P., 1968, L'Origine des Animaux domestiques en Palestine; (Publications de l'Institut de Préhistoire, Université de Bordeaux **6**); Bordeaux: Institut de Préhistoire.

Ducos, P., 1975, A new find of an equid metatarsal bone from Tell Mureibet in Syria and its relevance to the identification of equids from the early Holocene of the Levant. Journal of Archaeological Science **2**, 71–73.

Ducos, P., 1978, Tell-Mureybet (Syrie, IXe-VIIe Millenaires): Étude archéozoologique et problèmes d'écologie humaine: 1; Paris: Éditions du CNRS.

Ducos, P., 1986, The equid of Tell Muraibit, Syria. In: Meadow, R. H./Uerpmann, H.-P., eds., Equids in the Ancient World; (Beihefte zum Tübinger Atlas des Vorderen Orients, Reihe A **19/1**); Wiesbaden: Dr. Ludwig Reichert Verlag, 237–245.

Durante, S./Tosi, M., 1977, The aceramic shell middens of Ra's al-Hamra: a preliminary note. Journal of Oman Studies **3**, 137–172.

Eisenmann, V./Beckouche, S., 1986, Identification and discrimination of metapodials from Pleistocene and modern *Equus* wild and domestic. In: Meadow, R. H./Uerpmann, H.-P., eds., Equids in the Ancient World; (Beihefte zum Tübinger Atlas des Vorderen Orients, Reihe A **19/1**); Wiesbaden: Dr. Ludwig Reichert Verlag, 117–163.

Hoch, E., 1979, Reflections on prehistoric life at Umm an-Nar (Trucial Oman) based on faunal remains from the third millennium BC. In: Taddei, M., ed., South Asian Archaeology 1977; Naples: Istituto Universitario Orientale, Seminario di Studi Asiatici, 589–638.

Masry, A., 1974, Prehistory in Northeastern Arabia; the Problem of Interregional Interaction; Miami: Field Research Projects.

Meadow, R. H./Uerpmann, H.-P., 1986, eds., Equids in the Ancient World; (Beihefte zum Tübinger Atlas des Vorderen Orients, Reihe A **19/1**); Wiesbaden: Dr. Ludwig Reichert Verlag.

Phillips, C. S./Wilkinson, T. J., 1979, Recently discovered shell middens near Quriyat. Journal of Oman Studies 5, 107–110.

Postgate, J. N., 1986, The equids of Sumer, again. In: Meadow, R. H./Uerpmann, H.-P., eds., Equids in the Ancient World; (Beihefte zum Tübinger Atlas des Vorderen Orients, Reihe A **19/1**); Wiesbaden: Dr. Ludwig Reichert Verlag, 194–206.

Turnbull, P. F., 1981 ms, Faunal remains from Ras en-Naqb, southern Jordan: a preliminary report based on the 1979 collection; (in press).

Turnbull, P. F., 1986, Measurements of *Equus hemionus* from Palegawra Cave (Zarzian, Iraq). In: Meadow, R. H./Uerpmann, H.-P., eds., Equids in the Ancient World; (Beihefte zum Tübinger Atlas des Vorderen Orients, Reihe A **19/1**); Wiesbaden: Dr. Ludwig Reichert Verlag, 319–365.

Uerpmann, H.-P., 1981, The major faunal areas of the Middle East during the late Pleistocene and early Holocene. In: Préhistoire du Levant; Paris: Éditions du CNRS, 99–106.

Uerpmann, H.-P., 1982, Faunal remains from Shams ed-Din Tannira, a Halafian site in northern Syria. Berytus **30**, 3–52.

Uerpmann, H.-P., 1986, Halafian equid remains from Shams ed-Din Tannira in northern Syria. In: Meadow, R. H./Uerpmann, H.-P., eds., Equids in the Ancient World; (Beihefte zum Tübinger Atlas des Vorderen Orients, Reihe A **19/1**); Wiesbaden: Dr. Ludwig Reichert Verlag, 246–265.

Uerpmann, H.-P., 1987, The Ancient Distribution of Ungulate Mammals in the Middle East; (Beihefte zum Tübinger Atlas des Vorderen Orients, Reihe A **27**); Wiesbaden: Dr. Ludwig Reichert Verlag.

Vita-Finzi, C., 1982, Recent coastal deformation near the Strait of Hormuz. Proceedings of the Royal Society London **A 382**, 441–457.

Weisgerber, G., 1981, Mehr als Kupfer in Oman. Der Anschnitt **5–6**, 174–263.

Zarins, J., 1986, Equids associated with human burials in third millennium B.C. Mesopotamia: two complementary facets. In: Meadow, R.H./Uerpmann, H.-P., eds., Equids in the Ancient World; (Beihefte zum Tübinger Atlas des Vorderen Orients, Reihe A **19/1**); Wiebaden: Dr. Ludwig Reichert Verlag, 164–193.

Zeder, M.A., Undated ms., A preliminary report on the faunal material from Dosariyah, Abu Khamis, and Ain Qannas.

Hans-Peter Uerpmann, Institut für Urgeschichte, Universität Tübingen, Schloss, D-7400 Tübingen 1, Federal Republic of Germany

HIJLKE BUITENHUIS

Some Equid Remains from South Turkey, North Syria, and Jordan

Contents

1. Introduction

Western Asia, especially Syria, occupies a critical position in the geographical distribution of equids in the Near East during the Holocene. Here is the western-most spread of *Equus hemionus* and the northern-most distribution of *Equus* (*Asinus*) *africanus.* Also in the Early Holocene period the now extinct *Equus hydruntinus* may have been present in this area. From the point of view of the geographical distribution of equid species in the Near East it is important to establish which species lived in which period in this region. And not only the species have to be established. Hardly anything is known about the *Equus hemionus* subspecies that lived in Syria. It has been assumed that *Equus hemionus hemippus* was the representative of this subspecies, but now it is thought that *E.h.hemippus* as it was found in the 19th century AD in Syria may have been a remnant population and as such not representative of the original population of the area. Another problem arises in the later Holocene period. Man had domesticated different *Equus* species in different areas and spread the domestic animals all over the Near East, thereby breaking all natural geographical boundaries. To understand the human societies involved in the exploitation of *Equus,* it is necessary to establish exactly which species or breeds were used and domesticated by man. In this article equid remains from a number of archaeological sites in Southern Turkey, Northern Syria and Jordan are presented and analyzed. The analysis is mainly based on single measurement comparisons of published data on equid finds elsewhere in the Near East and on data collected during a visit to the Muséum National d'Histoire Naturelle (MNHN) in Paris, France. In addition, a number of equid remains from Northern Syria from recent times were collected and used in this analysis.

2. The sites

Nahr el Homr (CLASON/BUITENHUIS 1978) (Fig. 1) is a small campsite excavated in 1974 (BOERMA/ROODENBERG 1977). It dates from the Natufian (c. 8000 BC), and all faunal remains are from hunted animals. Equids form 85.0 % (193 fragments) of all remains.

Hayaz Höyük (Fig. 1) is a relatively small mound with remains from the Neolithic period (c. 6000 BC), the Chalcolothic period (c. 4400 BC), the Early Bronze Age (c. 3300-2500 BC) and the Early Middle Ages (c. 700–1200 AD) (ROODENBERG 1979–80). These remains have been described in a preliminary report (BUITENHUIS 1985) and in a dissertation (BUITENHUIS 1988). Only one fragment of a mandible of an equid was found among the Neolithic remains (less than 0.1 % of the total collection). Eight fragments (0.9 % or 3.0 % in weight) were found among the remains of the Chalcolithic. One cannot, however, rule out the possibility that these remains were an admixture from the Bronze Age. The equid remains from the Early Bronze Age form 1.0 % (7 fragments) or 4.2 % in weight of the total collection. In the Early Middle

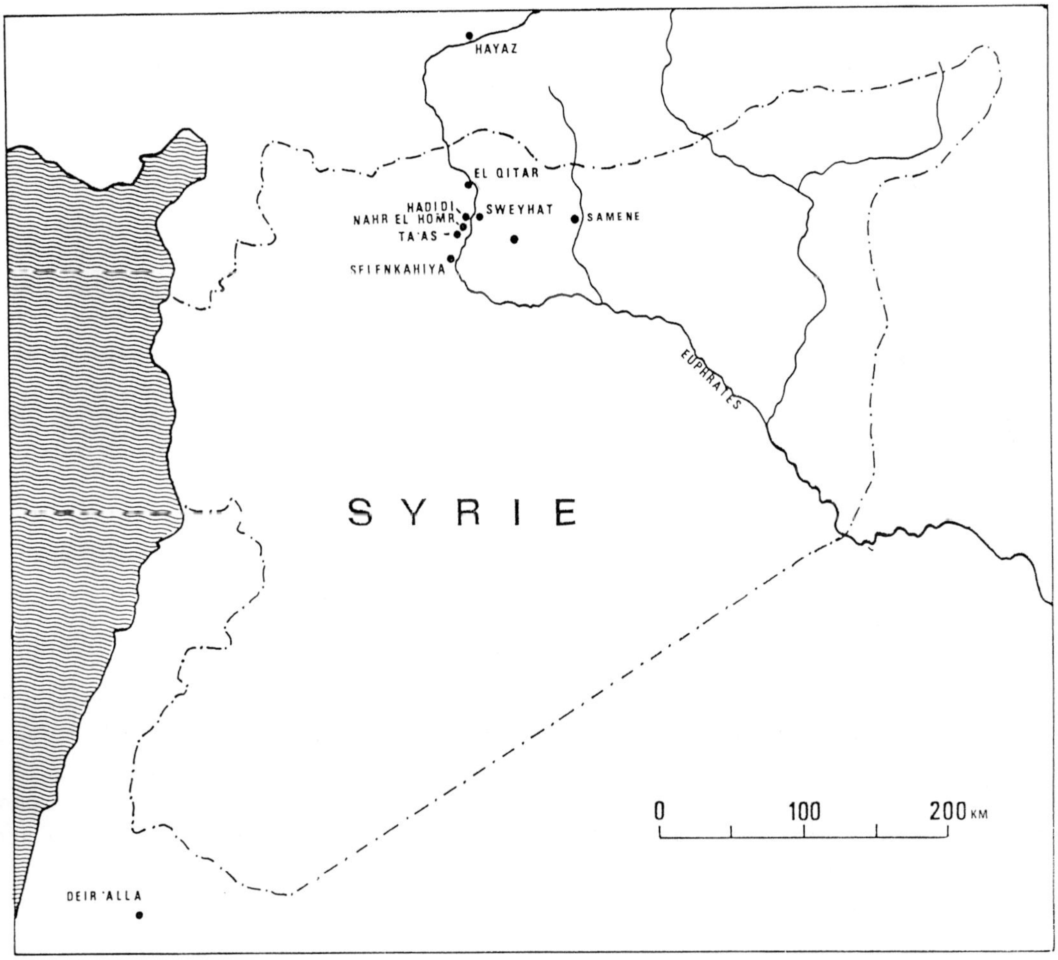

Fig. 1: Geographical location of the sites

Ages, 21 fragments of equids were found (1.4 % or 7.5 % in weight). All remains in all periods are typical settlement debris, broken and heavily fragmented. Hayaz Höyük lies in a relatively isolated location and never grew into an urban center. It is best seen as a small, self-supporting, farming community.

Sweyhat (Fig. 1) is a large mound in North Syria on the east bank of the Euphrates with extensive architectural remains from the Early Bronze Age (c. 2400–1900 BC) and remains from the Hellenistic (3rd century BC) and the Roman (4th century AD) periods. It was excavated by HOLLAND (1976, 1977), and a report on the faunal remains is published (BUITENHUIS 1983). Equid remains from the Early Bronze Age form 4.2 % (93 fragments) of the total collection. In the Hellenistic period, 11 fragments (2.8 %) were found and no equid remains were recognized among the Roman debris.

Hadidi (Fig. 1) is a large mound in North Syria on the west bank of the Euphrates, opposite Sweyhat, with extensive material from the Middle to Late Bronze Age (c. 1700–1300 BC) and from the Early Middle Ages (c. 800–1400 AD). The tell was excavated by V. D. LEEUW (1976) and DORNEMANN (1975). Only part of the material has been analyzed so far (BUITENHUIS 1979). Equid remains form 7.5 % (146 fragments) of the remains from the Bronze Age and 3.1 % (33 fragments) of the remains from the Early Middle Ages.

El Qitar (Fig. 1) lies on the west bank of the Euphrates c. 15 km north of Hadidi. The site was excavated by CULICAN/MCCLELLAN (1983–84, MCCLELLAN 1984–85). It is a heavily fortified urban center from the Late Bronze Age (c. 1500–1300 BC). Equid remains form 2.7 % (397 fragments) of the faunal remains.

Selenkahiya (Fig. 1) is a large mound excavated by VAN LOON (1977) on the west bank of the Euphrates c. 30 km south of Hadidi. The faunal remains have been studied by G. IJZEREEF. All material dates from the the Early Bronze Age (c. 2200–1900 BC).

Selenkahiya, Hadidi, Sweyhat, and El Qitar all lie along the Euphrates. They are urban centers with heavy defense systems in the Early to Late Bronze Ages and are part of a chain of trading settlements between the Anatolian plateau, Mesopotamia and the Mediterranean coast. It is known that this trade was carried out principally with asses used as pack animals (VAN LOON 1982).

Ta'as (Fig. 1) is a small village near Hadidi with material from the Middle Ages (c. 1200–1400 AD) (CLASON/BUITENHUIS 1978).

Deir 'Alla (Fig. 1) is a large urban center in the Jordan valley, excavated by FRANKEN/IBRAHIM (1977–78) and dating mainly from the Late Bronze Age. The faunal material is still being studied.

In addition to the above described material from excavations, the author and Dr. CLASON collected recent skeletal remains from equids in North Syria. A large group of bones was found on the outskirts of the village Ta'as. Other bones were found scattered between Hadidi and Selenkahiya. These have been given the designation 'North Syria' in the measurement tables. Some material was collected near Tell es-Samène in the Balikh valley (Fig. 1) during a survey in 1974.

3. The material

In the following presentation of the material, I refer to the excellent work of EISENMANN (1980, 1981, 1986) on the skull and jaw remains of the different equid species for my interpretations. Also the work of GROMOVA (1955), NOBIS (1971) and DAVIS (1980) are cited in this article when species characteristics are mentioned. The material described here, is stored in the Biologisch-Archeologisch Instituut, Groningen, except for the material from Hayaz Höyük, which is stored in the museum of Adiyaman, Turkey, and the material of Selenkahiya which is stored in the Instituut voor Pre- en Protohistorie, Amsterdam.

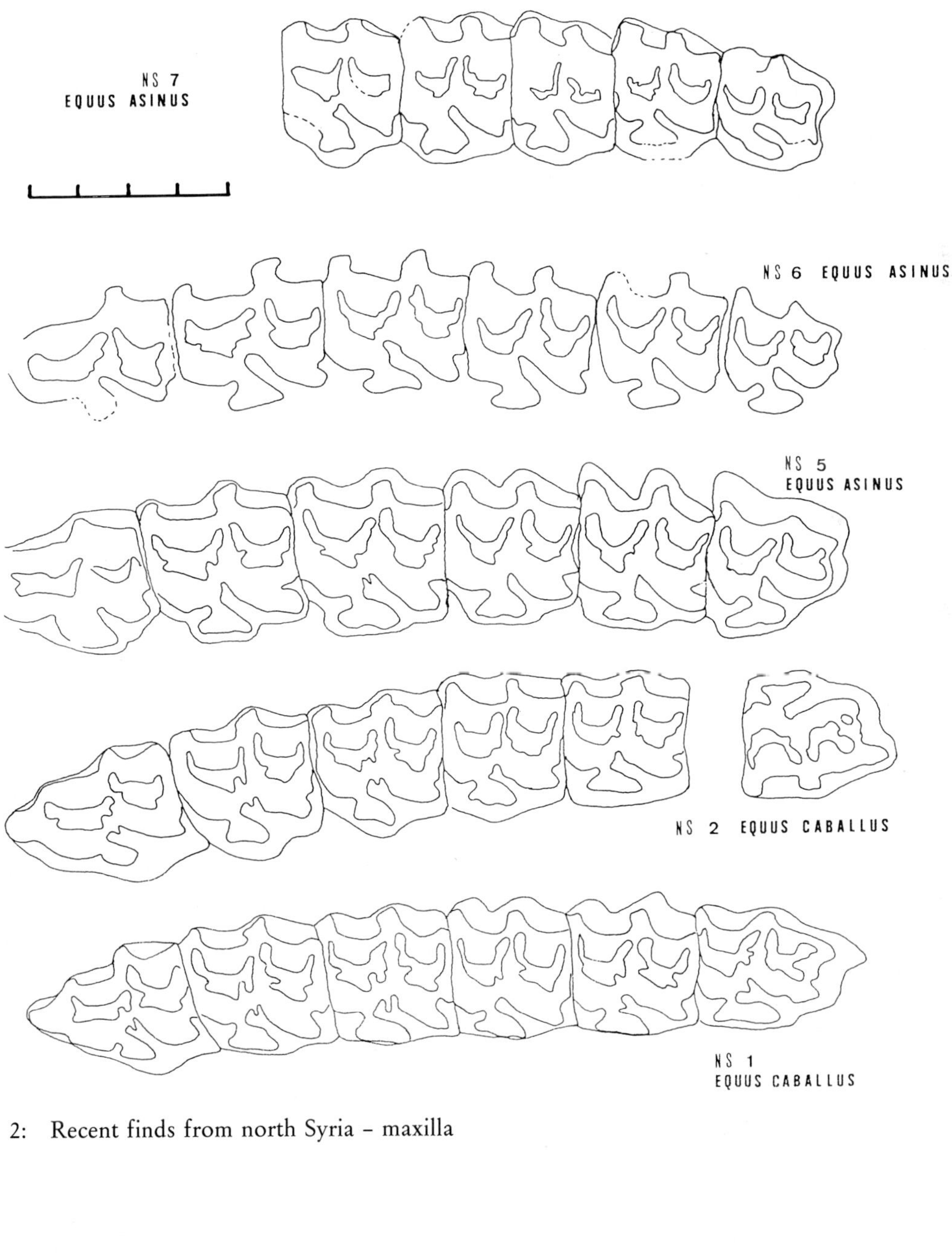

Fig. 2: Recent finds from north Syria – maxilla

3.1. Cranium

In Figure 2 are shown the enamel patterns of the maxillary teeth of recent material from North Syria. In Table 1 are given the measurements of the cranial material from which these patterns derive.

TABLE 1

Cranial dimensions of recent equids from North Syria.

A. Measurements in mm following von den Driesch (1976)

	species	cab NS 1	cab NS 2	cab Ta'asF
1.	total length	535.8	(535)	529
2.	condylobasal length	516.0	(520)	514
3.	basal length	495.7	(469)	(511)
3a.	basilar length	481		(499)
4.	short skull length	357.2	348.5	377.2
5.	basicranial axis	125.2	124.6	132.7
6.	basifacial axis	357.3		(378)
8.	viscerocranium length	323.5		
9.	upper neurocranium length	178.0	178.3	171.3
10.	facial length	377.5		357.7
11.	basion–facial crest	269.0	308.2	288.5
12.	facial crest–prosthion	238.8		244.4
13.	short lateral facial length	319.8		
14.	length braincase	195.8	193.5	193.6
15.	lateral facial length	380.0		351.3
17.	basion–staphylion	224.7	218.5	(236)
18.	median palatal length	270.5		275.5
18a.	palatal length	270.2		275.3
19.	dental length	303.7		
20.	lateral length premaxilla	(200.5)		191.8
21.	length diastema	105.1		95.2
31.	inner length orbital	62.0	60.5	(54.3)
32.	inner height orbital	53.5	55.0	(52.3)
33.	breadth mastoid	120.3	115.3	
34.	breadth occipital condyles	85.3	81.7	
35.	breadth across bases paroccipitals	112.0	(108)	
36.	breadth foramen magnum	40.5	32.7	
37.	height foramen magnum	37.0	36.6	
38.	breadth neurocranium	116.4	107.0	(114.4)
39.	smallest breadth frontals	(51.3)	(59)	
40.	smallest breadth between supraorbital foramina	132.0	146.7	147.0
41.	breadth skull	196.5	205.6	206.5
42.	smallest breadth orbitals	139.8	145.6	
43.	facial breadth (1)	145.4	(155)	
44.	facial breadth (2)	78.8	69.0	
45.	breadth "snout"	66.2		68.2
46.	breadth premaxilla	64.3		70.7
47.	smallest breadth diastemas	53.8	(58)	
48.	breadth palate	118.6	119.8	121.2
50.	height basion	103.5	98.8	(95.7)

B. Measurements in mm following EISENMANN (1980, 1986)

	species	cab NS 1	cab NS 2	cab Ta'asF
9.	length choanes	69.4		
10.	maximum width choanes	51.5	47.1	46.6
10b.	minimum width choanes	42.3	37.9	43.6
14.	zygomatic breadth	188.6	192.2	187.6
17b.	smallest breadth "snout"	44.0		39.3
19.	height infraorbitale	9.3	6.3	7.6
20.	height external earbone	13.5	13.7	
23.	length anterior ocular line	383.0	(385)	
24.	length posterior ocular line	195.5	197.5	
25.	height "snout"	(91.0)		
26.	facial height between P4 and M1	(118)	(114.5)	
27.	facial height behind M3	(128.5)	(129.2)	
28.	height cranium	(97.0)	(102)	

C. Measurements in mm following VON DEN DRIESCH (1976)

		cab NS 1	cab NS 2	as NS 5	as NS 6	as NS 7	as Ta'asF
22.	length tooth row	166.5	160.5	133.1	(133)		181.5
23.	length molar row	81.3	77.2	59.7	58.7		97.5
24.	length premolar row	89.7	85.4	75.5	73.2		86.7
25a.	length P 2	36.9	34.6		27.5		38.4
25a.	breadth P 2	22.6	23.1	21.5	21.1		24.4
26a.	length P 3	27.3	26.7	24.6	23.4	22.6	29.2
26b.	breadth P 3	26.6	28.0	23.4	23.0	26.0	25.3
27a.	length P 4	25.8	25.9	22.1	22.0	21.9	28.2
27b.	breadth P 4	27.0	26.8	24.1	23.8	27.2	26.0
28a.	length M 1	22.8	23.3	19.5	19.7	19.0	25.1
28b.	breadth M 1	25.8	26.4	22.0	23.4	26.5	25.6
29a.	length M 2	23.1	24.2	20.5	19.8	20.3	26.6
29b.	breadth M 2	24.2	23.8	21.9	23.0	24.8	23.2
30a.	length M 3	27.8	30.2	21.4	19.6		25.9
30b.	breadth M 3	22.8	21.5	20.7	19.9		23.1

(NS = North Syria, cab = *Equus caballus,* as = *Equus asinus*)

The identification of this cranium material to species (and also of the mandibles) is based on information from the local people, but has also been checked with reference to the identification characteristics given by EISENMANN (1980, 1981, 1986) for crania, maxillae and mandibles. Table 2 gives the measurements for the mandibles of which the enamel dental patterns are shown in Figure 3.

TABLE 2:

Mandibular dimensions of recent equids from North Syria.

A. Measurements in mm following von den Driesch (1976)

	species	cab NS 8	as NS 21	as NS 11	as NS 14	cab Ta'as
1.	length angle–infradentale	410				
2.	length condyle–infradentale	436				
3.	length gonion–M3	132.8				120.0
4.	length M3–infradentale	282				
5.	length P2–gonion	296				298.7
15.	length diastema	100.5	79.3			
16.	breadth "snout"	59.2	46.0			
17.	breadth I3–I3	58.1				
18.	smallest breadth diastemas		33.6			
19.	height vertical ramus	232.4				205.6
20.	middle height ramus	217.4				189.5
22a.	height behind M3	94.3				105.2
22b.	height behind P4	61.1		61.3	60.9	80.4

B. Measurements in mm following Eisenmann (1981, 1986)

2.	breadth ramus for M3	130.8				123.9
6.	length symphysis	81.5	60.3			

C. Measurements in mm following von den Driesch (1976)

	species	cab NS8	as NS21	spec NS25	as NS11	as NS14	cab Ta'asF
6.	length tooth row	165.2		133.0	134.7	131.2	182.6
7.	length molar row	79.7		66.0	69.6	64.8	85.8
8.	length premolar row	84.0		70.0	65.6	66.3	94.2
9a.	length P2	29.5	25.2			22.1	31.2
9b.	breadth P2	14.2	13.2			13.6	14.5
10a.	length P3	27.1	22.7		21.3	22.1	27.9
10b.	breadth P3	17.1	15.1		14.7	15.4	16.6
11a.	length P4	25.6			21.2	21.9	28.3
11b.	breadth P4	17.6			16.0	17.0	16.6
12a.	length M1	23.6			19.9	19.2	26.6
12b.	breadth M1	16.1			15.3	15.8	15.3
13a.	length M2	25.2			20.3	19.9	28.0
13b.	breadth M2	16.1			13.2	14.4	15.3
14a.	length M3	30.0			23.6	24.4	30.0
14b.	breadth M3	15.7			12.0	13.4	13.1

(NS = North Syria, cab = *Equus caballus*, as = *Equus asinus*)

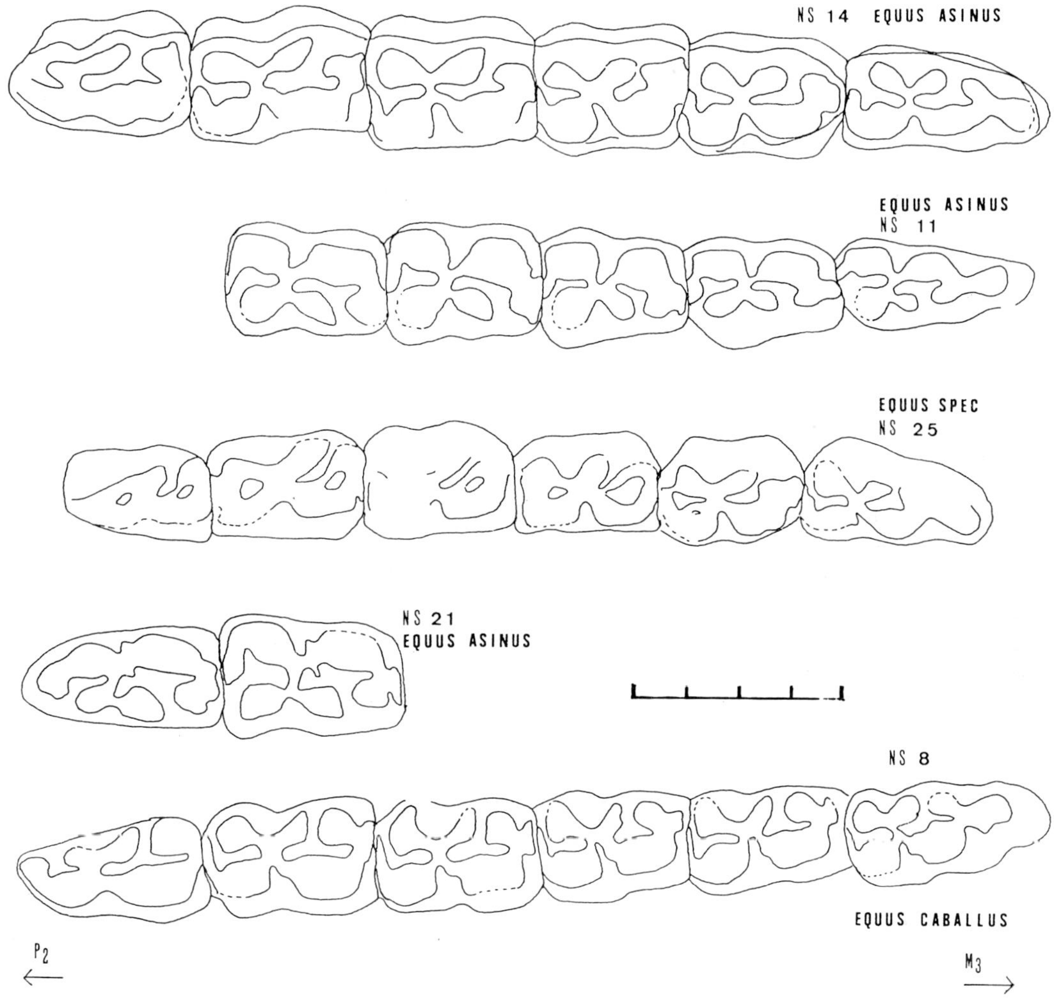

Fig. 3: Recent finds from north Syria – mandibula

In Figure 4 the enamel patterns of the cheekteeth found in Nahr el Homr are shown. It is clear that no caballine equids are included in this material. The *vallis interna* (for nomenclature, see Nobis 1971) of the mandibular cheekteeth of C1 130–140 show a pattern clearly characteristic of *Equus hemionus* in the angularity of the V-shape. The other three lower cheekteeth show a less clear pattern and could just as well be from an asinine equid. The maxillary cheekteeth all show a clearly symmetrical protocone, typical for hemiones and asses. No definite conclusions can therefore be drawn about the species to which these teeth belonged.

The material from Nahr el Homr is chronologically from the same population as the material found in nearby Mureybet. Ducos (1978, 1986) has analysed these remains and comes to the conclusion that the equids found in Mureybet are from a wild ass, *Equus africanus mureybeti.* This would mean that two distinct equid species lived in the same region at the same time. This is not completely impossible, but not probable. As we shall see later during the analysis of the postcranial elements there is a clear difference in size between on the one hand *Equus afri-*

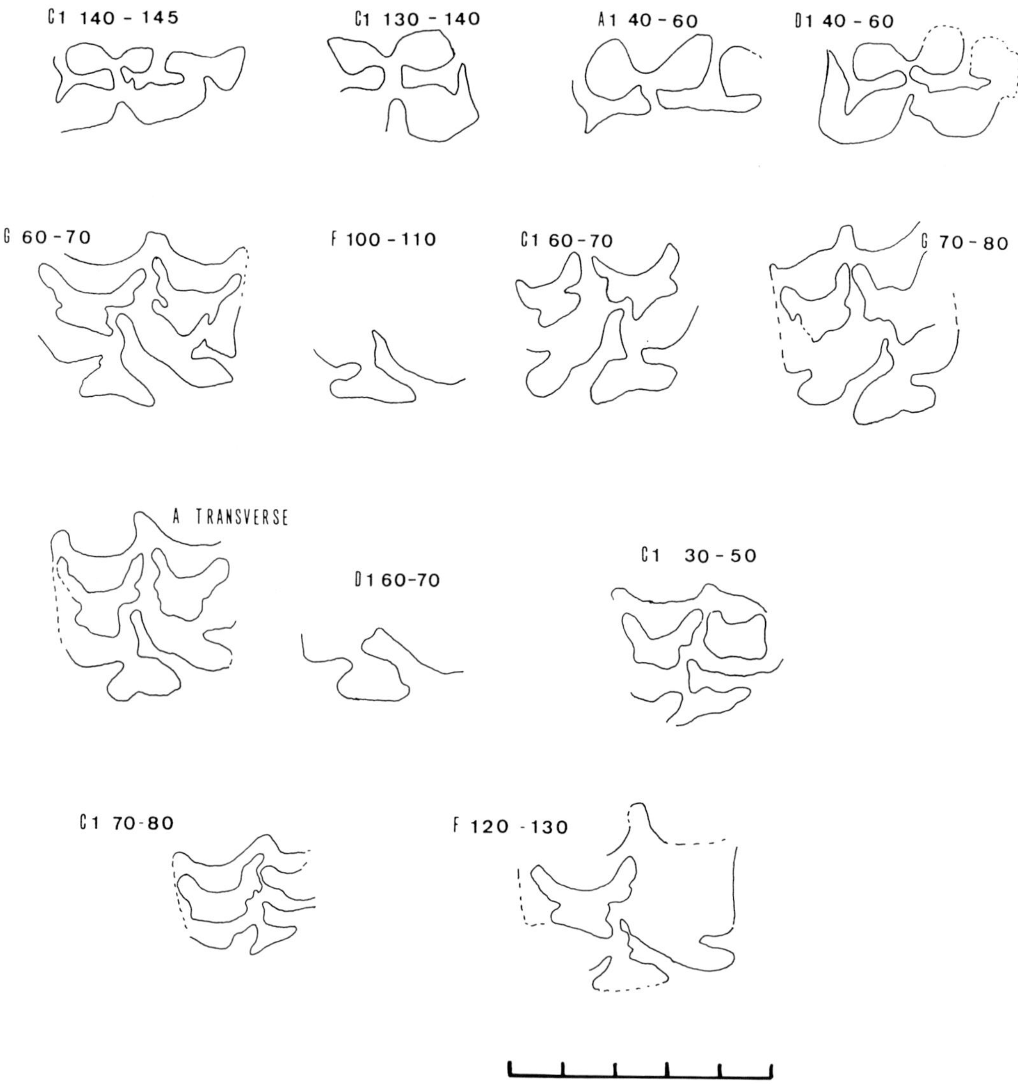

Fig. 4: Nahr el Homr 74

canus mureybeti and *Equus hemionus* and on the other hand between the former and *Equus africanus africanus.*

In Figures 5, 6, and 7 are shown the enamel patterns of the upper and lower cheekteeth found among the remains of Tell Hadidi. The protocones of 77S332, 74S243, 77S171 and 73A5 manifest very asymmetrical shapes, indicating the presence of horse, *Equus caballus.* The lower cheekteeth, except for 74O338, all show a relatively V-shaped *vallis interna* and can be considered asinine. The lower cheektooth 74O338 has a more U-shaped *vallis interna* and a higher penetration of the *vallis externa* between the para and entoconids. This tooth can therefore be considered to be from *Equus caballus.*

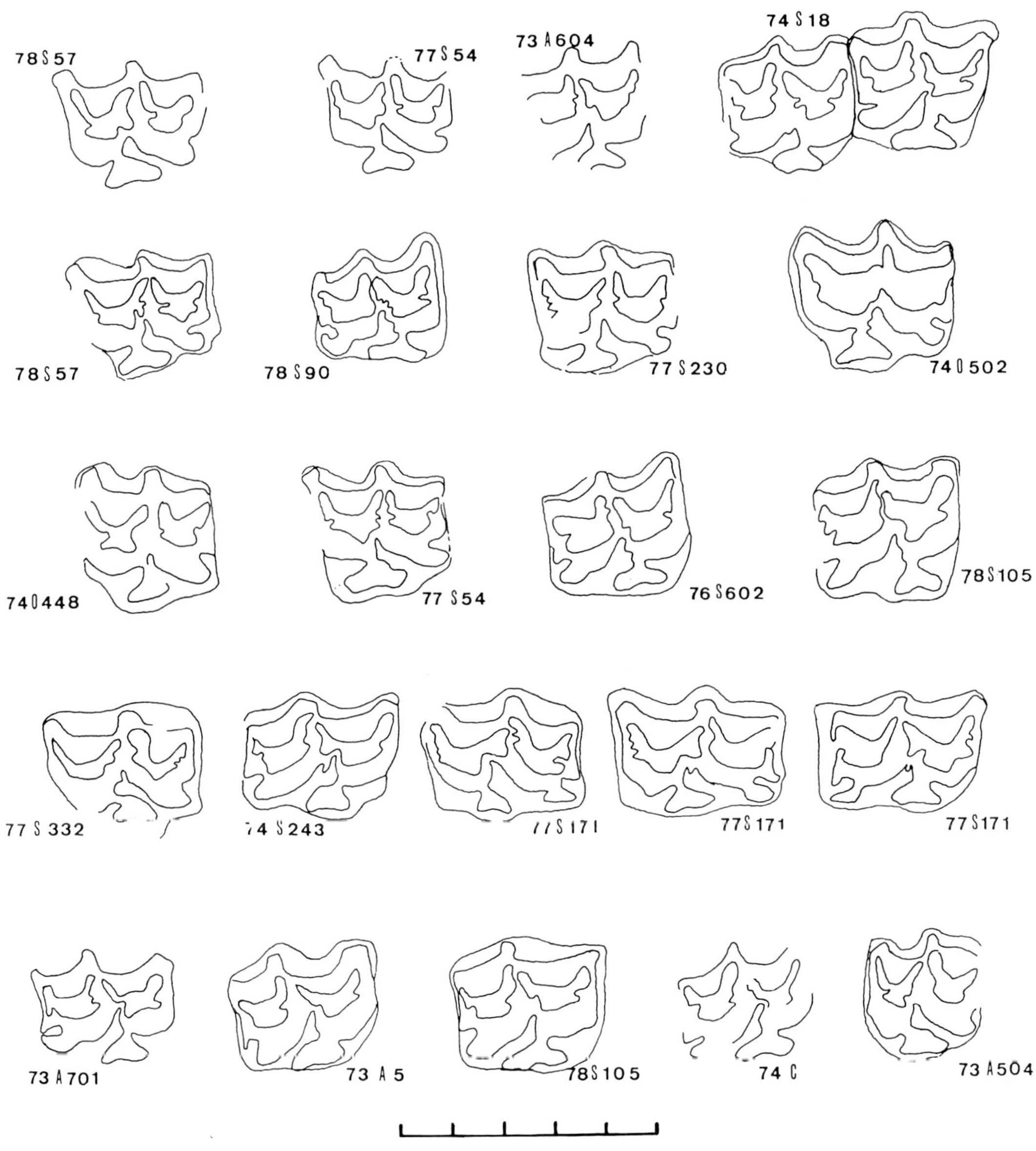
78S57
77S54
73A604
74S18
78S57
78S90
77S230
740502
740448
77S54
76S602
78S105
77S332
74S243
77S171
77S171
77S171
73A701
73 A 5
78S105
74 C
73A504

Fig. 5: Hadidi

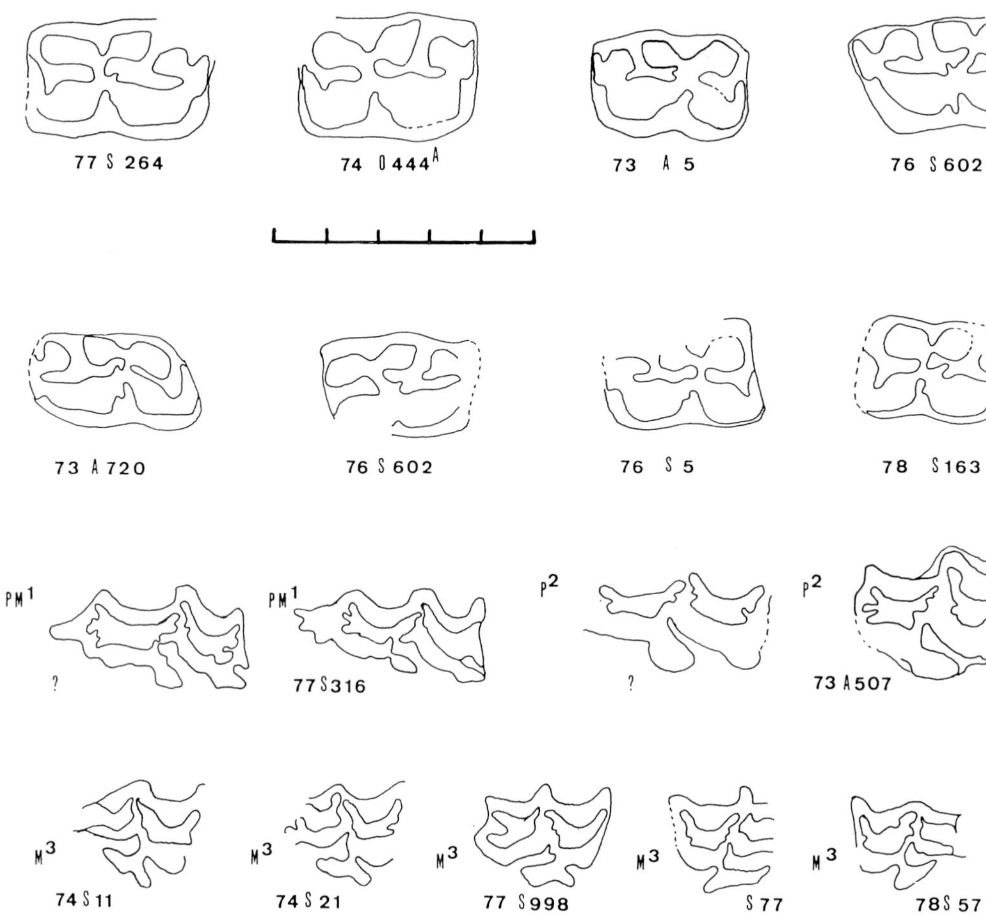
77 S 264
74 0 444 A
73 A 5
76 S 602
73 A 720
76 S 602
76 S 5
78 S 163
PM 1
?
PM 1
77 S 316
P 2
?
P 2
73 A 507
M 3
74 S 11
M 3
74 S 21
M 3
77 S 998
M 3
S 77
M 3
78 S 57

Fig. 6: Hadidi

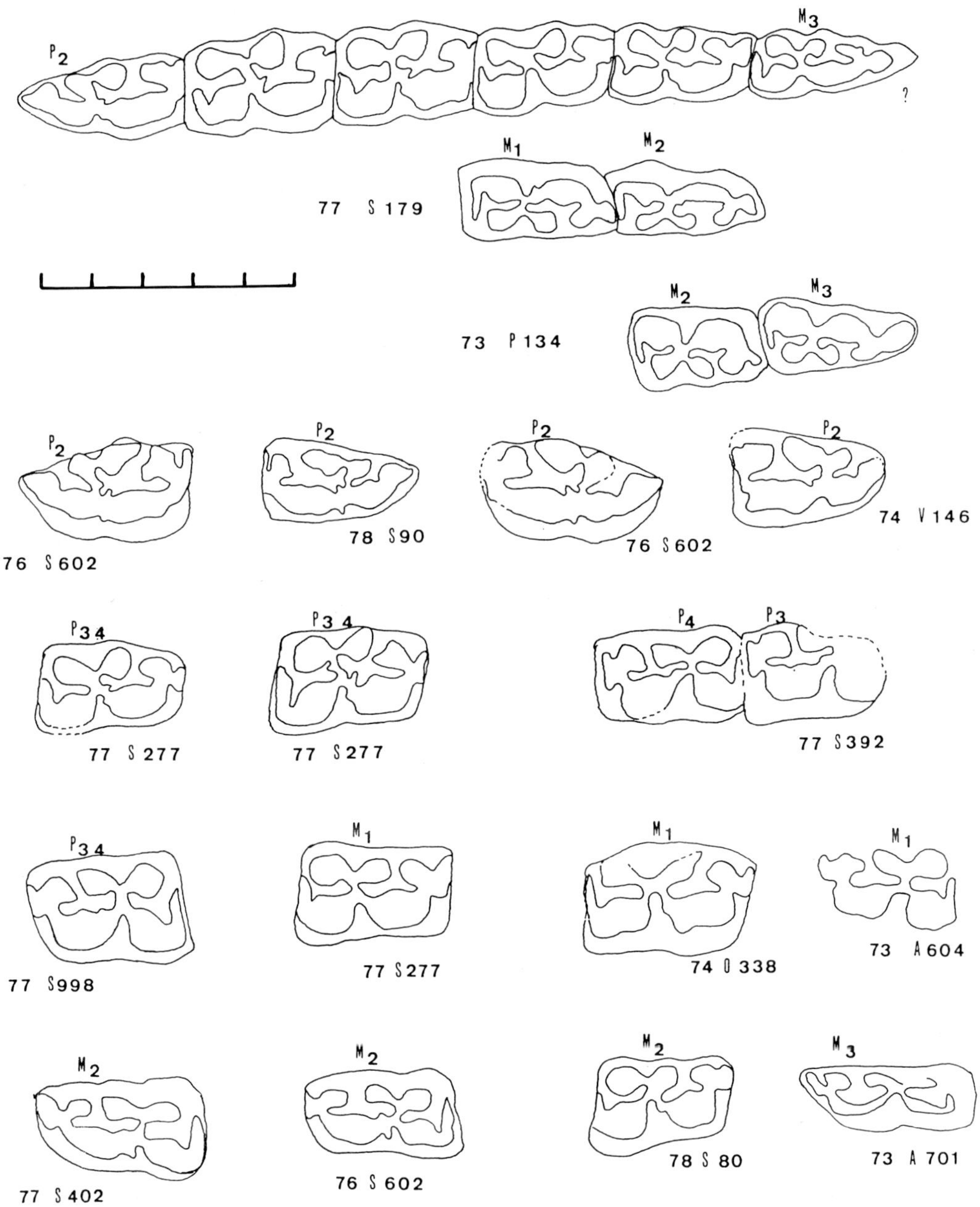

Fig. 7: Hadidi

In Figure 8 are given the enamel patterns of the cheekteeth from Medieval Ta'as. Teeth A144 and A214 can be considered caballine. The others must be asinine.

The enamel patterns of the teeth from Deir 'Alla (Fig. 9) show the same division between caballine and asinine. Only the three teeth AA305 with the asymmetrical protocones and clearly developed plis caballine can be from *Equus caballus.* All others are probably asinine.

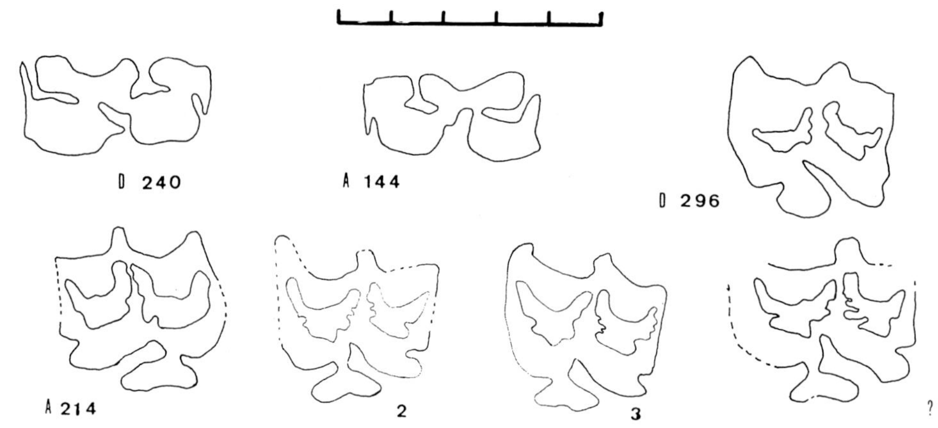

Fig. 8: Ta'as 72/73

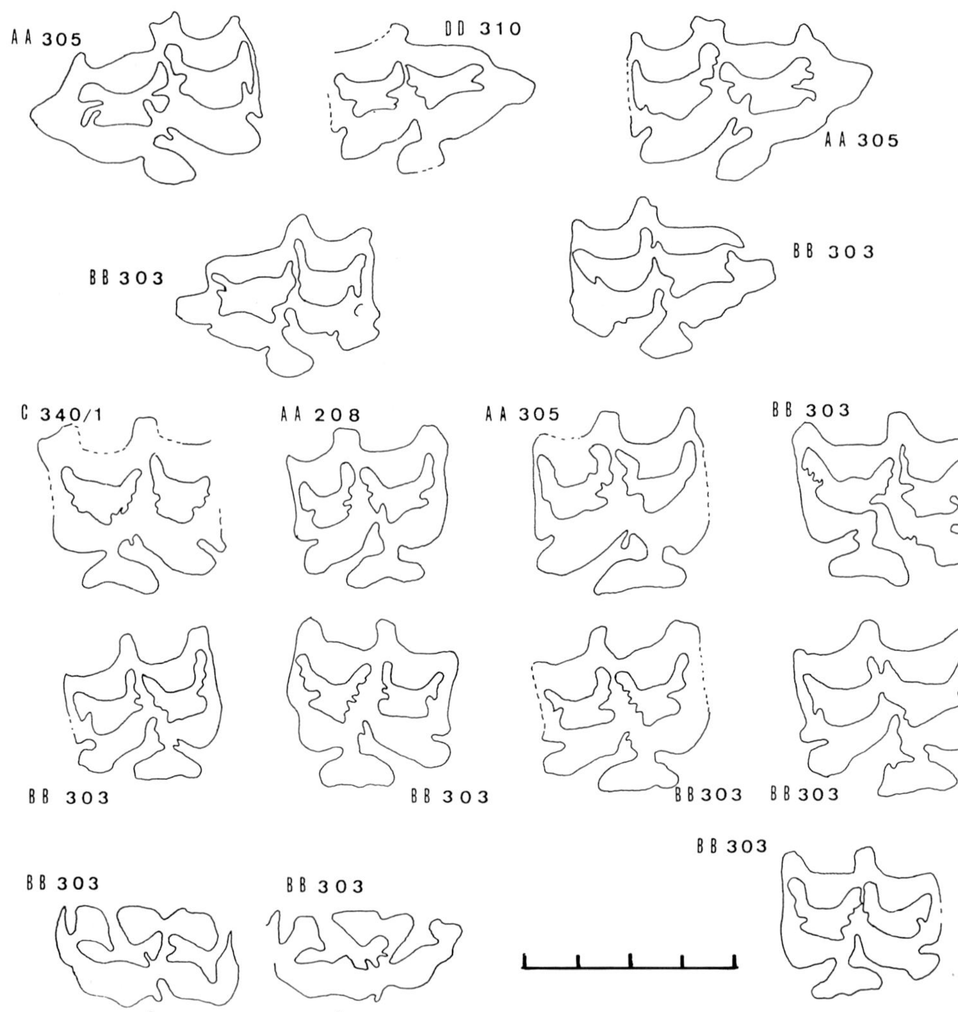

Fig. 9: Deir 'Alla

The enamel patterns of the equid teeth from the settlements of Hadidi, Ta'as, and Deir' Alla that are not caballine are identified as asinine. Although the patterns from asinine and hemione equids are rather similar, we must assume that most of these remains are from domestic animals. The importance of hunting in these settlements was minimal. If the teeth were mainly from hemiones, then hunting activities would have been much more important, which is not consistent with the overall pattern of animal exploitation in these periods in the Near East.

3.2. Postcranial skeletal elements

The postcranial remains of equids from excavations from later Holocene periods are usually poorly preserved when compared to the dental remains. Most of these postcranial remains were found among settlement debris, where they were probably dumped by man or brought by scavengers. Hardly any butchering marks are to be noted on these remains, and no articulated bones were found. It is most probable that the remains of equids are from animals that died of natural causes or because of accidents and were not butchered by man. The remains of recent equids collected in North Syria all show clear chewing and gnawing marks made by carnivores. Anyone who has ever travelled in the Near East will have observed that dead asses and horses are dragged out of the settlements and are there devoured by dogs, foxes, jackals, and other carnivores. In the process, a complete carcass is then scattered over a large area. It may also account for the fact that hardly any complete bones of equids were found among the archaeological remains.

The studies of GROMOVA (1955), NOBIS (1971), and especially EISENMANN (1979) have shown that identification of complete post-cranial elements to the species level (*Equus caballus, Equus asinus, or Equus hemionus*) is possible. The main factor in this identification is size. In the following analysis, I have tried to find parameters whereby also fragments of postcranial elements could be identified on the basis of size. For this purpose I have taken the measurements of the bones of equids from the excavated sites of Hayaz, Hadidi, Sweyhat, El Qitar, Deir 'Alla and Selenkahiya (see the measurement table) and compared individual measurements with those of published data from other archaeological settlements and with modern comparative material. The latter consists mainly of the elements collected from North Syria and of postcranial elements from *Equus hemionus* in the Muséum National d'Histoire Naturelle in Paris.

The reports on which I depended for comparative studies of archaeological material are: BOESSNECK 1970, 1976 (Thebes, Soleb); BOESSNECK/VON DEN DRIESCH 1976 (Norsuntepe); BÖKÖNYI 1973 (Umm Dabagiyah); CLUTTON-BROCK 1974 (Buhen); CLUTTON-BROCK/BURLEIGH 1978 (Abu Salabikh); COMPAGNONI 1978 (Shahr i-Sokhta); DAVIS 1976, 1980 (Arad, Israel); DENIZ/CALISLAR/ÖZGÜDEN 1964 (Sardis); VON DEN DRIESCH/AMBERGER 1981 (Ababra); DUCOS 1968, 1978 (Tel Gat, Mureybet); HECKER 1975 (Beidha); HERRE/RÖHRS 1954 (Osmankayasi); KOLB 1972 (Takht i-Suleiman); KRAUSS 1975 (Bastam); RAUH 1981 (Demircihüyük); TURNBULL 1967 (Palegawra).

In Figures 10–25, the individual measurements of breadth and depth of skeletal parts are compared. For each species the modern and archaeological material are shown separately. The data for *Equus hemionus* are a combination of the data for *Equus hemionus hemionus, Equus hemionus onager, Equus hemionus khur,* and *Equus hemionus kulan.* Data on these four subspecies come from the museum in Paris and data published by COMPAGNONI (1978).

Separating these four closely related subspecies would scatter the available data over too many groups. Also, we must take into account that modern comparative material like that from Paris is mainly from zoo specimens.

To give at least an idea of the total variation of the population of *Equus hemionus,* I have shown these four subspecies as one group. In the diagrams, blocks have been drawn whenever the sample included six or more specimens. The blocks consist of a vertical line from minimum to maximum and a horizontal line showing the arithmetic mean value. The little blocks above and below the mean value show the extent of plus and minus one standard deviation (for samples). The data from the archaeological deposits of Hadidi, Selenkahiya, El Qitar, Sweyhat, Ta'as, Hayaz, and Deir'Alla are presented without a species identification in the diagrams. However, identifications based on the diagrams have been given in the measurement tables. In almost all diagrams, *Equus ferus* is mentioned. By this term is meant the material from Norsuntepe that Boessneck/von den Driesch (1976) presented as wild horses. Furthermore, the name *Equus africanus* is used for the wild asses, while *Equus asinus* is reserved in this analysis for domesticated asses. The material from Mureybet is always presented separately.

Scapula (Figure 10): The smallest length of the *collum* (SLC) and the greatest length of the *processus articularis* (GLP) are used here to separate the species. These two measurements together with the length (LG) and breadth (BG) of the glenoid cavity are the most commonly available dimensions for archaeological material. The latter two measurements, however, show too much overlap in size to be useful. Therefore only the SLC and GLP are presented. As one can see, there is a large overlap between the measurements for *Equus hemionus* and those for *Equus asinus. Equus caballus* however, is clearly distinguishable from the other species. This pattern can be observed for all postcranial elements. The bones from Hadidi, Sweyhat, and El Qitar are mainly from domestic asses. The two largest scapulae from Sweyhat and the two

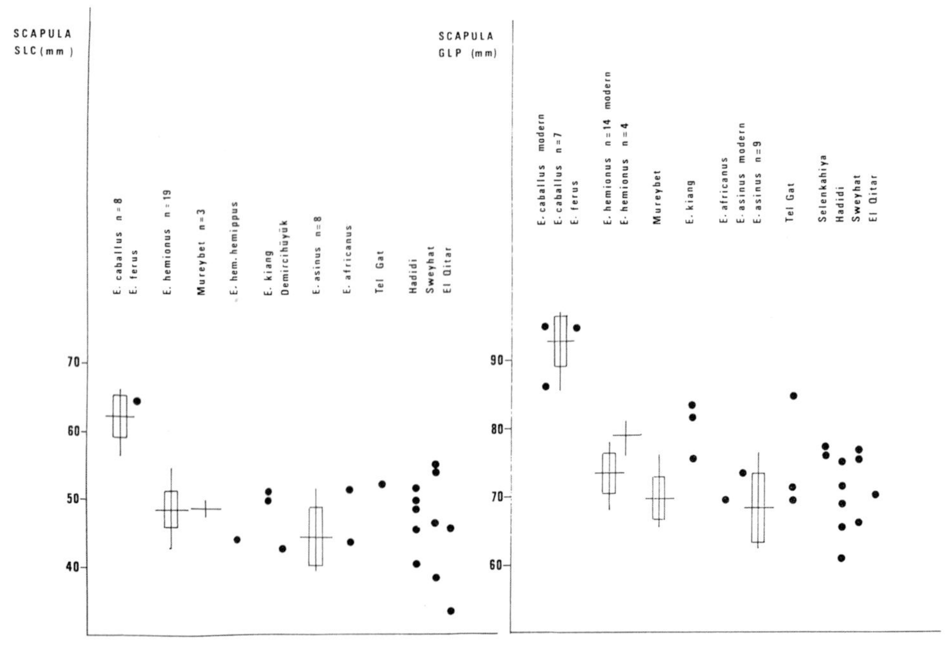

Fig. 10: Scapula, measurements

scapulae from Selenkahiya, however, reveal a size difference with respect to *Equus asinus.* These bones are clearly as large as *Equus hemionus* but one factor is missing in these diagrams. Mules and hinnies, the sterile cross-breeds of horses and asses, have the same size bones as hemiones. It is therefore impossible to separate these two groups only on the basis of size.

One more observation must be made. DUCOS (1968) identified the remains from Tel Gat (in Palestine) as *Equus asinus palestinae.* It is clear from these diagrams that at least the largest scapulae from Tel Gat cannot have been from an ass, but must have been from a hemione or mule.

Humerus (Figure 11): The greatest breadth of the distal end (Bd) and the breadth of the trochlea(BT) are the most frequently encountered measurements for humeri in the archaeozoological literature. They are here presented as individual measurements. The BT dimension, in particular, shows a marked separation between the three main species. We can also observe here the fact that *Equus asinus* from recent times, in these diagrams mainly represented by asses collected in North Syria, are clearly smaller than the asses from archaeological contexts. As with the scapulae we again have some large humeri in the collections from Hadidi and Sweyhat that could be some species other than *Equus asinus.*

Radius (Figures 12–14): For the radius, the greatest breadth of the proximal end (Bp) and the greatest length of the *facies articularis proximalis* (BFp) have been measured together with the corresponding dimensions for the distal end (repectively Bd and BFd). The same observations as with scapula and humerus can be made, but we can now also observe a general trend toward large animals among the remains of Hadidi and Sweyhat, the remains from Hayaz being generally somewhat smaller.

In Figure 14 it is evident that the greatest length (GL) also provides a means for distinguishing between the species. In the second part of Figure 14, GL is shown in relation to the smallest

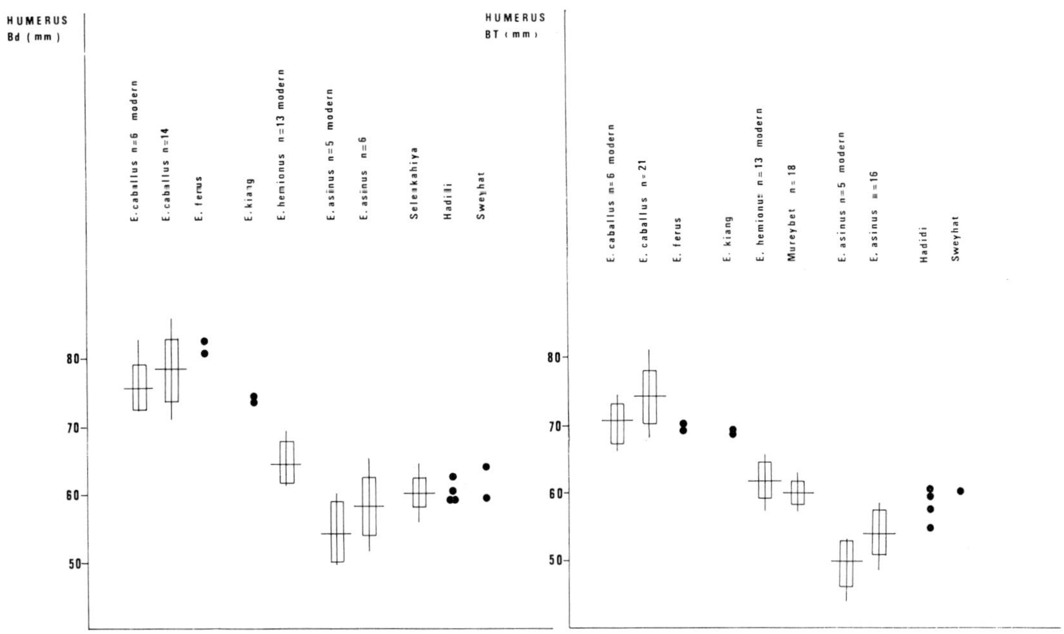

Fig. 11: Humerus distal, measurements

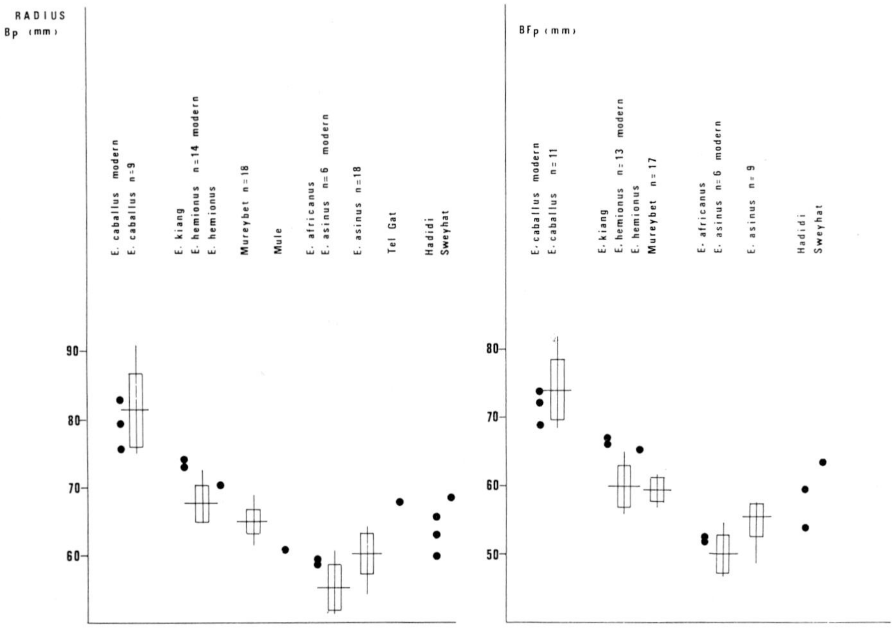

Fig. 12: Radius proximal, measurements

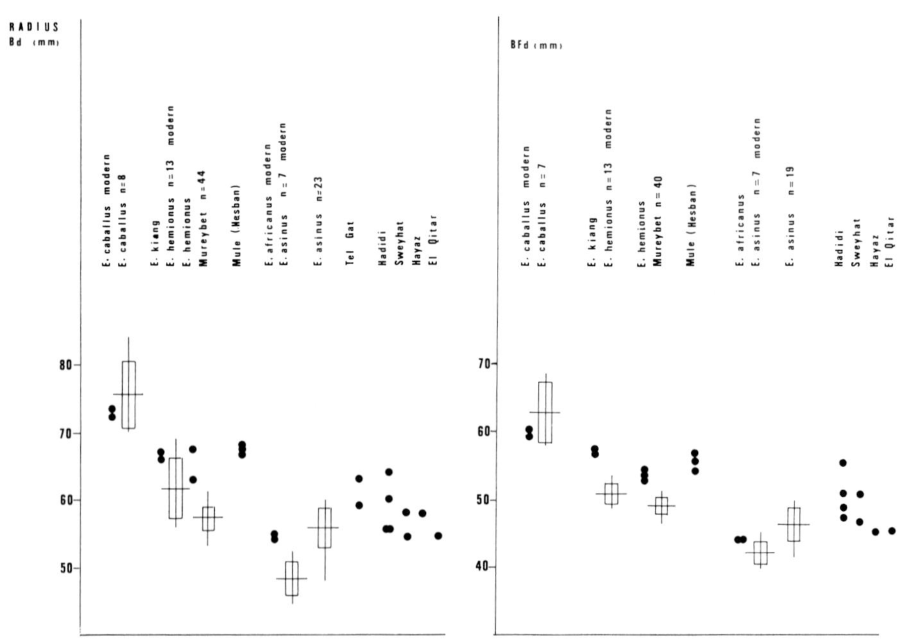

Fig. 13: Radius distal, measurements

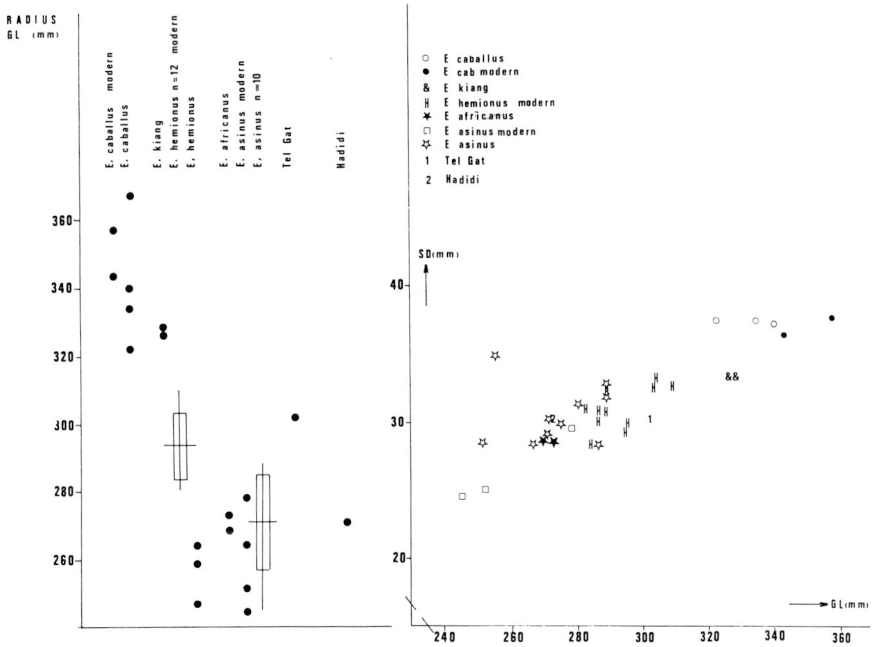

Fig. 14: Radius, length measurements

breadth of the diaphysis (SD). This illustrates the often used slenderness index (GL:SD). We can observe that, with this index, the greatest influence on the observed variation is due to GL and that SD barely affects the index. This observation has been tested for all bones and all width-depth indices. No indices could be found that indicate a difference in the relation between the breadth and depth of bone parts for the principle species.

Ulna (Figure 15): Although no archaeological material is available from the sites under study I have included these diagrams to have a complete set for analysis of all major long bones. Numbers of bones presented here are small, but the same patterns are true for the ulna as for the above described bones.

Metacarpus (Figures 16–17): Already EISENMANN (1979) has shown that the metapodial bones can be separated quite easily as to species when they are complete. In Figures 16–17, the breadth and depth measurements of the proximal parts (Bp and Dp) and of the distal parts (Bd and Dd) have been plotted individually. In Figure 16 we can see that *Equus caballus* does occur among the material from Hayaz Höyük and Deir' Alla and that a few of the larger bones might be from mules or hemiones. This holds true also for the material from Tel Gat (DUCOS 1968). In Figure 17, the distal ends of metacarpi are compared. (In using these diagrams it is, of course, necessary to be absolutely certain about the identification as to metacarpus or metatarsus.) No horse remains were found among the material of Selenkahiya, Hadidi, and El Qitar, but the largest bones of these sites must have been from animals other than asses.

Pelvis (Figure 18): Although pelvis bones are often found, it is hardly ever possible to measure anything except the length of the acetabulum (LA), which is represented here in Figure 18. The large size variation among asses makes this measurement less useful for identification purposes.

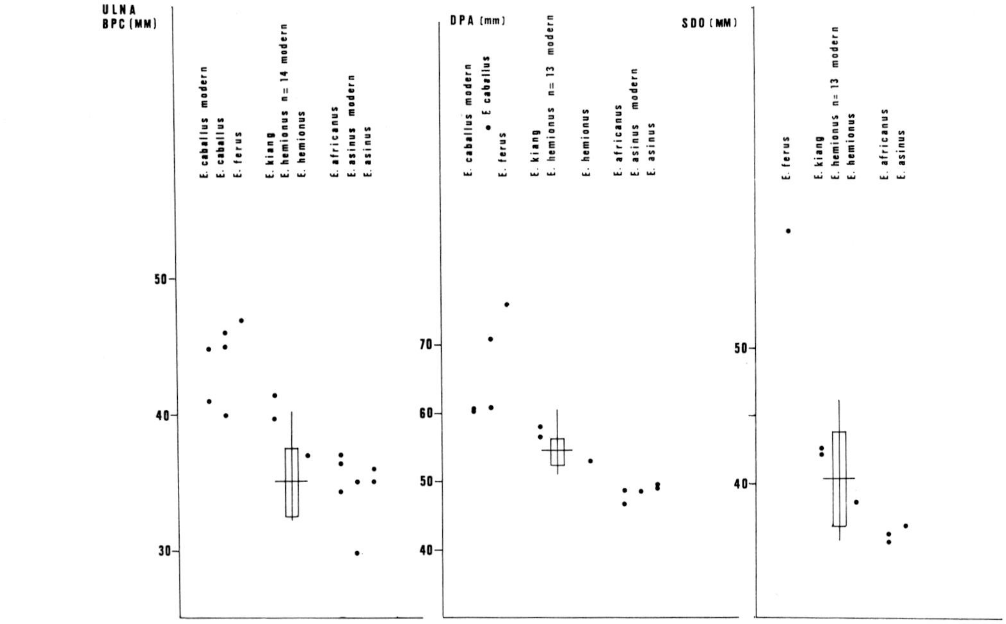

Fig. 15: Ulna, measurements

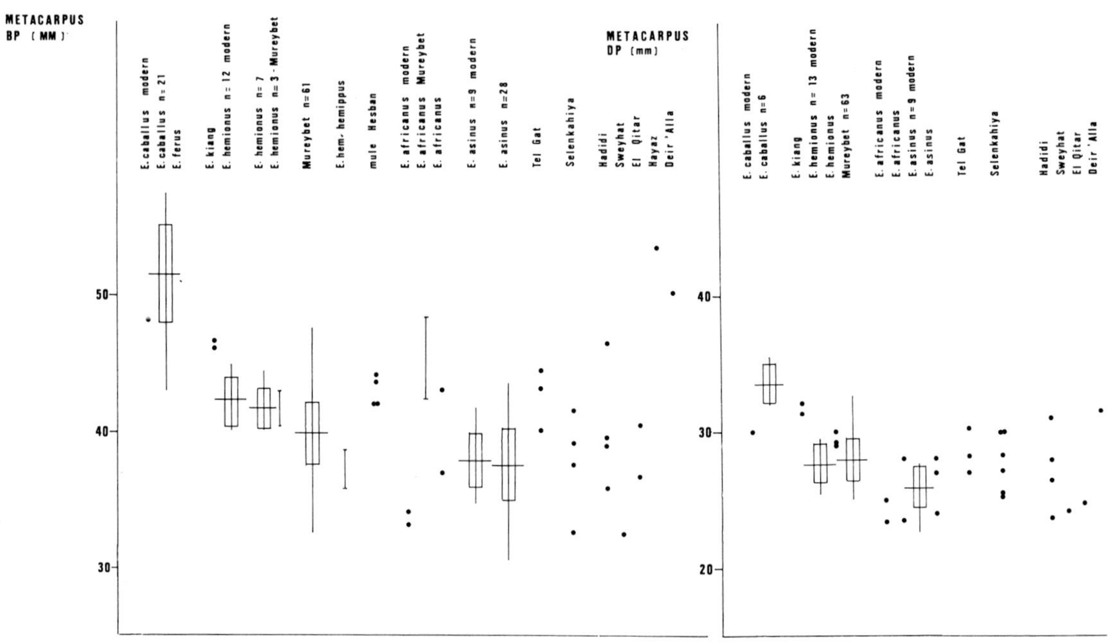

Fig. 16: Metacarpus III proximal, measurements

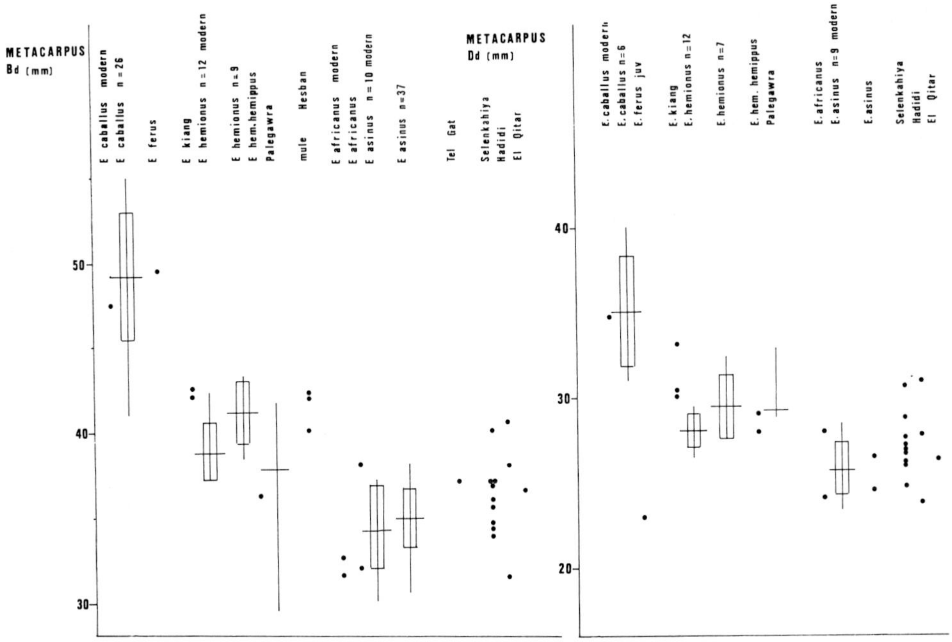

Fig. 17: Metacarpus III distal, measurements

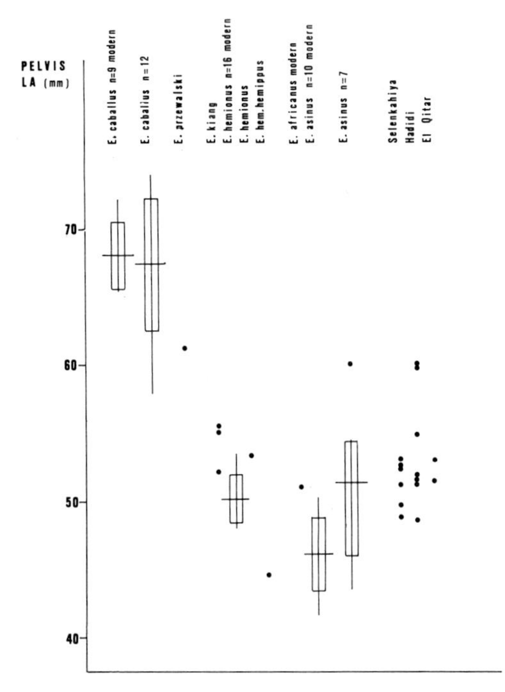

Fig. 18: Pelvis, measurements

Femur (Figure 19): The greatest breadth of the proximal end (Bp) and the greatest depth of the *caput* (DC) are presented here as those skeletal part are often found during excavations. The DC, in particular, shows great variation in size with a very small *caput femoris* among the remains from El Qitar and rather large ones, possibly mules or hemiones, from Selenkahiya and Hadidi. The greatest breadth of the distal end (Bd) is less well represented and also shows less difference in size between the species.

Tibia (Figures 20–21): The greatest breadth of the proximal end (Bp) again highlights a few bones from Selenkahiya and Hadidi that fall outside the size range for *Equus asinus.* The smallest breadth of the diaphysis (SD) shows the same pattern. Here we can see an even more marked difference between the asses from recent times and asses from archaeological context. The distal end of the tibia is a frequently found element among faunal remains. Both the greatest breadth (Bd) and the greatest depth (Dd) are presented here. Again we find clear evidence for the presence of horse, *Equus caballus,* among the remains from Hadidi and Ta'as. The other remains show again that a number of bones are larger than can reasonably be expected for *Equus asinus.* The differences in these diagrams are quite marked.

Talus (Figures 22–23): The greatest height (GH), greatest breadth (GB), the breadth of the *facies articularis distalis* (BFd) and the length of the medial part of the trochlea (LmT) are shown here. The GH and LmT (Figure 22), in particular, reveal hardly any overlap between the measurements for *Equus hemionus* and *Equus asinus* (modern domestic or wild). A few horse bones from Hayaz and Deir 'Alla are revealed and also a relatively high number of bones from individuals larger than the largest known *Equus asinus* in this analysis. Again we see that the ass elements from Hayaz are generally smaller than those from the urban centers of Selenkahiya, Hadidi, Sweyhat, and El Qitar.

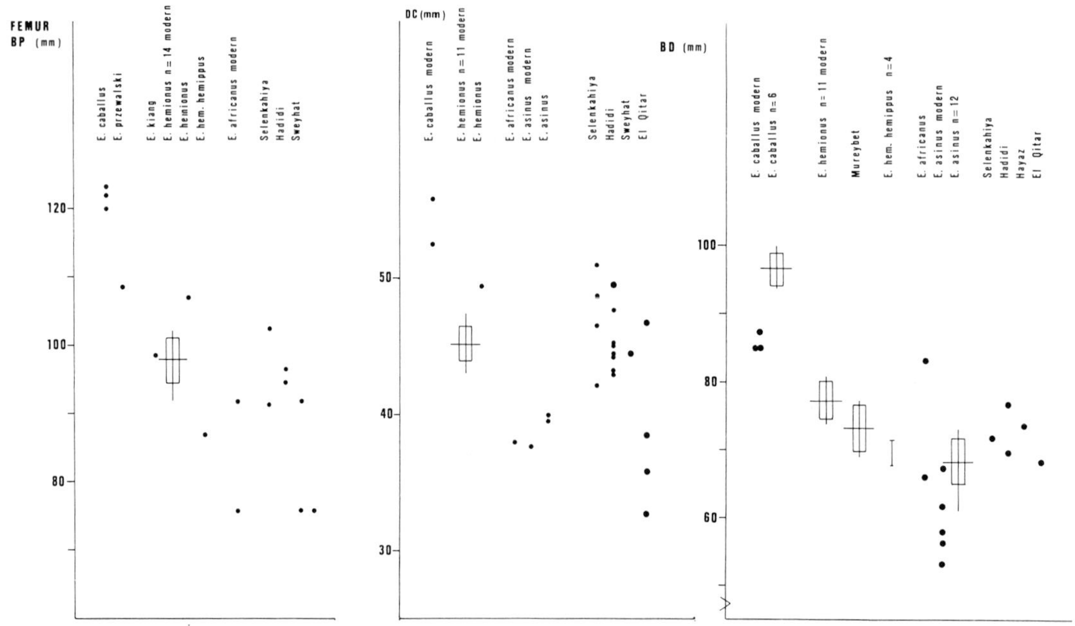

Fig. 19: Femur, measurements

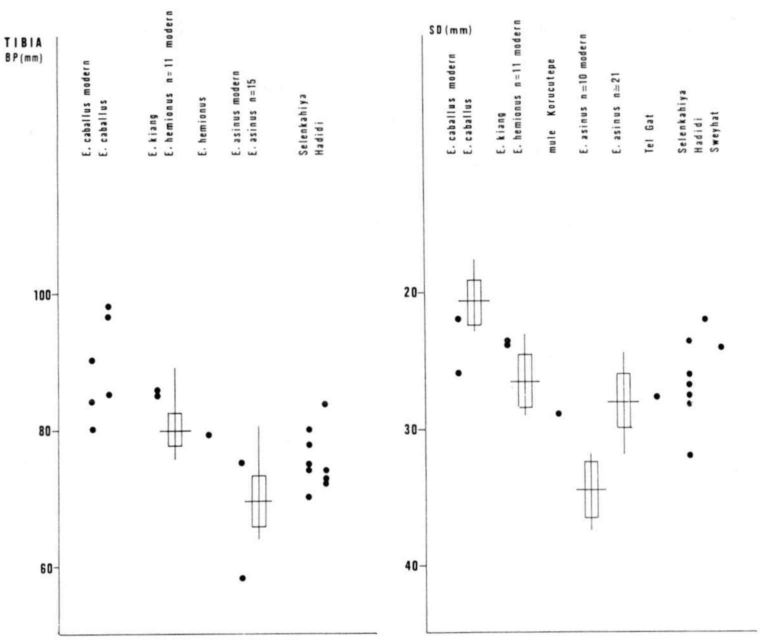

Fig. 20: Tibia proximal and shaft, measurements

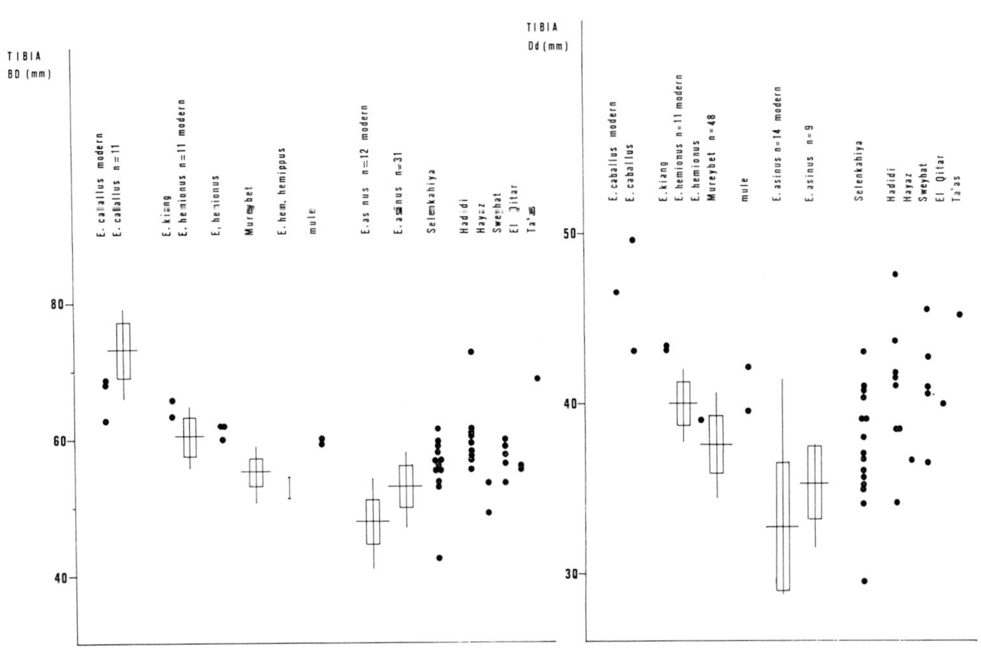

Fig. 21: Tibia distal, measurements

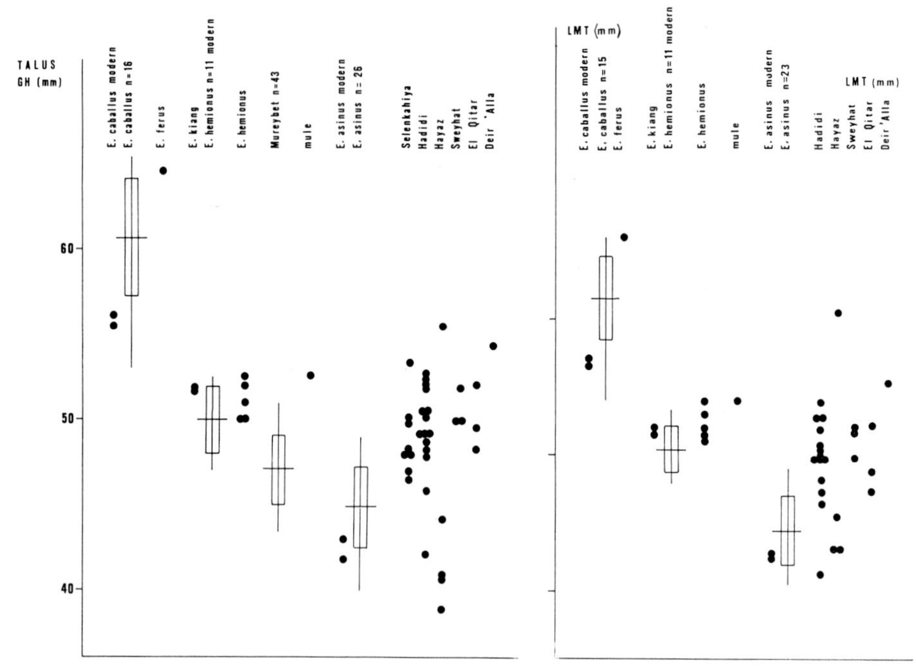

Fig. 22: Talus, length measurements

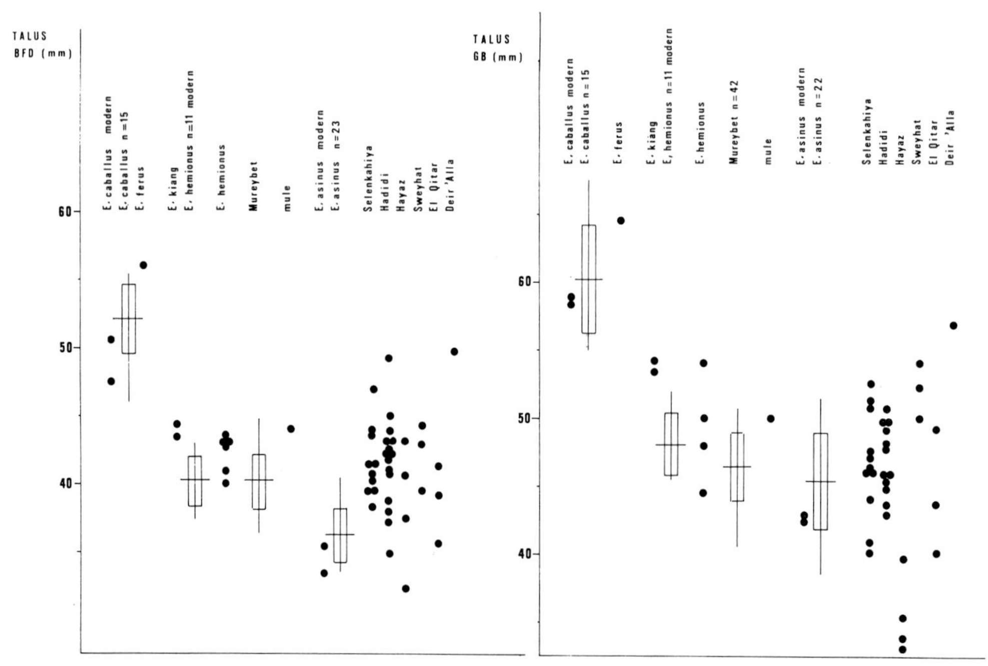

Fig. 23: Talus, breadth measurements

Calcaneum (Figure 24): The greatest length (GL) and breadth (GB) are presented here. They show the same organization of size for the different species as do the earlier mentioned bones. It is remarkable that, while in the case of all other measurements for other bones the size of the animals of Mureybet lies between *Equus hemionus* and *Equus asinus,* the GB of the calcanea of the animals from Mureybet is much smaller than that for *Equus asinus.* A difference in measuring might account for this discrepancy.

Metatarsus (Figure 25): For the metatarsus only the greatest breadth (Bp) and greatest depth (Dp) of the proximal end are shown in this analysis. The same pattern as observed for the other bones holds true here as well. Ideally it would be worthwhile to establish the patterns for the distal end of the metatarsi and to compare these with the distal end of the metacarpi, thus permiting one to separate these bones. However, it is remarkable that, while loose distal ends of metacarpi are quite often recognized, this is not true for metatarsi.

In the measurement tables are also given measurements for the phalanges 1, 2 and 3 found in archaeological contexts in the sites discussed above. They have not been analysed in the same fashion, mainly because there is hardly any comparative material and all identifications of archaeological materials are handicapped by the fact that there is a great deal of variation between the anterior and posterior bones and their placement is often difficult and subjective. Unless most of these phalanges can be placed with any certainty, rather than only some of them as is now the case, there is no point in attempting to work out any size variation. [But see DIVE/EISENMANN, this volume.]

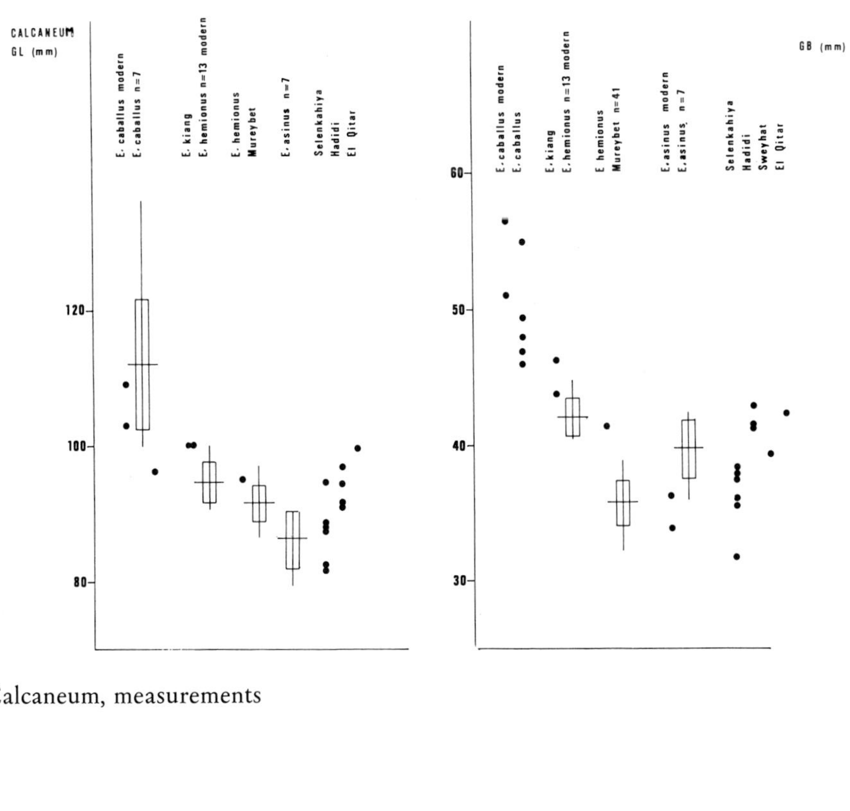

Fig. 24: Calcaneum, measurements

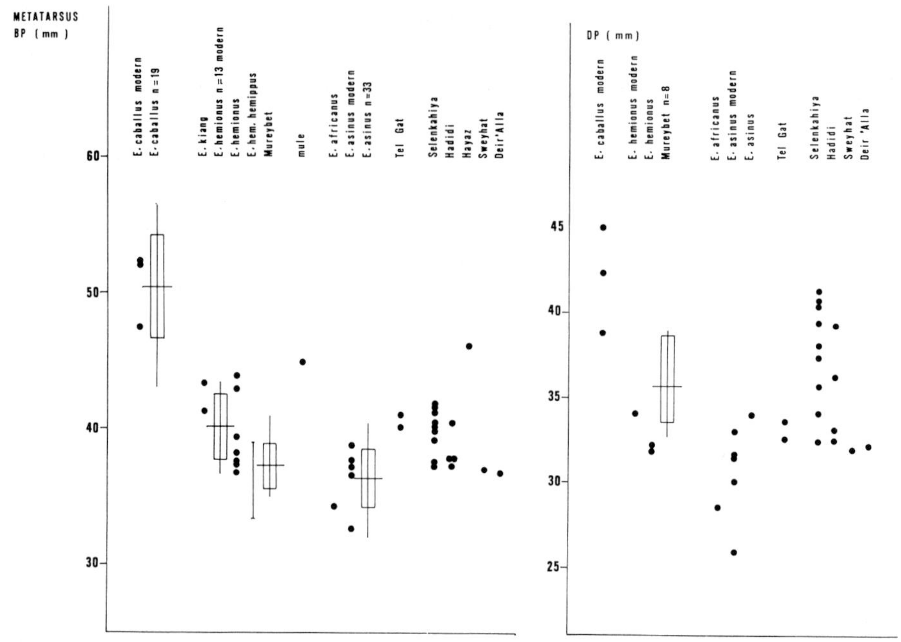

Fig. 25: Metatarsus III proximal, measurements

4. Conclusion

In this article, material is presented from several sites located along the Euphrates in South Turkey and North Syria and from one settlement in the Jordan valley. An attempt has been made to use breadth and depth dimensions of postcranial elements to separate and identify the different *Equus* species. It has been shown that, although difficulties do exist, the species identification of fragments from archaeological sites can be undertaken. The main problem that remains is to establish criteria for the identification of the cross-breeds, principally from *Equus caballus* × *Equus asinus.* A thorough anatomical analysis of skeletons of these cross-breeds must be undertaken in the near future.

In this analysis an attempt has also been made to identify and interpret the material from Syria, Turkey, and Jordan. The species names given to the fragments in the measurement tables are based on the diagrams presented here. We have seen that for Hayaz, a small farming community, a few horses were found and that the majority of the equid bones are from small asses. The material from the urban centers of Selenkahiya, Sweyhat, Hadidi, and El Qitar show a very consistent pattern. A few horse remains are found as are quite a number of clearly recognizable asses. But a considerable number of fragments, about one-third of all non-caballine remains, are from animals larger than the asses from archaeological contexts in all other places and periods. It is unlikely that these were all remains from hunted animals (*Equus hemionus*), although a few bones might have been. Considering the trade that took place between these cities and other areas where asses and mules were used, it is most probable that the majority of these remains are from horse × ass hybrids. These cross-breeds had a high value as

they were less vulnerable and less demanding than horses but had even greater carrying abilities than asses.
Seeing that within a relatively short period in a small area all settlements show this characteristic occurrence of large non-caballine equids in a similar cultural context, it can be concluded that in the Bronze Age in North Syria mules and/or hinnies were deliberately bred on a large scale to meet the needs of developing trade.
The question of whether the hemione that lived in Syria can be equated with *Equus hemionus hemippus* cannot be satisfactorily solved. There are not enough finds of wild equids from North Syria, except for the material of Mureybet, to establish if and when hemiones lived in this area and if they can be separated from other hemiones on the basis of their size.

5. Acknowledgements

I would like to express my thanks for the help and cooperation that Dr. F. Poplin of the Laboratoire d'Anatomie comparée gave me during my study of the equid material in Paris. Also I would like to thank Dr. IJzereef, now of the Rijksdienst voor Oudheidkundig Bodemonderzoek, Amersfoort, for his kind permission to use and publish the equid data from Selenkahiya. The English text was corrected by Mrs. S. v. Gelder-Ottway and Richard Meadow.

6. Literature

Boerma, J. H. K./Roodenberg, J. J., 1977, Une deuxième industrie épipaléolithique sur le Nahr el Homr. Palaeohistoria **19**, 8–17.

Boessneck, J., 1970, Ein altägyptisches Pferdeskelett. Mitteilungen des Deutschen Archäologischen Instituts Abteilung Kairo **26**, 43–47.

Boessneck, J., 1976, Tell el-Dab'a III. Die Tierknochenfunde 1966–1969; Wien: Österreichische Akademie der Wissenschaften.

Boessneck, J./Driesch, A. V. D., 1976, Pferde im 4./3. Jahrtausend v. Chr. in Ostanatolien. Säugetierkundliche Mitteilungen **24**(2), 81–87.

Bökönyi, S., 1973, The fauna of Umm Dabagiyah: a preliminary report. Iraq **35**(1), 9–11.

Buitenhuis, H., 1979, The faunal remains from Tell Hadidi. In: Kubasiewicz, M., ed., Archaezoology; Szczecin: Agricultural Academy, 164–175.

Buitenhuis, H., 1983, The animal remains of Tell Sweyhat, Syria. Palaeohistoria **25**, 131–144.

Buitenhuis, H., 1985, Preliminary report on the faunal remains of Hayaz Höyük from the 1979–1983 seasons. Anatolica **12**, 61–74.

Buitenhuis, H., 1988, Archeozoölogisch Onderzoek Langs de Midden-Eufraat (Archaeozoological Investigations along the Middle Euphrates); Dissertation, Rijksuniversiteit, Groningen.

Clason, A. T./Buitenhuis, H., 1978, A preliminary report on the faunal remains of Nahr el Homr, Hadidi and Ta'as in the Tabqa dam region in Syria. Journal of Archaeological Science **5**(1), 75–83.

Clutton-Brock, J., 1974, The Buhen horse. Journal of Archaeological Science **1**(1), 89–100.

Clutton-Brock, J./Burleigh, R., 1978, The animal remains from Abu Salabikh, a preliminary report. Iraq **40**, 89–100.

Compagnoni, B., 1978, The bone remains of *Equus hemionus* from Shahr-i Sokhta. In: Meadow, R.H./Zeder, M.A., eds., Approaches to Faunal Analysis in the Middle East; (Peabody Museum Bulletin 2); Cambridge (MA): Peabody Museum, Harvard University, 105–118.

Culican, W./McClellan, T.L., 1983–84, El Qitar: First season of excavations 1982–83. Abr-nahrain **22,** 29–63.

Davis, S.J., 1976, Mammal bones from the Early Bronze Age city of Arad, northern Negev, Israel: some implications concerning human exploitation. Journal of Archaeological Science **3,** 153–164.

Davis, S.J., 1980, Late Pleistocene and Holocene equid remains from Israel. Zoological Journal of the Linnean Society **70,** 289–312.

Deniz, E./Calislar, T./Özgüden, T., 1964, Osteological investigations of the animal remains recovered from the excavations of ancient Sardis. Anatolia **8,** 49–56.

Dornemann, R.H., 1975, M.P.M. Euphrates Valley expedition, 1974: excavations at Tell Hadidi. Lore **25**(1), 2–23.

Driesch, A. von den, 1976, A Guide to the Measurement of Animal Bones from Archaeological Sites; (Peabody Museum Bulletin **1**); Cambridge (MA): Peabody Museum, Harvard University.

Driesch, A. von den/Amberger, G., 1981, Ein altbabylonisches Eselskelett vom Tell Ababra, Iraq. Bonner Zoologische Beiträge **32**(1–2), 67–74.

Ducos, P., 1968, L'Origine des Animaux domestiques en Palestine; (Publications de l'Institut de Préhistoire, Université de Bordeaux **6**); Bordeaux: Institut de Préhistoire.

Ducos, P., 1978, Tell-Mureybet (Syrie, IXe–VIIe Millenaires): Étude archéozoologique et problèmes d'écologie humaine: 1; Paris: Éditions du CNRS.

Ducos, P., 1986, The equid of Tell Muraibit, Syria. In: Meadow, R.H./Uerpmann, H.P., eds., Equids in the Ancient World; (Beihefte zum Tübinger Atlas des Vorderen Orients, Reihe A **19/1**); Wiesbaden: Dr. Ludwig Reichert Verlag, 237–245.

Eisenmann, V., 1979, Les métapodes d'*Equus sensu lato* (Mammalia, Perissodactyla). Géobios **12**(6), 863–886.

Eisenmann, V., 1980, Les Chevaux (*Equus sensu lato*) fossiles et actueles: crânes et dents jugales supérieures; (Cahiers de Paléontologie); Paris: Éditions du CNRS.

Eisenmann, V., 1981, Étude des dents jugales inférieures des *Equus* (Mammalia, Perissodactyla) actuels et fossiles. Palaeovertebrata **10** (3–4), 127–226.

Eisenmann, V., 1986, Comparative osteology of modern and fossil horses, half-asses, and asses. In: Meadow, R.H./Uerpmann, H.-P., eds., Equids in the Ancient World; (Beihefte zum Tübinger Atlas des Vorderen Orients, Reihe A **19/1**); Wiesbaden: Dr. Ludwig Reichert Verlag, 67–116.

Franken, H.J./Ibrahim, M.M., 1977–78, Two seasons of excavations at Tell Deir'Alla, 1976–78. Annals of the Department of Antiquities of Jordan, Amman **22,** 57–228.

Gromova, V., 1955, L'histoire des chevaux (genre *Equus*) de l'ancien monde: première partie: revue et description des formes. Annales du Centre d'Études et de Documentation Paléontologiques **13,** 1–202. (French Translation of Gromova, V., 1949, Istorija loshadej (roda *Equus*) v Starom Svete, Chast' 1, Obzor i opisanie form; Trudy Paleontologitcheskovo Instituta Akademii Nauk S.S.S.R. **17**(1), 1–373).

Hecker, H.M., 1975, The faunal analysis of the primary food animals from Pre-Pottery Neolithic Beidha (Jordan); Dissertation, Columbia University, New York.

Herre, W./Röhrs, M., 1958, Die Tierreste aus den Hethitergräbern von Osmankayasi. In: Bittel, K., *et al.,* Die Hethitischen Grabfunde on Osmankayasi; Bogazköy-Hattusu 2; (Wisenschaftliche Veröffentlichung der Deutschen Orient-Gesellschaft **71**), 60–80.

Holland, T.A., 1976, Preliminary report on excavations at Tell es-Sweyhat, Syria, 1973–4. Levant **8,** 36–75.

Holland, T.A., 1977, Preliminary report on excavations at Tell es-Sweyhat, Syria, 1975. Levant **9,** 36–68.

Kolb, R., 1972, Die Tierknochenfunde vom Takht-i Suleiman in der iranischen Provinz Aserbaidschan (Fundmaterial der Grabung, 1969); Dissertation, Universität München, Munich.

KRAUSS, R., 1975, Tierknochenfunde aus Bastam in Nordwest-Azerbaidjan/Iran (Fundmaterial der Grabungen 1970 und 1971); Dissertation, Universität München, München.

LEEUW, S.E. v.d., 1976, Studies in the technology of ancient pottery; Dissertation, Rijksuniversitat, Amsterdam.

LOON, M.N. van, 1977, Preliminary results of the 1974 and 1975 excavations at Selenkahiya near Meskene, Syria. Annual of the American Schools of Oriental Research **44,** 97–112.

LOON, M.N. VAN, 1982, Hammân et-Turkmân on the Balikh: Background and first results of the University of Amsterdam's 1981 excavation. Akkadica **27,** 30–45.

MCCLELLAN, T.L., 1984–85, El Qitar: second season of excavations, 1983–84. Abr-nahrain **23,** 39–72.

NOBIS, G., 1971, Vom Wildpferd zum Hauspferd; (Fundamenta, Reihe B, Band **6**); Köln: Böhlau Verlag.

RAUH, H., 1981, Knochenfunde von Säugetieren aus dem Demircihüyük (Nordwestanatolien); Dissertation, Ludwig-Maximiliens-Universität, München.

ROODENBERG, J.J., 1979–80, Premiers resultats des recherches archéologiques à Hayaz Höyük. Anatolica **7,** 3–19.

TURNBULL, P.F., 1967, Bones of Palegawra. Bulletin (Field Museum of Natural History) **38**(9), 4–5.

Hijlke Buitenhuis, Biologisch-Archaeologisch Instituut, Poststraat 6, 9712 Groningen, The Netherlands

APPENDIX

Equus asinus

Scapula

Site: Unit etc.	1	2	3	4	5	6	7	8
HADI:74-3107	--	--	--	37.4	--	--	--	--
HADI:A604	--	--	44.6	39.2	--	--	--	--
HADI:77-3130	47.7	65.5	44.0	--	--	--	--	--
HADI:C134117	39.5	60.8	40.4	35.6	--	--	--	--
HADI:F122	45.0	--	--	--	--	--	--	--
HADI:74-314	--	--	44.6	36.9	--	--	--	--
HADI:G10-35	--	--	44.1	33.9	--	--	--	--
HADI:74-350	49.0	71.3	46.5	--	--	--	--	--
HADI:M119	--	69.0	45.0	40.4	--	--	--	--
QITA:XK11-30	45.0	69.9	45.3	35.6	--	--	--	--
QITA:Y238-25	32.7	--	--	--	--	--	--	--
SWEY:04N044	37.9	66.4	42.5	34.4	--	--	--	--
SWEY:04F111	46.0	--	--	35.2	--	--	--	--

Measurements 1–8.

1 - SLC (smallest length of collum)
2 - GLP (greatest length of proc. art.)
3 - LG (length of facies art.)
4 - BG (breadth of facies art.)

Humerus

Site: Unit etc.	1	2	3	4	5	6	7	8
HADI:75-S39	59.4	54.7	26.7	--	--	210.6	--	--
HADI:76-S1078	59.3	57.2	26.8	--	--	--	--	--
HADI:74-0378	--	--	28.2	--	--	--	--	--
SELE:	60.7	--	--	--	--	--	--	--
SELE:	58.0	--	29.1	--	--	--	--	--
SELE:	60.0	--	--	--	--	--	--	--
SELE:	58.0	--	--	--	--	--	--	--
SELE:	57.2	--	--	--	--	--	--	--
SELE:	--	--	24.4	--	--	--	--	--
SELE:	60.5	--	--	--	--	--	--	--
SELE:	56.0	--	--	--	--	--	--	--
SELE:	58.8	--	--	--	--	--	--	--
SELE:	58.1	--	--	--	--	--	--	--
SELE:	--	--	29.7	--	--	--	--	--
SELE:	60.8	--	--	--	--	--	--	--
SELE:	59.2	--	30.0	--	--	--	--	--
SELE:	61.0	--	28.0	--	--	--	--	--
SWEY:01C141	59.5	--	--	--	--	--	--	--

Measurements 1–8.

1 - Bd (distal breadth)
2 - BT (breadth of trochlea)
3 - SD (smallest breadth of diaphysis)
6 - GLC (greatest length from caput)

Equus asinus

Radius

Site: Unit etc.	1	2	3	4	5	6	7	8
	Measurements							
HADI:74-S40	59.8	--	--	--	--	--	53.6	--
HADI:77-S295	--	--	60.0	--	--	--	--	51.4
HADI:75-S39	63.0	30.5	--	271.0	--	--	--	--
HADI:73-A504	--	--	55.7	--	--	--	--	47.8
HADI:77-S386	--	--	55.7	--	--	--	--	49.3
HAYA:GG0034	--	--	57.7	--	--	--	--	45.6
QITA:YK38-17	--	--	54.8	--	--	--	--	45.6
SWEY:03GN16	--	--	54.5	--	--	--	--	47.2
SWEY:04F111	--	--	58.3	--	--	--	--	51.4

1 - Bp (proximal breadth)
2 - SD (smallest breadth of diaphysis)
3 - Bd (distal breadth)
4 - GL (greatest length)
7 - BFp (breadth of prox. facies art.)
8 - BFd (breadth of dist. facies art.)

Metacarpus III

Site: Unit etc.	1	2	3	4	5	6	7	8
	Measurements							
HADI:77-S418	39.5	27.9	25.8	37.9	27.8	212.7	--	--
HADI:77-S341	39.0	26.5	--	--	--	--	--	--
HADI:77-S405	35.8	23.7	22.7	31.4	23.8	175.4	--	--
QITA:YL37-90	36.7	--	--	--	--	--	--	--
QITA:XK12-63	40.4	24.8	25.7	36.4	26.3	182.3	--	--
SELE:	41.5	28.5	--	--	--	--	--	--
SELE:	--	--	--	37.0	26.0	--	--	--
SELE:	--	--	--	36.8	28.8	--	--	--
SELE:	--	--	--	34.6	26.2	--	--	--
SELE:	--	--	--	36.0	27.6	--	--	--
SELE:	--	--	--	35.6	26.8	--	--	--
SELE:	32.6	25.3	--	--	--	--	--	--
SELE:	--	--	--	37.0	27.0	--	--	--
SELE:	39.1	27.2	--	--	--	--	--	--
SELE:	37.6	25.7	24.0	33.8	24.7	176.6	--	--
SELE:	--	--	--	34.2	27.2	--	--	--
SWEY:01C141	32.5	24.2	19.9	--	--	--	--	--

1 - Bp (proximal breadth)
2 - Dp (proximal depth)
3 - SD (smallest breadth of diaphysis)
4 - Bd (distal breadth)
5 - Dd (distal depth)
6 - GL (greatest length)

Equus asinus

Pelvis (Acetabula)

Site: Unit etc.	1	2	3	4	5	6	7	8
HADI:74-S8	51.8	49.1	--	--	--	--	--	--
HADI:74-S93	51.6	48.7	--	--	--	--	--	--
HADI:74-S24	48.5	46.1	--	--	--	--	--	--
HADI:73-A604	52.3	49.6	--	--	--	--	--	--
QITA:YH38-54	51.4	--	--	--	--	--	--	--
QITA:XK13-88	53.0	50.6	--	--	--	--	--	--
QITA:YL37-90	--	46.3	--	--	--	--	--	--
SELE:	49.6	--	--	--	--	--	--	--
SELE:	52.6	--	--	--	--	--	--	--
SELE:	52.4	--	--	--	--	--	--	--
SELE:	48.9	--	--	--	--	--	--	--
SELE:	53.0	--	--	--	--	--	--	--
SELE:	51.1	--	--	--	--	--	--	--

(Measurements: columns 1–8)

1 - LA (length of the acetabulum) 2 - LAR (length of the acet. on rim)

Femur

Site: Unit etc.	1	2	3	4	5	6	7	8
HADI:75-S198	--	44.6	--	--	--	--	--	--
HADI:74-S31	--	44.4	--	--	--	--	--	--
HADI:74-V122	--	--	--	69.3	--	--	--	--
HADI:74-S17	--	43.2	--	--	--	--	--	--
QITA:YH37-97	--	35.9	--	--	--	--	--	--
QITA:XK11-92	--	--	--	68.0	--	--	--	--
QITA:YJ37-75	--	32.9	--	--	--	--	--	--
QITA:YH38-80	--	38.5	--	--	--	--	--	--
SELE:	--	--	--	71.6	--	--	--	--
SELE:	91.5	42.1	--	--	--	--	--	--
SWEY:04F1014	91.8	44.4	--	--	--	--	--	--
SWEY:04M017	73.4	--	--	--	--	--	--	--

(Measurements: columns 1–8)

1 - Bp (proximal breadth) 2 - DC (depth of the caput)
4 - Bd (distal breadth)

Tibia

Site: Unit etc.	1	2	3	4	5	6	7	8
HADI:78-S193	--	--	57.4	--	--	--	--	--
HADI:74-S35	--	--	58.0	38.4	--	--	--	--
HADI:74-M355	72.5	--	--	--	--	--	--	--
HADI:73-0109	72.0	--	--	--	--	--	--	--
HADI:74-S40	--	--	57.6	38.5	--	--	--	--

(Measurements: columns 1–8)

continued

1 - Bp (proximal breadth) 3 - Bd (distal breadth)
4 - Dd (distal depth)

Equus asinus

Tibia

Site: Unit etc.	1	2	3	4	5	6	7	8
	continued							
HADI:73-A505	--	--	55.7	34.2	--	--	--	--
HADI:73-A	74.2	--	--	--	--	--	--	--
HAYA:FG0239	--	--	53.5	36.7	--	--	--	--
HAYA:JE0055	--	--	49.0	--	--	--	--	--
QITA:YH37-91	--	--	55.9	--	--	--	--	--
QITA:XK11-26	--	--	56.0	40.0	--	--	--	--
SELE:	--	33.1	56.8	36.7	--	--	--	--
SELE:	--	--	56.8	36.0	--	--	--	--
SELE:	--	--	--	35.2	--	--	--	--
SELE:	--	--	55.6	34.0	--	--	--	--
SELE:	74.0	--	--	--	--	--	--	--
SELE:	70.0	--	--	--	--	--	--	--
SELE:	--	28.0	42.5	29.4	--	--	--	--
SELE:	--	31.8	53.5	--	--	--	--	--
SELE:	--	32.4	53.1	37.0	--	--	--	--
SELE:	--	--	55.6	--	--	--	--	--
SELE:	--	33.8	--	35.0	--	--	--	--
SELE:	--	--	--	35.6	--	--	--	--
SWEY:04K013	--	--	57.8	45.5	--	--	--	--
SWEY:04N044	--	35.8	56.8	41.0	--	--	--	--
SWEY:01C133	--	--	53.6	36.6	--	--	--	--

Measurements

1 - Bp (proximal breadth)
2 - SD (smallest breadth of diaphysis)
3 - Bd (distal breadth)
4 - Dd (distal depth)

Astragalus

Site: Unit etc.	1	2	3	4	5	6	7	8
HADI:74-0449	49.1	49.8	--	--	45.9	42.5	--	--
HADI:D+	42.1	41.2	--	--	49.2	35.0	--	--
HADI:73-D	45.8	40.4	--	--	44.9	37.3	--	--
HADI:74-S8	49.1	50.0	--	--	49.8	42.3	--	--
HADI:73-B	48.3	49.3	--	--	45.3	49.3	--	--
HADI:77-S220	49.2	47.3	--	--	43.8	38.8	--	--
HADI:73-A207	48.8	48.2	--	--	43.0	38.1	--	--
HADI:74-M355	49.3	49.8	--	--	48.2	42.2	--	--
HADI:77-S268	47.9	51.8	--	--	47.9	40.8	--	--
HAYA:DF0013	38.8	43.1	--	--	39.8	43.1	--	--
HAYA:EF0060	40.9	39.2	--	--	37.8	33.7	--	--
HAYA:JE0069	41.0	43.3	--	--	40.7	33.0	--	--
HAYA:KE0029	44.2	45.5	--	--	--	--	--	--
QITA:XK11-30	48.4	48.7	--	--	40.1	35.7	--	--
QITA:XK10-02	49.6	47.2	--	--	43.7	39.2	--	--
SELE:	46.5	--	--	--	41.0	38.3	--	--
	continued							

Measurements

1 - GH (greatest height)
2 - LmT (medial length of trochlea)
5 - GB (greatest breadth)
6 - BFd (breadth of dist. facies art.)

Equus asinus

Astragalus

Site: Unit etc.	Measurements 1	2	3	4	5	6	7	8
	continued							
SELE:	47.0	--	--	--	44.0	39.3	--	--
SELE:	48.0	--	--	--	46.1	40.8	--	--
SELE:	--	--	--	--	40.1	47.0	--	--
SELE:	48.3	--	--	--	47.6	40.2	--	--
SELE:	48.0	--	--	--	46.0	39.6	--	--
SELE:	--	--	--	--	46.2	--	--	--
SELE:	--	--	--	--	47.1	41.4	--	--

1 - GH (greatest height)
5 - GB (greatest breadth)
6 - BFd (breadth of dist. facies art.)

Calcaneus

Site: Unit etc.	Measurements 1	2	3	4	5	6	7	8
SELE:	82.5	31.9	--	--	--	--	--	--
SELE:	87.5	37.5	--	--	--	--	--	--
SELE:	87.9	38.4	--	--	--	--	--	--
SELE:	--	38.0	--	--	--	--	--	--
SELE:	88.4	36.1	--	--	--	--	--	--
SELE:	81.7	35.6	--	--	--	--	--	--

1 - GL (greatest length)
2 - GB (greatest breadth)

Metatarsus III

Site: Unit etc.	Measurements 1	2	3	4	5	6	7	8
DEIR:D-	37.0	32.2	23.9	34.5	26.5	229.0	229.2	--
SELE:	39.4	32.5	23.7	36.4	30.8	--	223.8	--
SELE:	37.8	38.1	--	--	--	--	--	--
SELE:	37.5	34.2	24.3	--	--	--	--	--
SWEY:03GN16	37.3	31.9	--	--	--	--	--	--

1 - Bp (proximal breadth)
2 - Dp (proximal depth)
3 - SD (smallest breadth of diaphysis)
4 - Bd (distal breadth)
5 - Dd (distal depth)
6 - GLl (greatest lateral length)
7 - GL (greatest length)

Equus caballus

Humerus

Site: Unit etc.	Measurements 1	2	3	4	5	6	7	8
HADI:74-S67	--	--	36.6	--	--	--	--	--

3 - SD (smallest breadth of diaphysis)

Metacarpus III

Site: Unit etc.	Measurements 1	2	3	4	5	6	7	8
DEIR:B152/60	50.1	31.4	28.9	--	--	--	--	--
HAYA:HE0007	53.5	--	--	--	--	--	--	--

1 - Bp (proximal breadth) 2 - Dp (proximal depth)
3 - SD (smallest breadth of diaphysis)

Pelvis (Acetabula)

Site: Unit etc.	Measurements 1	2	3	4	5	6	7	8
HADI:73-F128	59.8	54.5	--	--	--	--	--	--
HADI:78-S3	60.0	--	--	--	--	--	--	--

1 - LA (length of the acetabulum) 2 - LAR (length of the acet. on rim)

Tibia

Site: Unit etc.	Measurements 1	2	3	4	5	6	7	8
HADI:74-S52	83.5	--	--	--	--	--	--	--
HADI:77-S247	--	37.9	72.6	47.6	--	--	--	--
TAAS:C124	--	--	68.6	45.1	--	--	--	--

1 - Bp (proximal breadth) 2 - SD (smallest breadth of diaphysis)
3 - Bd (distal breadth) 4 - Dd (distal depth)

Equus caballus

Astragalus

Site: Unit etc.	Measurements 1	2	3	4	5	6	7	8
DEIR:AA305	54.4	55.1	--	--	56.9	49.8	--	--
HAYA:KE0021	55.6	60.5	--	--	--	--	--	--

1 - GH (greatest height)
2 - LmT (medial length of trochlea)
5 - GB (greatest breadth)
6 - BFd (breadth of dist. facies art.)

Metatarsus III

Site: Unit etc.	Measurements 1	2	3	4	5	6	7	8
HAYA:KE0015	46.3	--	--	--	--	--	--	--
HAYA:EG0071	--	--	--	55.8	--	--	--	--

1 - Bp (proximal breadth)
4 - Bd (distal breadth)

Phalanx 2 anterior or posterior

Site: Unit etc.	Measurements 1	2	3	4	5	6	7	8
DEIR:B233/1	55.3	43.7	47.6	50.3	33.2	--	--	--
DEIR:B152/21	49.3	41.3	47.6	49.6	30.5	--	--	--

1 - Bp (proximal breadth)
2 - SD (smallest breadth of diaphysis)
3 - Bd (distal breadth)
4 - GL (greatest length)
5 - Dp (proximal depth)

Phalanx 3 anterior or posterior

Site: Unit etc.	Measurements 1	2	3	4	5	6	7	8
DEIR:B23312	67.4	57.4	41.8	78.2	46.4	--	--	--
DEIR:B152/20	--	--	34.3	69.0	47.5	--	--	--
DEIR:AA218	64.0	58.2	43.0	77.5	48.1	--	--	--

1 - GL (greatest length)
2 - Ld (dorsal length)
3 - HP (heigth at proc. extensorius)
4 - GB (greatest breadth)
5 - BF (breadth of facies art.)

Equus sp.

Scapula

Site: Unit etc.	1	2	3	4	5	6	7	8
	Measurements							
HADI:M355	--	74.7	46.8	39.2	--	--	--	--
HADI:74-338	50.8	--	--	--	--	--	--	--
SELE:	55.0	77.0	49.3	43.6	--	--	--	--
SELE:	--	76.0	47.1	42.6	--	--	--	--
SWEY:03GH18	53.5	76.3	47.5	41.2	--	--	--	--
SWEY:0G015	54.2	75.3	45.2	40.9	--	--	--	--

1 - SLC (smallest length of collum)
2 - GLP (greatest length of proc. art.)
3 - LG (length of facies art.)
4 - BG (breadth of facies art.)

Humerus

Site: Unit etc.	1	2	3	4	5	6	7	8
	Measurements							
SELE:	61.5	--	--	--	--	--	--	--
SELE:	61.5	--	--	--	--	--	--	--
SELE:	61.4	--	--	--	--	--	--	--
SELE:	64.5	--	--	--	--	--	--	--
SELE:	64.0	--	--	--	--	--	--	--
SELE:	62.3	--	--	--	--	--	--	--
SELE:	63.0	--	--	--	--	--	--	--
SELE:	61.5	--	--	--	--	--	--	--
HADI:74-M355	62.9	60.2	--	--	--	--	--	--
HADI:73-A6	61.0	59.1	--	--	--	--	--	--
SWEY:09A110	64.2	60.2	--	--	--	--	--	--

1 - Bd (distal breadth)
2 - BT (breadth of trochlea)

Radius

Site: Unit etc.	1	2	3	4	5	6	7	8
	Measurements							
HADI:78-345	65.9	--	--	--	--	--	59.4	--
HADI:73-A506	--	--	63.8	--	--	--	--	56.0
SWEY:04A024	68.5	--	--	--	--	--	63.2	--

1 - Bp (proximal breadth)
3 - Bd (distal breadth)
7 - BFp (breadth of prox. facies art.)
8 - BFd (breadth of dist. facies art.)

Equus sp.

Metacarpus III

Site: Unit etc.	1	2	3	4	5	6	7	8
			Measurements					
HADI:77-S219	46.4	30.9	28.0	40.5	30.9	201.3	--	--
SELE:	--	30.4	--	--	--	--	--	--
SELE:	--	--	--	40.0	30.6	--	--	--
SELE:	--	30.5	--	--	--	--	--	--

1 - Bp (proximal breadth)
2 - Dp (proximal depth)
3 - SD (smallest breadth of diaphysis)
4 - Bd (distal breadth)
5 - Dd (distal depth)
6 - GL (greatest length)

Phalanx 1, anterior

Site: Unit etc.	1	2	3	4	5	6	7	8
			Measurements					
HADI:77-S316	42.2	25.3	34.4	72.6	30.4	--	--	--
HADI:73-A6	35.0	23.8	33.4	70.6	27.8	--	--	--
HADI:74-S60	39.7	--	--	--	29.0	--	--	--
HADI:74-S40	39.5	23.1	32.0	66.8	28.1	--	--	--
HADI:74-S101	33.7	21.9	29.5	68.7	25.7	--	--	--
HADI:73-A707	40.3	26.3	35.4	76.3	30.4	--	--	--
HADI:77-S200	39.4	24.6	--	--	28.9	--	--	--
HADI:77-S343	40.7	23.4	32.2	69.4	29.3	--	--	--

1 - Bp (proximal breadth)
2 - SD (smallest breadth of diaphysis)
3 - Bd (distal breadth)
4 - GL (greatest length)
5 - Dp (proximal depth)

Pelvis (Acetabula)

Site: Unit etc.	1	2	3	4	5	6	7	8
			Measurements					
HADI:74-S8	54.8	--	--	--	--	--	--	--

1 - LA (length of the acetabulum)

Femur

Site: Unit etc.	1	2	3	4	5	6	7	8
			Measurements					
HADI:74-S10	--	49.6	--	--	--	--	--	--
HADI:77-S85	--	47.8	--	--	--	--	--	--
HADI:73-P157	--	--	--	76.5	--	--	--	--

continued

2 - DC (depth of the caput)
4 - Bd (distal breadth)

Equus sp.

Femur

Site: Unit etc.	Measurements 1	2	3	4	5	6	7	8
	continued							
HADI:74-S42	94.8	45.1	--	--	--	--	--	--
HADI:74-S3	--	45.3	--	--	--	--	--	--
HADI:73-O109	96.7	43.3	--	--	--	--	--	--
HAYA:KE0029	--	--	30.0	73.2	--	--	--	--
SELE:	--	49.0	--	--	--	--	--	--
SELE:	102.5	--	--	--	--	--	--	--
SELE:	--	51.0	--	--	--	--	--	--
SELE:	--	46.6	--	--	--	--	--	--

1 - Bp (proximal breadth)
2 - DC (depth of the caput)
3 - SD (smallest breadth of diaphysis)
4 - Bd (distal breadth)

Tibia

Site: Unit etc.	Measurements 1	2	3	4	5	6	7	8
HADI:73-A704	--	--	60.3	43.7	--	--	--	--
HADI:74-S30	--	--	59.3	41.6	--	--	--	--
HADI:76-S721	--	--	61.1	41.0	--	--	--	--
SELE:	--	36.3	59.8	41.0	--	--	--	--
SELE:	75.0	--	61.5	40.3	--	--	--	--
SELE:	--	--	--	43.0	--	--	--	--
SELE:	--	--	57.0	40.9	--	--	--	--
SELE:	--	--	59.4	39.1	--	--	--	--
SELE:	--	--	56.0	39.1	--	--	--	--
SELE:	--	--	58.0	38.0	--	--	--	--
SWEY:03GH18	--	--	59.3	42.7	--	--	--	--
SWEY:020083	--	--	59.9	40.6	--	--	--	--

1 - Bp (proximal breadth)
2 - SD (smallest breadth of diaphysis)
3 - Bd (distal breadth)
4 - Dd (distal depth)

Astragalus

Site: Unit etc.	Measurements 1	2	3	4	5	6	7	8
HADI:73-A707	50.7	52.7	--	--	49.8	43.1	--	--
HADI:73-A508	52.2	52.8	--	--	45.3	40.6	--	--
HADI:74-S9	50.2	49.8	--	--	50.8	41.8	--	--
HADI:74-O448	52.8	53.9	--	--	--	43.2	--	--
HADI:74-O413	52.5	--	--	--	--	45.0	--	--
HADI:74-S68	52.0	50.5	--	--	--	43.9	--	--
	continued							

1 - GH (greatest height)
2 - LmT (medial length of trochlea)
5 - GB (greatest breadth)
6 - BFd (breadth of dist. facies art.)

Equus sp.

Astragalus

Site: Unit etc.	1	2	3	4	5	6	7	8
	continued							
HADI:74-S18	50.6	50.7	--	--	--	--	--	--
QITA:YL38-25	52.1	52.1	--	--	49.3	41.2	--	--
SELE:	50.2	--	--	--	50.8	41.3	--	--
SELE:	53.4	--	--	--	52.6	43.5	--	--
SELE:	49.9	--	--	--	51.3	44.0	--	--
SWEY:03F022	50.0	51.6	--	--	52.2	44.3	--	--
SWEY:03GH18	52.0	52.0	--	--	54.1	43.0	--	--
SWEY:020085	50.0	49.8	--	--	50.0	39.5	--	--

Measurements columns 1–8.

1 - GH (greatest height)
2 - LmT (medial length of trochlea)
5 - GB (greatest breadth)
6 - BFd (breadth of dist. facies art.)

Calcaneus

Site: Unit etc.	1	2	3	4	5	6	7	8
SELE:	94.5	36.1	--	--	--	--	--	--
SWEY:020085	--	39.4	--	--	--	--	--	--
HADI:74-M355	91.0	41.6	41.5	--	--	--	--	--
HADI:78-S76	96.9	41.4	42.4	--	--	--	--	--
HADI:74-S48	94.3	42.9	43.0	--	--	--	--	--
HADI:78-S110	91.8	--	43.5	--	--	--	--	--
QITA:YH38-45	99.6	42.4	--	--	--	--	--	--

1 - GL (greatest length)
2 - GB (greatest breadth)
3 - GD (greatest depth)

Metatarsus III

Site: Unit etc.	1	2	3	4	5	6	7	8
QITA:YH38-53	--	--	26.4	41.1	29.7	--	239.4	--
SELE:	41.8	37.4	28.6	--	--	--	--	--
SELE:	41.5	40.8	23.7	35.7	28.3	--	248.5	--
SELE:	40.6	40.4	29.2	39.8	29.2	--	256.9	--
SELE:	40.1	39.4	29.1	--	--	--	--	--
SELE:	42.0	35.7	--	--	--	--	--	--
SELE:	40.5	41.3	25.9	--	--	--	--	--
SELE:	--	--	--	--	31.6	--	--	--
SELE:	--	--	24.7	36.5	30.7	--	--	--

1 - Bp (proximal breadth)
2 - Dp (proximal depth)
3 - SD (smallest breadth of diaphysis)
4 - Bd (distal breadth)
5 - Dd (distal depth)
7 - GL (greatest length)

Equus sp.

Phalanx 1, posterior

	Measurements							
Site: Unit etc.	1	2	3	4	5	6	7	8
HADI:78-S68	41.0	26.0	37.3	--	--	--	--	--
HADI:74-S72	36.1	20.9	29.1	64.0	26.2	--	--	--
HADI:74-O339	35.5	22.2	33.0	75.5	29.7	--	--	--
HADI:74-S57	43.0	--	--	--	--	--	--	--
HADI:73-A714	42.7	25.9	34.1	74.4	32.1	--	--	--
HADI:77-S344	42.7	--	32.8	73.0	30.9	--	--	--
HADI:78-S115	43.8	25.3	33.6	72.9	32.6	--	--	--

1 - Bp (proximal breadth)
2 - SD (smallest breadth of diaphysis)
3 - Bd (distal breadth)
4 - GL (greatest length)
5 - Dp (proximal depth)

Phalanx 1 anterior or posterior

	Measurements							
Site: Unit etc.	1	2	3	4	5	6	7	8
HADI:74-S93	--	23.8	31.1	--	--	--	--	--
HADI:77-S182	37.2	23.0	32.1	69.0	27.0	--	--	--
HADI:77-S409	--	20.1	31.1	--	--	--	--	--
HADI:73-A6	--	26.1	30.8	--	--	--	--	--
HADI:73-A504	35.5	23.2	32.4	--	--	--	--	--
HADI:78-S156	39.4	23.8	33.6	74.6	28.0	--	--	--
HADI:74-S67	40.5	24.8	32.9	72.0	30.0	--	--	--

1 - Bp (proximal breadth)
2 - SD (smallest breadth of diaphysis)
3 - Bd (distal breadth)
4 - GL (greatest length)
5 - Dp (proximal depth)

Phalanx 2 anterior or posterior

	Measurements							
Site: Unit etc.	1	2	3	4	5	6	7	8
HADI:73-A707	40.1	38.8	39.2	37.2	25.0	--	--	--
HADI:78-S118	39.2	35.2	35.1	40.5	25.7	--	--	--
HADI:74-O339	36.4	31.7	34.4	38.0	24.1	--	--	--
HADI:74-S67	37.8	30.7	34.3	37.8	25.2	--	--	--
HADI:73-B	37.3	34.4	35.2	38.1	24.7	--	--	--
HADI:77-S219	29.7	25.0	26.8	30.9	19.7	--	--	--
HADI:74-S81	35.2	30.1	--	36.0	24.8	--	--	--

1 - Bp (proximal breadth)
2 - SD (smallest breadth of diaphysis)
3 - Bd (distal breadth)
4 - GL (greatest length)
5 - Dp (proximal depth)

Equus sp.

Phalanx 3 anterior or posterior

Site: Unit etc.	Measurements 1	2	3	4	5	6	7	8
HADI:77-S321	--	40.3	31.0	44.0	35.8	--	--	--
HADI:78-S98	47.9	38.7	27.9	52.4	35.6	--	--	--
HADI:74-S67	38.3	37.3	29.1	45.1	30.5	--	--	--
HADI:77-S343	--	35.9	28.9	37.4	34.0	--	--	--
HADI:78-S202	47.7	--	34.5	53.7	46.7	--	--	--
HADI:73-A504	--	41.7	--	--	--	--	--	--

1 - GL (greatest length)
2 - Ld (dorsal length)
3 - HP (heigth at proc. extensorius)
4 - GB (greatest breadth)
5 - BF (breadth of facies art.)

Key to site abbreviations:

DEIR : Deir' Alla
HADI : Tell Hadidi
HAYA : Hayaz Hüyük
QITA : El Qitar
SELE : Selenkahiya
SWEY : Sweyhat
TAAS : Ta'as

ALLAN S. GILBERT

Equid Remains from Godin Tepe, Western Iran: an interim summary and interpretation, with notes on the introduction of the horse into Southwest Asia

Contents

1. Introduction

Ongoing analysis of the faunal remains from Godin Tepe has, to date, yielded a small but informative collection of equid bones. A description of these bones follows, together with preliminary interpretive remarks that will place the remains into cultural perspective as it is presently understood.

Godin Tepe is located at an elevation of c. 1600 m in the mountainous Kangavar valley of western Iran, just west of the highest Zagros thrusts rimming the Persian plateau (Figure 1). The site encompasses some 32 m of occupational sediments representing at least ten cultural phases beginning c. 4500 BC (YOUNG 1969, YOUNG/LEVINE 1974). Almost all of the faunal material recovered from four seasons of excavations (carried out between 1967 and 1973 by the Royal Ontario Museum under the direction of T. CUYLER YOUNG, JR.) comes from a 700 sq m deep sounding sunk into the mound summit which extensively sampled remains of Godin periods III, IV, and V. To date, fauna from the eastern part of the sounding has been analyzed (GILBERT 1979). This analyzed fraction includes 24,416 bone fragments, 15,282 (69 %) of which were identified to taxon, including almost entirely sheep, goats, and cattle with minor suid and equid components. Remains of wild fauna were found to be rare, the most commonly represented being red deer, but with only 121 bones. The number of equid bones plus loose teeth is 337, this amounting to only 2 percent of the identified sample.

Godin V is the earliest occupation represented by analyzed fauna. The sounding uncovered a single architectural complex, remodeled several times, that was enclosed by an oval wall the original dimensions of which approached 33 by 21 m. The excavators date this Godin V complex between 3200 and 3000 BC and interpret it as an outpost maintained by a foreign pre-

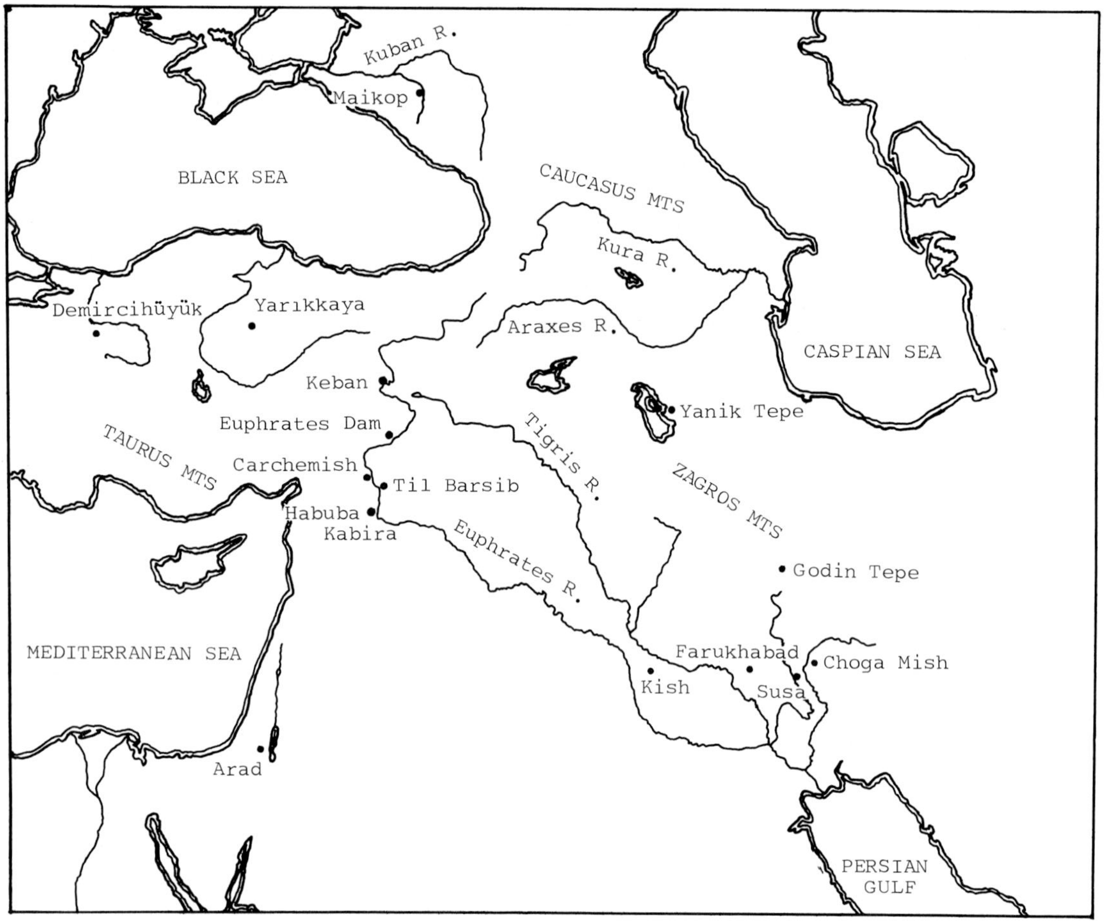

Fig. 1: Map of the Near East showing the location of sites mentioned in the text.

sence from lowland Mesopotamia (Young 1985, Badler in preparation). As determined by other test pits, the outpost was placed at the mound summit amidst houses of a village continuously occupied into Godin V times by the previous local culture of Godin VI (Weiss/Young 1975). The outpost appears to have formed one link in a long-distance trade network for the procurement and transport of goods and raw materials into Mesopotamia from the Persian plateau and beyond. As the compound was kept quite clear of debris during its use, few bones were recovered. The presence of a probable cooking hearth within one room suggests that domestic as well as administrative and commercial activities took place there, and thus it is assumed that rubbish generated by food preparation and consumption was discarded outside the excavated area.

Godin IV marked an intrusion by peoples from the north who disrupted plateau-lowland connections through the central Zagros mountains at about 3000 BC. Called the "Yanik culture" by Dyson (1973: 697–699) after its initial unearthing by Burney at Yanik Tepe in Iranian Azerbaijan and the "Early Trans-Caucasian culture" by Burney because it origins are traceable to Transcaucasia (Burney/Lang 1971: 52–53), the new Godin occupants covered the summit of the mound with structures of uncertain function, although there are suggestions of industrial activities in the eastern part of the deep sounding and of public or ceremonial activities in the west (YOUNG/WEISS 1974: 208–209). The present faunal sample was removed from the eastern, industrial-looking sector.

A hiatus of possibly lengthy duration separated the abandonment of Godin IV from the start of Godin III, which is an impressively long-lived series of complex building levels spanning the interval between 2600 and 1500 BC. Recent stratigraphic analysis (HENRICKSON 1984) has subdivided the period into six phases and documented the level-to-level changes in the ceramic repertory (also, HENRICKSON 1985, 1986). Expansion of the site at this time to its greatest size (15 hectares as presently preserved after heavy erosion of the north side) signals a period of relative stability and prosperity lasting through 7 m of cultural deposits.

2. The Equid Remains

2.1. Recovery Distributions and Taphonomy

Distributions of recovered fauna cannot be properly analyzed and interpreted without considering the generative agencies potentially responsible for producing them. Prior to recent interest in taphonomy, faunal quantification was usually a straightforward counting procedure designed to elucidate slaughter patterns under the assumption that the recovered sample distributions were adequately representative of death assemblage populations. It is now widely acknowledged that this is often not the case (e.g., GILBERT/SINGER 1982) and that burial processes frequently modify archaeological materials, creating a configuration that reflects its origins only when such transformations are recognized and their modifications corrected.

Succeeding paragraphs attempt to articulate the jumble of information relevant to equid remains and, in so doing, clarify the circumstances of their deposition.

2.2. Bone Distributions

The breakdown of equid remains by period is given in Table 1. Only 6 bones (2 percent of the collection as a whole) are of unknown date with an equal number coming from sealed Godin V deposits. Numbers of pieces securely dated to Godin IV and all phases of Godin III are 30 (9 %) and 142 (42 %) respectively. Although mixing of materials between Godin III and Godin IV was minimal, substantial contamination occurred between Godin IV and Godin V with 146 equid specimens (43.3 %) falling into this uncertain category. The temporal distribution of equid remains among the three periods depends, therefore, upon how one deals with the large mixed component.

Excavated lots were assigned a "mixed" status based upon association with mixed ceramics. Lots designated "IV/V" did contain both Godin IV and V pottery in varying proportions, but there is reason to suspect that most of the accompanying bone is of Godin IV date. First, the rarity of faunal remains recovered from the oval compound of Godin V prompts the notion that any internal garbage accumulations were very likely removed to other areas of the site.

TABLE 1

Equid remains from Godin Tepe (as of 1985): distribution by Period.

Period:	III:						III/IV	IV	IV/V	V	?	Total
Subphase:	1	2	3	4	5	6						
Mandible		2		4				3	9			18
Teeth, Maxilla		2		4		1		1	7	1	1	17
Teeth, Mandible		6	2	1		1		1	7			18
Teeth, Incisors		1	5			1			3			10
Atlas									1			1
Axis									1			1
Cervical Vert.			2			1			2			5
Sacrum								1				1
Scapula				1		2		2	7			12
C.Humerus				1								1
D.Humerus		1		1	1	1		2	4	1	1	12
P.Radius		3	1	2		1	1		6			14
D.Radius				1		1		3	7			12
P.Ulna		1		1			1		2			5
Carpals				4		6		1	12			23
P.Metacarpal II		1		1		1			1			4
C.Metacarpal III									1			1
P.Metacarpal III			1	2		2			4			9
D.Metacarpal III									1			1
Innominate		1		2			1	3	10			17
P.Femur				1					3			4
D.Femur						1			3			4
Femur Shaft		1		1		3		2	2			9
Patella						1		1	1			3
C.Tibia											1	1
P.Tibia		1		2		1			1			5
D.Tibia		2		1		1		2	6			12
Astragalus				1		1		1	7	1	1	12
Calcaneum		1	2	1		1			7		1	13
Tarsals			1	5		1		1	16			24
P.Metatarsal II							1					1
C.Metatarsal III				1					1			2
P.Metatarsal III							1		3			4
D.Metatarsal III		2				1						3
P.Metatarsal IV							1	1				2
D.Metapodial III		1		1		1						3
Metapodial Shaft									1			1
Phalanx 1		3	2	4		2	1	4	4	1		21
Phalanx 2		1	1	6		1			4	2		15
Phalanx 3		2	1	4		1		1	2		1	12
P.Sesamoid				2								2
D.Sesamoid				1		1						2
TOTAL	0	32	18	56	1	35	7	30	146	6	6	337

C. = Complete P. = Proximal D. = Distal

TABLE 2

Equid remains from Godin Tepe (as of 1985):
distribution of bones by findspot context and period*

Deposit Code**	III: 1	III: 2	III: 3	III: 4	III: 5	III: 6	III/IV	IV	IV/V	V	Total	%
12 undisturbed surface, open area										2	2	.4
16 collapsed 2nd storey floor deposit		2									2	.4
21 trash accumulation on occupation floor		2				5			5	3	15	3.0
22 trash in pit or well, clear boundaries			1								1	.2
23 amorphous trash, no clear boundaries				1		7		13	106	2	129	25.8
26 post-occupation trash deposit				16				1			17	3.4
28 kiln, hearth, oven, other container								13	44		57	11.4
34 amorphous bricky fill, no associated walls			5	15		5		7	7		39	7.8
35 bricky fill below associated walls		8		2		4		1	13	4	32	6.4
36 feature removal: mat'l of feature		5		1		3	1	10	42		62	12.4
38 baulk removal						1			6		7	1.4
39 dump				1							1	.2
40 unknown		1			1	1		3			6	1.2
41 clean-up									2		2	.4
42 non-bricky fill with associated walls									1		1	.2
47 mixed fill/brick, no associated walls		3	3	35		17	6	2	8		74	14.8
48 mixed fill/brick with associated walls		15		4		3					22	4.4
52 pot, drain, etc.: contents of								1	13		14	2.8
53 pebble layer			10	7							17	3.4
TOTALS	0	36	19	82	1	46	7	51	247	11	500	100.0

* Totals reflect numbers greater than bone counts because findspots for each bone may be characterized by more than one context code.

** Characterizations according to the Deposit Code devised by the Ohio State University/University Museum Expedition to Tal-e Malyan. See SUMNER (1978).

Such actions would make sense for a limited, enclosed space that had to be kept open for other business. In contrast to expectations for fauna, the abrupt abandonment of the Godin V compound, presumably in the face of advancing Yanik culture groups, appears to have left much Godin V ceramic *in situ.* Smashed and flattened under the collapse of compound structures or submerged beneath newly formed overburden, this pottery remained shallowly buried during the initial Godin IV occupation. Thus, the digging of pits and modifying of pre-existing walls during Godin IV times could easily have brought up earlier ceramic and incorporated it within the extensive middens that excavators found radiating outward from the mound's summit. In this way, heavy contamination of essentially Godin IV lots in the deep sounding by earlier faunal remains seems less likely than the obvious mixing of pottery would imply.

Second, the depositional contexts of Godin IV lots and mixed IV/V lots are quite similar. In Table 2, the findspots of equid bones, separated into functional categories, are displayed by period.[1] Because deposits could occasionally be described by more than one context code, the number of findspots (500) exceeds the number of specimens recovered (337), an imbalance that results from allotting each context code equal weight. A comparison of the tabulations for Godin IV, IV/V, and V shows that context types containing substantial amounts of Godin IV bone are also well represented in IV/V, especially amorphous trash deposits (#23), kilns, hearths, and ovens, etc. (#28), and feature removals (#36). Contexts containing Godin V equids do not correspond as closely to those of IV/V, and they are too few to sustain confidence in the comparison.

In sum, although some Godin V bone may be present within the mixed IV/V material, it seems more probable that the bulk of that latter fauna is of Godin IV origin. If this is so, then the distribution of equid remains by period yields relatively high figures for Godin IV.

There are at least four possible explanations for an increase in equid remains during the Yanik (Godin IV) occupation. The first is simply the chance that sample error has overemphasized them. As more of the fauna from the site is analyzed, it will become clear whether the pattern of recovery deviates from that described above or remains consistent. Little else can now be said about this possibility.

A second possible explanation anticipates an increase in equid finds as a result of a generally higher discard rate for bones of all species during the Yanik phase.

Third, one might account for more equid bones with the assumption that preferential retention for specific secondary uses increased their chances of within-site burial and thus skewed recovery distributions away from proportions more closely reflecting original slaughter ratios.

Last, it is possible that the increase in equids during Godin IV does preserve a record of greater equid utilization, which would have involved a higher kill-off and probably, but not necessarily, more frequent consumption of their flesh.

To evaluate the last three possibilities, Tables 3 and 4 were constructed. They enable comparison of raw counts, numbers of specimens showing polish (evidence of secondary use), and recovery rates per unit volume of earth excavated. Data are provided for equids and for all species combined. In Table 3A, three ways of allocating finds from mixed IV/V lots to IV and V are examined: 90%/10%, 80%/20%, and 70%/30%. There is no way to know what the true

[1] Deposit codes devised by the Ohio State University/University Museum project at Tal-e Malyan to characterize findspots were utilized by the Godin researchers (SUMNER 1978), and so they appear here with acknowledgement and thanks to William SUMNER.

TABLE 3

Equid remains from Godin Tepe (as of 1985): statistics per period and per unit volume of site earth for equids and all fauna, with estimated allocations of Godin IV/V lots to Godin IV and V.

A: *With estimated allocations to Godin IV and V*

Period	Approximate Volume of Site Earth	Species	Godin IV/V: 90%/10% n	%	bones/m³	Godin IV/V: 80%/20% n	%	bones/m³	Godin IV/V: 70%/30% n	%	bones/m³
III	2200 m³	Equids	142	42.1	.065	142	42.1	.065	142	42.1	.065
	(7.25 m deep)	All Fauna	10,099	66.1	4.590	10,099	66.1	4.590	10,099	66.1	4.590
IV	390 m³	Equids	161	47.8	.413	147	43.6	.377	132	39.2	.338
	(1.50 m deep)	All Fauna	3,580	23.4	9.179	3,411	22.3	8.746	3,241	21.2	8.310
V	710 m³	Equids	21	6.2	.030	35	10.4	.049	50	14.8	.070
	(3.00 m deep)	All Fauna	379	2.5	.534	548	3.6	.772	718	4.7	1.011
TOTAL	3300 m³	Equids	337	100.0	.102						
	(11.75 m deep)	All Fauna	15,282	100.0	4.631						

B: *Frequencies as excavated*

Period			Equids n	%	bones/m³	All Fauna n	%	bones/m³			
III			142	42.1	.065	10,099	66.1	4.590			
III/IV			7	2.1	.003	71	0.5	.027			
IV			30	8.9	.077	2,053	13.4	5.264			
IV/V			146	43.3	.133	1,697	11.1	1.543			
V			6	1.8	.008	209	1.4	.294			
? or UNASSIGNED			6	1.8	.002	1,153	7.5	349			
TOTAL			337	100.0	.100	15,282	100.0	4.631			

division was, nor whether such percentages were uniform across all bone types or species, but it is presumed for this analysis that between 70 and 90 percent of the mixed fauna is made up of Godin IV material. In Table 3B, the same information is displayed for the original, unmodified stratigraphic designations. Because recently updated stratigraphic assignments generously provided by R.C. Henrickson and V.R. Badler have lent greater precision to the dating of Godin III and V lots, it is likely that fauna from as yet unassigned loci (Table 3B) will further increase the counts for Godin IV (Henrickson 1984, Badler in preparation).

The columns headed "n" in Table 3A show plainly that more bones were deposited during Godin III than during Godin IV, so that the substantial number of equid finds from the Yanik culture phase cannot be explained as the result of any absolute increase in faunal deposition.

TABLE 4

Equid remains from Godin Tepe (as of 1985):
distribution of polished specimens by species and period.

	Sheep/Goats			Cattle			Pigs			Equids		
	no polish	polish	%*	no polish	polish	%	no polish	polish	%	no polish	polish	%
Godin III	3,691	1,733	32.0	868	352	28.9	79	21	21.0	100	12	10.7
Godin IV only	1,110	380	25.5	107	21	16.4	43	9	17.3	24	5	17.2
Godin IV plus IV/V	2,018	657	24.6	190	30	13.6	69	19	21.6	151	9	5.6
Godin V	122	14	10.3	23	0	0.0	1	1	50.0	0	5	0.0
Total Sample	6,337	2,530	28.5	1,196	402	25.2	169	46	27.2	276	23	7.7

* percent polished

Over 66 percent of all fauna recovered comes from Godin III against only 21–24 percent estimated for Godin IV, but the percentages of equid bones assignable to each period are comparable: 42 percent for Godin III and 39–48 percent for Godin IV estimates. Godin III deposits accumulated over the course of a millennium, doubtless with interruptions due to the on-and-off cycles of precinct decay and renovation. The duration of the Godin IV occupation is not known with certainty, but it is fixed between termini at c. 3000 and 2600 BC and cannot have been as long as Godin III. Roughly judging on a per-year basis, the stream of Godin IV equid bones entering the site deposits must have been elevated, as finds equivalent in number to Godin III accumulated within the eastern deep sounding over a period lasting less than half as long. Although there is some risk in extrapolating these deep sounding counts site-wide, the present sample does contradict explanation 2 posited above and suggests that increased deposition of equids during Godin IV was likely not a result of greater faunal deposition in general.

Without allocating fauna of mixed loci to Godin IV and V, it is still evident from Table 3B that Godin III contains most of the bones but not necessarily most of the equids. Again, aside from the *caveat* of presuming present distributions to be representative of the entire site, these data suggest the same preliminary conclusions.

The patterns and frequency of polish on recovered bones were recorded in order to gain some control over secondary utilization. Bone use, it was thought, would leave some physical evidence (mostly polish) provided that the duration of use lasted long enough for the formation of visibly detectable traces. Polish occurred in many gradations, from faint and barely recognizable to quite shiny. In most cases, it was clear that the gloss developed as a consequence of wear against the hand and that its intensity was probably a factor of handling time (Gilbert/Steinfeld 1977, Gilbert 1979: 179ff.). Some skeletal parts included many glossy specimens that were easily identifiable as implements. Others, however, have no apparent technological application but may tentatively be referred to as tokens, gaming pieces, souvenirs, amulets, or toys. The percentages offered in Table 4 demonstrate that equid remains were less polished in general than the rest of the collection. Aside from the anomalously high polished bone ratio recorded for equids in unmixed Godin IV lots, the more common sheep, goat, cattle, and pig bones display substantially higher polish percentages. This makes it fairly certain, barring sample error, that they tended to be retained more often for secondary uses than those of equids. Based upon the analysis of polished bone percentages in the full Godin assemblage, purposeful retention or retrieval for secondary uses was judged to be a major determinant of within-site bone deposition (Gilbert 1979: 232 ff.). Although the distribution of polish on equid bones conforms to the bovine pattern, the number of polished specimens remains low. This finding suggests that equid bone could substitute for the more generally available bovine bone in whatever tasks such material served, but also that technological variables seem to be less important factors in their entry into the Bronze Age town. The evidence better suggests that simple discard of carcass parts and meal remnants, or leftovers from canine scavenging, accounts in large part for their deposition. Explanation 3 (perferential retention of equid bones) thus cannot be sustained.

A note of caution must be sounded regarding the potential effects on equid bone counts from Godin IV of other secondary uses for bone. As the analyzed portion of the Godin IV fauna derives from the eastern, industrial-looking, sector of the deep sounding, it must be admitted that as yet unidentified activities involving technological use of equid carcass products might have artifically boosted recovery of their bones. Further study of the Godin IV archaeological remains, however, must precede any commitment to a linkage between elevated number of equid finds and an hypothesized manufacturing orientation.

For the moment, explanation 4 is left as the most promising. In the absence of evidence suggesting otherwise, it assumes that more frequent slaughter of equids led to the increased deposition of their remains during Godin IV. The preliminary conclusion then is that the distribution reflects greater kill-off, and possibly also consumption, of equids.[2]

[2] Two additional comments about Table 3 are instructive. First, bone concentrations per unit volume of site earth for each period seem to be related to the kinds of deposit from which the bones principally derive. (Volumes are approximate and were estimated from sections and plans from the eastern part of the deep sounding only, i.e., that part of the excavations from which the present sample was retrieved.) The concentrations for all fauna in Table 3A show very high values for Godin IV (8.31 to 9.18 bones per cubic meter) compared to Godin III (4.59 bones per cubic meter). The concentration of equid bone is also high in Godin IV, with figures calculated for this Yanik occupation (between 0.34 and 0.42 bones per cubic meter) being at least five times greater than those determined for Godin III (0.07 bones per cubic meter) or Godin V (between 0.03 and 0.07 bones per cubic meter). Such high densities suggest more frequent discard of bone into the steadily aggrading site surface of Godin IV. Deposit codes from Table 2 reveal that much of the bone-bearing sediment of Godin IV represented midden accumulation, a likely place to find domestic or industrial refuse and fairly high ratios of bone to earth. Conversely, most of the fauna from Godin III derives from deposits characterized as bricky fill with or without associated walls. The presence of architectural structures and materials has seemingly contributed to the "thinning out" of embedded artifactual debris, thus reducing, comparatively, the concentration of bone in these layers.

Second, it is curious that in both Tables 3A and 3B Godin V, like the combined Godin IV and IV/V samples, contained a higher percentage of the total equid sample than it contained of the total identified collection. Admittedly a poor indicator of the potential importance of equids during a period that has left precious few zooarchaeological data, it is nevertheless pertinent to consider the possibility that equids served a useful function connected with the Uruk outpost. Because percentages must

Summing up, the distribution of equid remains suggests heavy exploitation during Godin IV, and the nature of this exploitation may have been both domestic and industrial. Equid remains are proportionately rarer in Godin III deposits. Godin V data hint at the importance of equids but there are not enough finds from this period to supply more than a suggestion. As so few bones show traces of secondary use, equid remains appear to have been principally meal remains or pieces of butchering units discarded directly upon consumption.

2.3. Ageing

Age-related information from the equid sample was obtainable using epiphyseal union data. Frequencies of fused/unfused skeletal parts and survivorship estimates based upon them are displayed in Table 5. In Table 5A, the bones are separated into three groups (A, B, and C) according to closeness of age at fusion. The fusion intervals, representing those of modern domestic equids, are taken from SILVER (1969). The ancient ossification schedules were probably not identical to those offered here for comparison, and thus only an approximate idea of ages at death of the individuals represented by the recovered bones is intended.

Although Godin V has yielded insufficient data for comparison, the pattern plotted for all phases of Godin III (n = 73, with 14 indeterminate) compares favorably with that resulting from the lumping of Godin IV with mixed IV/V bones (n = 93 with 11 indeterminate). In Table 5B, percentages show almost all individuals in survived stages A and B (their first two years?) with a noticeable dropoff in survivorship in C (3–3.5 years?).

There is, however, something of a difference in stage C survivorship between the two temporal units, with the percentage being higher for Godin IV and IV/V than for Godin III. If this divergence was brought about by post-mortem biases, then the pattern may have resulted from differential attrition of the stage C diagnostics. As such diagnostic elements are generally the cancellous epiphyses of late-fusing skeletal parts, their survival is subject to greater variability because of lower resistance, and thus higher susceptibility to damage and destruction, than other, more compact, skeletal parts. Elimination of immature or partially ossified elements would create the impression of greater survivorship, and an older-looking assemblage such as that reflected by the Godin IV sample might result.

For the moment, it is not possible to know how much of the pattern of epiphyseal union might be due to bias. With the data at hand, there is no way to evaluate the possible destruction of bones from older individuals for grease or collagen glue extraction, although such practices may have been part of domestic or industrial activities conducted during Godin IV. Furthermore, the effects of canine scavenging, which often acts selectively on bone recovery distributions by completely destroying elements if they are small or spongy, are difficult to assess for the larger ungulates. For example, a low frequency of gnaw marks on the softer cancellous epiphyses of Godin caprines was considered an indication that such elements, always rarer than the more compact bone ends, survived only if the town dogs overlooked them (Gilbert 1979: 264). Gnaw marks on the bones of larger ungulates were more evenly distributed across spongy and compact elements, however, suggesting that bovine and equid bones were large

necessarily add up to unity, Godin III not surprisingly contained less of the total equid sample than it did of the total collection. Little or no significance may attach to these statistics, but it is of interest that the Godin V pattern should resemble in this respect that of Godin IV.

TABLE 5

Equid Remains from Godin Tepe (as of 1985):
frequencies of fused/unfused epiphyseal elements and survivorship estimates

A: Frequencies

	Age at Fusion (SILVER 1969)	n fused/unfused/?		
		III	IV & IV/V	V
Fusion Group A				
p. metacarpal III	foetal	5/0	5/0	0/0
p. metatarsal III	foetal	1/0	4/0	0/0
d. phalanx 1	foetal	10/0/ 1	8/0	1/0
d. phalanx 2	foetal	9/0	4/0	2/0
		25/0/ 1	21/0	3/0
Fusion Group B				
p. phalanx 2	9–12 mo.	9/0	4/0	2/0
p. phalanx 1	13–15 mo.	8/0/ 3	8/0	0/0/1
d. humerus	15–18 mo.	2/1/ 1	4/0/2	1/0
p. radius	15–18 mo.	6/0/ 1	6/0	0/0
d. metacarpal III	15–18 mo.	0/0	2/0	0/0
d. metatarsal III	16–20 mo.	4/0	1/0	0/0
d. scapula	20 mo.	0/0/ 3	9/0	0/0
d. tibia	20–24 mo.	2/2	7/1	0/0
acetabulum	18–24 mo.	0/0/ 3	7/0/6	0/0
		31/3/11	48/1/8	3/0/1
Fusion Group C				
calcaneum	36 mo.	3/1/1	4/2/1	0/0
p. humerus	36–42 mo.	0/1	0/0	0/0
p. femur	36–42 mo.	0/1	0/3	0/0
d. femur	36–42 mo.	1/0	3/0	0/0
p. tibia	36–42 mo.	3/1	1/0	0/0
p. ulna	42 mo.	0/1/1	0/0/2	0/0
d. radius	42 mo.	1/1	9/1	0/0
		8/6/2	17/6/3	0/0

B: Survivorship percentages calculated from Table 5 A

	Fusion Group		
	A Foetal	B (1–2 Years?)	C (3–3.5 Years?)
Godin III	100%	91%	57%
Godin IV & IV/V	100%	98%	74%

TABLE 6

Equid remains from Godin Tepe (as of 1985):
distribution of gnawed specimens by species and period.

	Sheep/Goats			Cattle			Pigs			Equids		
	intact	gnawed	%*	intact	gnawed	%	intact	gnawed	%	intact	gnawed	%
Godin III	4,270	1,147	21.2	876	344	28.2	72	28	28.0	76	36	32.1
Godin IV	1,064	422	28.4	102	26	20.3	36	16	30.8	19	10	34.5
Godin IV plus IV/V	1,911	755	28.3	171	48	21.9	65	23	26.1	114	46	28.8
Godin V	116	19	14.1	23	0	0.0	2	0	0.0	5	0	0.0
Total Sample	6,799	2,046	23.1	1,177	419	26.3	158	56	26.2	207	92	30.8

* percent gnawed

enough to resist such attacks regardless of bone structure (Gilbert 1979: 264–266). Table 6 demonstrates that equid remains revealed slightly more gnawed specimens than the other major taxa, which could mean not only similar survival rates for compact and spongy elements but also the possibility that dogs may have brought much of the equid bone into the site in the first place.

Taking the epiphyseal union data at face value would leave Godin III and IV with similar slaughter profiles differing mostly in the somewhat younger-looking profile for Godin III. Thus, as noted, the kill-off of equids appears to have occurred at rather advanced ages (even if the actual ages are not precisely knowable). During Godin III, 57 percent of the specimens diagnostic of age group C were c. 3.5 years or older at the time of death. The individuals they represent might have been appreciably older than 3.5 years, and thus it is perhaps better to imagine only 43 percent dying prior to the fourth year of life. For Godin IV and IV/V, the percentage is an even lower 26 percent dying young. If the animals in question possessed longer fusion intervals than today's domesticates, then such delayed ossification would push ages at death yet higher. As the ageable equid elements were few in number, one cannot eliminate random variability as a potential causal agent. In spite of this ever-present possibility, however, one suspects that the high survivorship registered for both Godin III and Godin IV and IV/V in fusion group B is likely to be a real effect of preference for animals greater than about two years of age. The divergence in fusion group C may stem from myriad factors, including a poor or skewed sample. These results are not inconsistent with an interpretation in which mostly mature animals, used principally for service functions, finally complete their active life as food.

2.4. Species

Osteological separation of the equids poses serious problems stemming from the uncertainties of overlapping and variable skeletal criteria. The Godin collection is noteworthy in this regard because it shows some blending of the diagnostic traits generally used to identify species.

The recovered bones and teeth can almost all be identified as *Equus hemionus* (probably *E.h.onager*) and *E.caballus.* The hemione remains make up about 94 percent of the equid sample. Their osteometric size range and cheektooth cusp patterns conform quite closely to expectations for hemiones, and their central Zagros provenience falls well within the traditional zoogeographic distribution of recent onagers. Several specimens, however, exhibit a size range or dental characteristics that support a caballine identification. Twenty-one limb elements, all dated to Godin III, seem rather clearly to be caballine, and two specimens from Godin IV/V can be identified as caballine more tentatively. Four specimens may be from *E.asinus,* but with so few examples, it is not possible to be more positive about their identification.

Generally speaking, it is unclear whether true breeding genotypes exclusively are reflected in this breakdown of identified taxa or whether at least some of the specimens come from hybrid animals.

Maxillary teeth all appear to have come from hemiones. In occlusal or buccal view they reveal buccal surfaces with concave intercolumnar channels curving with a smooth transition into the columns (STEHLIN/GRAZIOSI 1935: 6, BÖKÖNYI 1972: 14). The protocones are variable in length, but they are usually less than half of the oral-aboral length of the P^3, P^4, M^1, or M^2 and are never truly long (STEHLIN/GRAZIOSI 1935: 6–7, BÖKÖNYI 1972: 15). In one case, a P^3 and P^4 from the same left jaw (datable to Godin IV/V) gave ratios of 14.1/26.0 and 12.2/26.3, respectively, for the protocone length/tooth length (in mm). In all cases, the *pli caballin* is absent or

TABLE 7

Equid remains from Godin Tepe (as of 1985):
distribution of hemione and caballine diagnostics on mandibular cheekteeth

			SINUS LINGUALIS				
			sharp V-shape	rounded V-shape	U-shape	?	TOTAL
SULCUS EXTERNUS	Godin III	shallow		1			1
		deep	2	3	○	2	7
		?				2	2
	Godin IV, V & IV/V	shallow		1		1	2
		deep	1	1	(1)	1	4
		?				3	3
		TOTAL	3	6	1	9	19

Rectangles: expected pattern for hemiones
Circles: expected pattern for caballines

weakly developed, and the one measurable M^3 (from Godin IV/V) yielded dimensions comparable to equid teeth recovered from the Zagros site of Palegawra Cave that were identified as *E.hemionus* (Turnbull/Reed 1974: 113) and that were hunted during the Epipaleolithic Zarzian occupation of that site (c. 12,000 BC: Solecki/Solecki 1983)

Mandibular cheekteeth reveal a less clear pattern. In all undamaged examples except one, an angular or slightly rounded V-shaped valley (*sinus lingualis*) exists between the metaconid and metastylid. This hemione trait, however, appears frequently together with the caballine trait of the metalophid fold (*sulcus externus*) between the protoconid and hypoconid penetrating deeply, virtually touching the enamel boundaries of the metaflexid and entoflexid valleys and even abutting the *sinus lingualis* (Bökönyi 1972: 15). In only three out of the thirteen teeth in which the feature is clearly preserved does the metalophid fall short. Table 7 shows the distribution of mandibular teeth with respect to *sinus lingualis* shape and *sulcus externus* depth. Only two teeth fall within the rectangles enclosing the typical hemione combination of traits, and only one tooth falls within the circles circumscribing the anticipated caballine diagnostics. Excluding the broken or otherwise uncertain specimens, there are two teeth from Godin IV/V and five from Godin III that display a "blending" of hemione and caballine characteristics, namely a V-shaped *sinus lingualis* appearing with a deeply penetrating *sulcus externus.*

A cursory survey of intact hemione mandibles from osteological collections of the American Museum of Natural History in New York demonstrated that a certain variability in *sulcus externus* penetration is to be expected, especially in the cases of the P_4 and M_1, which tended to have somewhat greater fold depth than other cheekteeth. In no case, however, did penetration extend clear across the tooth to the opposite enamel border as is observed on some Godin specimens. Davis (1980) has found similar cusp patterns in the lower cheekteeth of equids from Israel and identified them as belonging to the extinct *Equus hydruntinus.* This species, found mostly in Middle Paleolithic contexts in the north of Israel, is described by Davis as "zebroid" because such patterns are characteristic of the zebras as well.

Concerning other features of the Godin mandibular cheekteeth, the hypoconulid is variable in size. The metaconid-metastylid borders show the flattened sections that are reportedly characteristic of both hemiones and caballines (Zarins 1976: 58–59), but the double knot itself tends to be short and rather symmetrical, more in the manner of *Equus asinus* or *E.hemionus* (Ducos 1968: 27, Bagtache/Hadjouis/Eisenmann 1984: 609–610).

Three of the mandibular cheekteeth with "blended" characteristics are illustrated in Figure 2 (a, b, c). Each was recovered from deposits of Godin III: 2 (c. 1700–1600 BC), and thus they are quite late in the Bronze Age sequence at Godin. Each exhibits a long *sulcus externus* (usually accompanied by a steep-sided groove in the cementum material within the fold, deepening outward) that bisects the fossettes and makes contact at one or more points on the lingual side. The tooth of Figure 2b possesses an open V-shaped *sinus lingualis* while its *sulcus externus* pushes between the arms of the entoflexid and metaflexid folds and terminates tangential to the V near its vertex. In Figure 2e and 2f, the normal hemione cusp pattern is represented as it appears on teeth from Godin IV and Godin III: 3 (c. 1900–1700 BC), respectively.

The problem of limited sample size, which potentially affects most of the faunal statistics for equids, is unrelated to the configuration of features observed on the occlusal tooth surface. As contrasting traits appear on the same specimen, either the diagnostics are inadequate to the task of discriminating equid species because of extreme variability, or the pattern reveals an interweaving of traits that may signal the presence of equid hybrids. Zarins (1976: 58–59) has already drawn attention to equid teeth from Old Elamite levels at Farukhabad, identified as

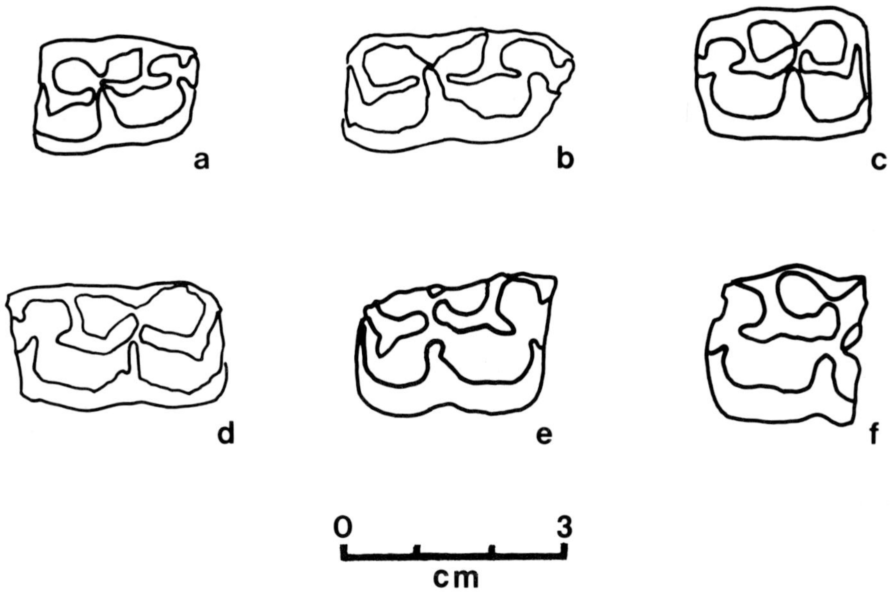

Fig. 2: Equid tooth enamel fold patterns (occlusal view): a–c: Godin III:2 specimens with "blending" of hemione and caballine traits; d: Godin IV/V specimen with caballine characteristics; e–f: Godin IV and III:3 specimens with hemione characteristis; scale: 1:1.

Equus asinus or *E.hemionus* that reveal deeply penetrating sulci. The curious zebrine/*E.hydruntinus* appearance of some teeth further raises the intriguing possibility that the Godin equids exhibit an ancient genetic configuration for deep sulci masked since the Ice Age.

If such mixed character patterns did indeed arise from interbreeding rather than simple variability, then the practice of hybridizing must have begun by at least the early third millennium BC. To be certain, one would need to know what cannot for the moment be known, namely, which phenotypic expressions of dental characteristics would dominate or blend in a cross between the original caballine and hemione genotypes available during the Bronze Age.

Twenty-one bones are identified as *Equus caballus* on the basis of size and robustness. These elements are listed in Table 8 and illustrated in Plate 1. It can immediately be observed that not only were the bones all from deposits of Godin III, phase 4 (c. 2100–1900 BC), but they were recovered from only three excavation lots. When first faced with the material, it seemed possible that all 21 bones were related, perhaps being parts of the same individual and having entered the site sediments in a similar manner if not at the same time. Exploring this idea led to the discovery that most of the horse bones found their way into a pit midden, although uncertainty about the bounds of the locus in question might mean that sediments outside of the hole were included during excavation with those from within the hole. Two bones were preserved not far away as part of the roadside garbage that ultimately became embedded within collapsed and weathered debris from surrounding structures.[3]

[3] Information about the findspots was kindly supplied by Dr. R. C. Henrickson in a letter dated 5 September 1984. Pertinent details include the points paraphrased as follows:

TABLE 8

Equid remains from Godin Tepe (as of 1985):
limb elements of *Equus caballus.*

Elements	Lot	Level		Measurements (mm.)*
right forelimb				
3rd carpal	AA2.193	III:4		GB 44.1
left forelimb				
2nd, 3rd, 4th carpals	AA2.185	III:4	?	-
prox. metacarpal II + III	AA2.185	III:4		Dp 37.5
right hindlimb				
dist. tibia (unfused)	AA2.193	III:4	?	Bd 72.5; Dd 49.5
astragalus	AA2.193	III:4		GH 64.0; LmT 61.4; GB 65.5; BFd 54.3
central, 1st+2nd, 3rd, 4th tarsals	AA2.185	III:4		GB 53.6 (central); GB 52.6 (3rd)
complete metatarsal	AA2.185	III:4	?	Bp 56.5; SD 33.0; Bd 54.5; Dd 42.5; GLl 263.3; GL 271.2
left hindlimb				
central tarsal	AA2.193	III:4		GB 54.0
digits				
phalanx 1	AA2.193	III:4		GL 89.0; L 84.7; Dd 27.0; Dp 37.0
phalanx 1	A2.172	III:4		GL 91.2; L 87.4
phalanx 2	A2.172	III:4		GL 53.0; L 44.2; Bp 51.3; Dp 34.6; SD 43.3; Bd 47.3; Dd 28.2; BFp 45.1
phalanx 3	AA2.193	III:4		GL 58.0; Ld 51.1; HP 37.0
prox. sesamoid (axial)	AA2.185	III:4		-
prox. sesamoid (peripheral)	AA2.185	III:4		-
dist. sesamoid	AA2.193	III:4		GB 44.8

* Measurement definitions as in von den Driesch (1976). Some error in measurements here reported is likely due to cracking and surface warping, but deviation from original values will probably not exceed 1-2 mm and only on elements designated '?'.

1. Lot AA2.185 lay immediately above AA2.193, both lots representing in large measure, but not necessarily exclusively, the fill of a pit (Pit 482) c. 3 m in diameter sunk into the road area 4005 in the extreme southeast corner of the deep sounding. The position of the road and adjoining buildings can be seen on the site plan labeled III: 4A in Figure 3.
2. Pottery from both pit lots indicates a dating within Godin III: 4, and the top of the pit (i.e., the upper boundary of AA2.185) was apparently covered by a pebble layer marking basal Godin III: 3.
3. Pit contents were described in field notes as "alternating layers of pebbles and loose wash" and "some painted sherds" in lot AA2.185 and "very soft ashy fill with ash and trash lenses sloping downwards" and including pottery in "great quantity, mostly fire blackened buff coarse ware" in lot AA2.193.
4. The pit was dug into the road area 4005 probably quite late in Godin III: 4. It penetrated into an earlier structure, Room 4001, that had been part of a previously existing building on the opposite side

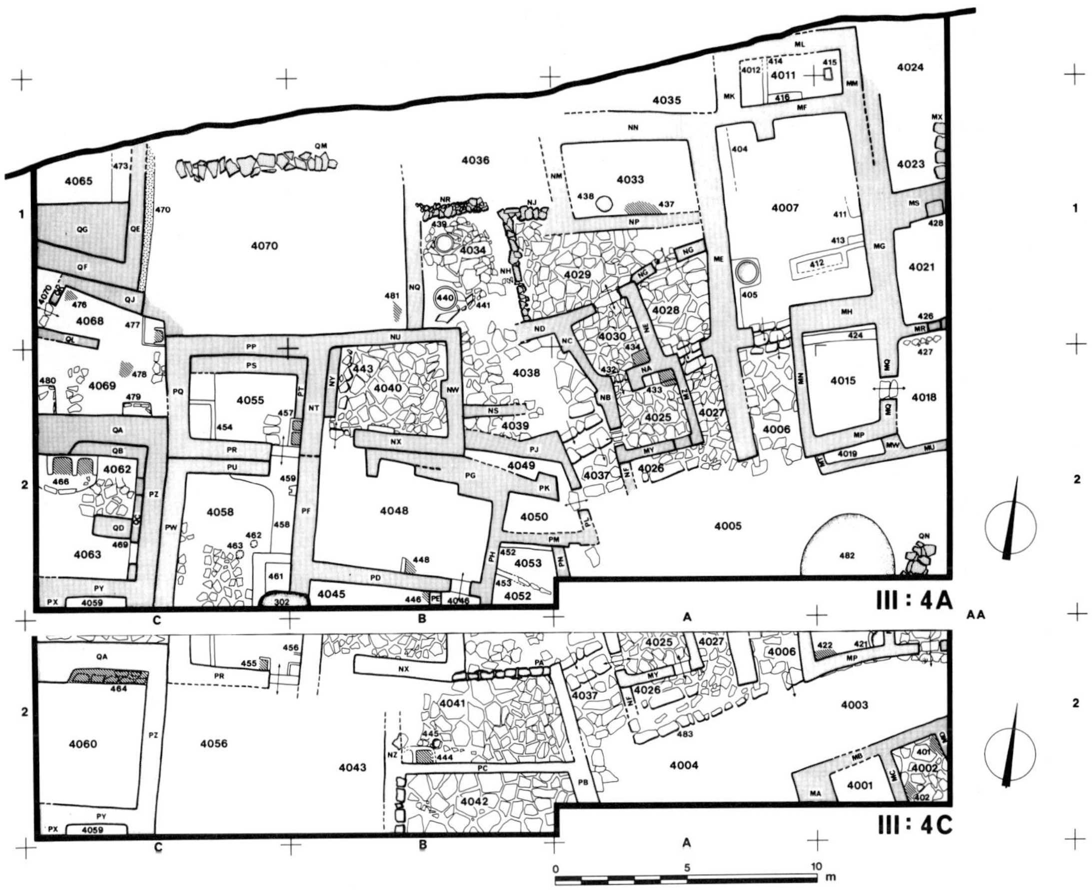

Fig. 3: Site plan of the deep sounding with remains of Godin III:4A and 4C showing location of Pit 482.

Discoloration of the horse bones varies with their findspot, and the post-depositional effects seem to agree with the functional contexts suggested by the field descriptions. Elements from AA2.185 (Plate 1 above centimeter scale) are whitened and crumbly with substantial calcitic encrustration. The metacarpals, carpals, and tarsals are, in fact, each cemented together by the mineral precipitate. The seven bones from AA2.193 (Plate 1 below the centimeter scale), lower

of the road (see plan labeled III: 4C in Figure 3). The pit removed virtually all of the contents of Room 4001 and parts of its walls, and its infilling was ongoing for an indeterminate length of time during the rest of phase 4.

5. Lot A2.172 was a layer of clayey, bricky collapse and pebbles located within road area 4005, several meters north and west of Pit 482. Also dated to Godin III: 4 on the basis of associated pottery, it appears again to be stratigraphically late in the layer.

6. Although similarly positioned stratigraphically, lot A2.172 is not necessarily connected to lots AA2.185 and 193, which certainly are related as upper and lower levels of the nearby rubbish pit.

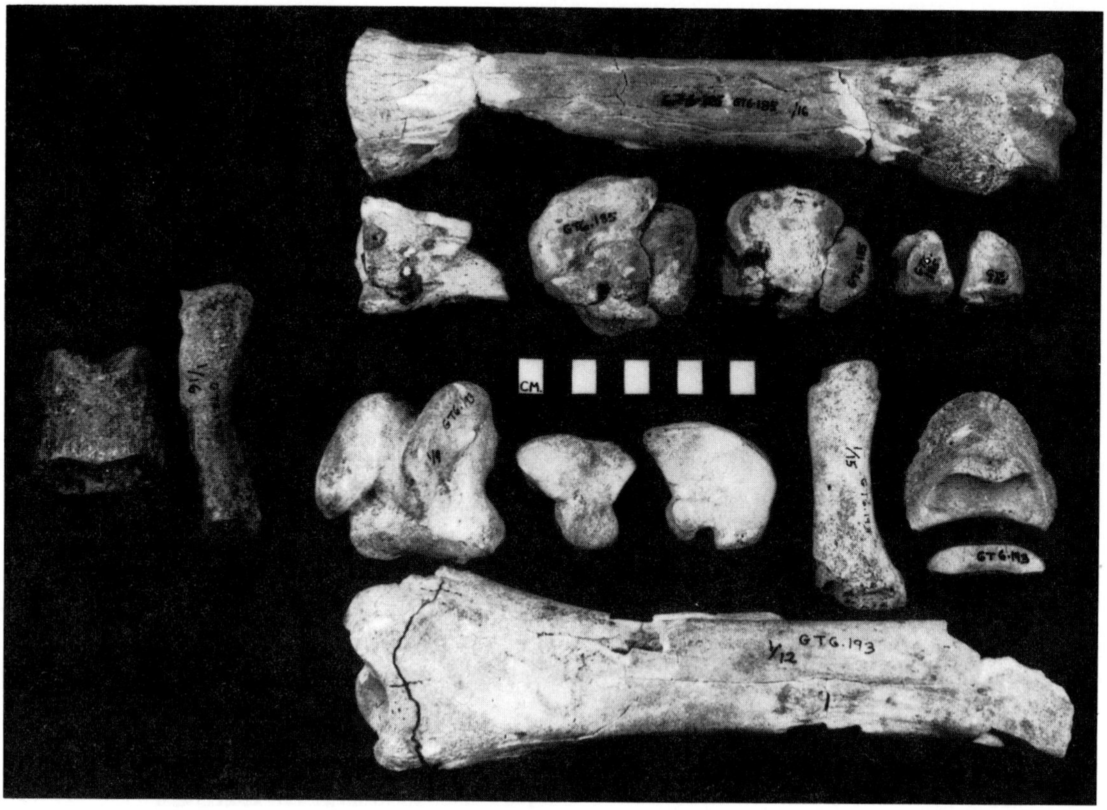

Plate 1: Remains of *Equus caballus* recovered from Pit 482 and road area of Godin III:4.

in the pit fill, are also whitened and crumbly but with noticeably less encrustation.[4] The two phalanges from lot A2.172 are, by contrast, darkened with a fine, earthy particulate staining that is apparently the result of burial within a medium fine enough to penetrate into the osseous fabric. These last bones remain quite dense, and both reveal signs of canine gnawing.

The fragmentary state of most of these caballine specimens makes it difficult to tell whether they conjoin and thus derive from one carcass, but it is very likely that they do. The complete tarsal joint of AA2.185, frozen in almost perfect alignment, seems certain to articulate with the complete metatarsal, but it also fits the astragalus from AA2.193 thus suggesting that all the right hindlimb finds might originally have been articulated and that separation of the pit lots is rather more arbitrary than real. Joins between the astragalus and distal tibia of AA2.193 and the first and second phalanges of A2.172 are less certain, but possible. As there are no duplicate skeletal parts or size differences to indicate the presence of more than one individual, it is

[4] Conceivably, organic residues at high concentrations within the pit may have lowered the pH which, in turn, would have brought about increased leaching and chemical erosion of the bones. Weak acidity may also have mobilized the calcium carbonate that subsequently reprecipitated around artifacts in the upper pit levels.

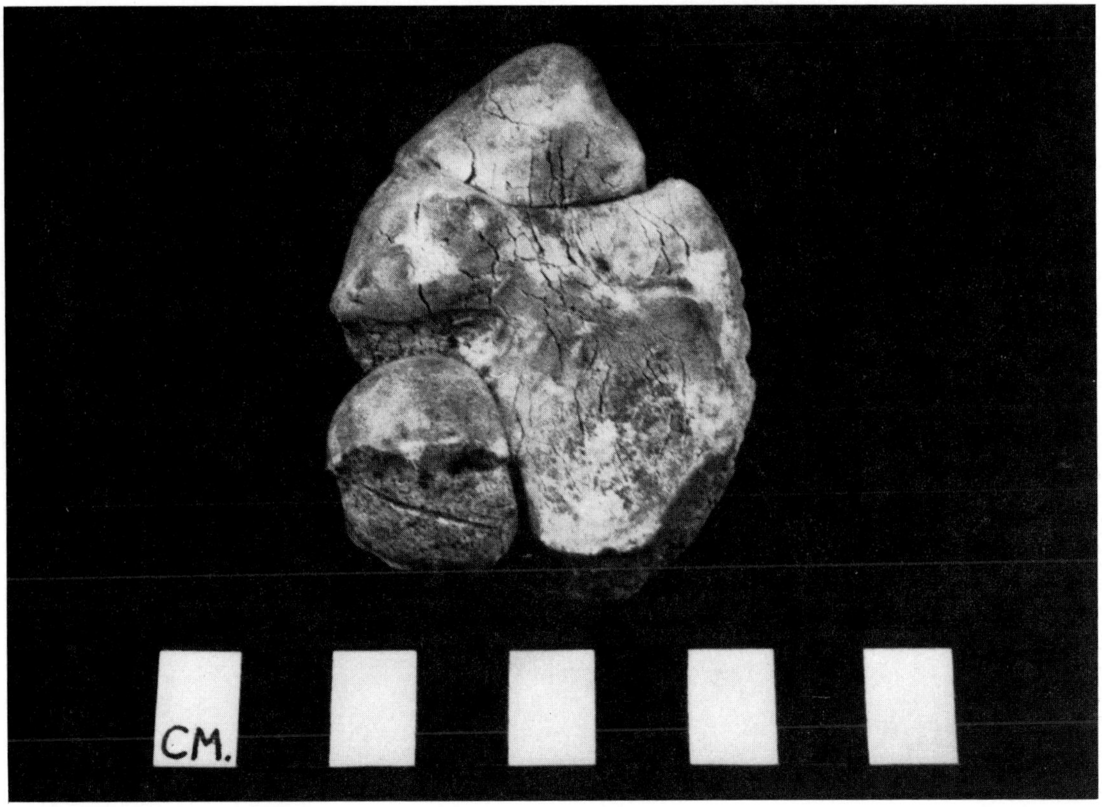

Plate 2: Distal row of left carpus of *Equus caballus* from Godin III:1, proximal view; second, third, and fourth carpals naturally cemented during burial; cut mark on medial periosteal surface of second carpal visible.

probable that the remains represent a single depositional episode involving discarded parts of a single carcass. Estimating from epiphyseal closure and based on modern fusion schedules, the horse would perhaps have been between 1.5 and 2 years old with fused proximal first and second phalanges, a recently fused distal metatarsal, and an unfused distal tibia. Skeletal part distribution and butchering mark evidence might help account for the presence of the horse bones in the Godin III sediments. Parts of each limb were recovered, but only of the lower elements that support almost no musculature. These footbones appear to have been removed deliberately from the limbs, as cut marks are frequent and conspicuously concentrated at the carpo-metacarpal and tarso-metatarsal joints.[5]

[5] Cut marks on the second and third carpals of the cemented carpus from AA2.185 suggest the severing of the medial ligament holding this joint together. A wide, sharp edged slit with weaker subparallel repeat slashes crosses the medial periosteal surfaces of both bones in a dorso-volar direction (Plate 2). The neighboring proximal metacarpals were cut in the same direction on the distal side of the articulation. The single right third carpal from AA2.193 bears cut marks in the identical position, implying that the same procedure was applied to the opposite member. A few tendon sheaths for muscles pass

Plate 3: Right tarsus of *Equus caballus* from Godin III:4, dorso-lateral view; central, first and second, third, and fourth tarsals naturally cemented during burial; cut marks on fourth tarsal visible.

The presence of footbones alone suggests that sinew recovery may have been the reason for their presence in the site. This does not mean that musculature was not also stripped from the carcass and utilized, but only that separation of the feet and their transport into the excavated area, probably from someplace outside of the site, appears likely to have been for the purpose of freeing and retrieving the tendons. Missing parts of the feet were either discarded elsewhere, carried off by dogs, or broken up for other secondary uses. In this regard, it is pertinent to mention that, in addition to being gnawed, both first phalanges were split longitudinally

nearby, but the cuts seem more likely to have been intended to split the joint, separating the meatless but sinew-bearing footbones.

The tarsal joint from AA2.185 reveals short, shallow nicks dorso-volarly across the lateral periosteal surface of the fourth tarsal just distal to its surface of articulation with the distal extremity of the calcaneum (Plate 3). The marks seem well placed to sever the long lateral ligament of the hock near where it attaches to the fourth tarsal and second metatarsal. The cuts may also have snipped the lateral extensor tendon that passes atop the ligament, but such action would not have parted the tendon at its origin but rather somewhere in the middle. Thus, these marks might have been intended either to split the joint or to remove sinew, or both.

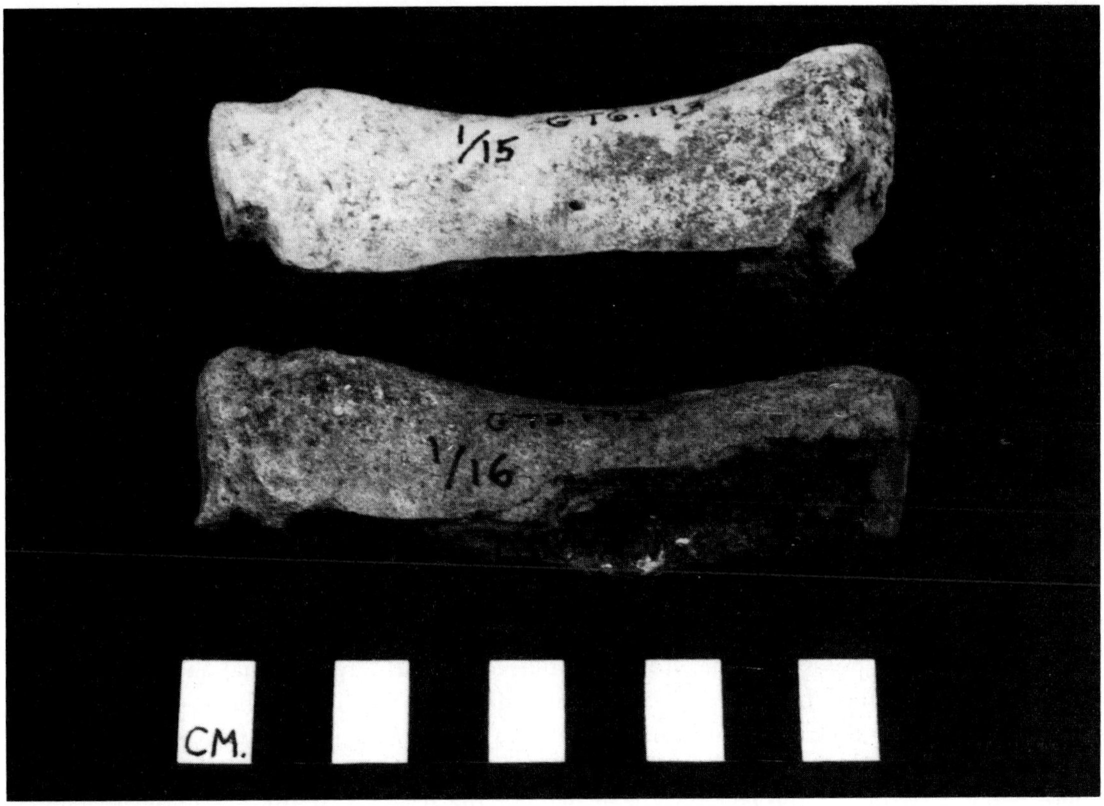

Plate 4: Two first phalanges from *Equus caballus* of Godin III:4, dorsal view; both specimens split along sagittal plane.

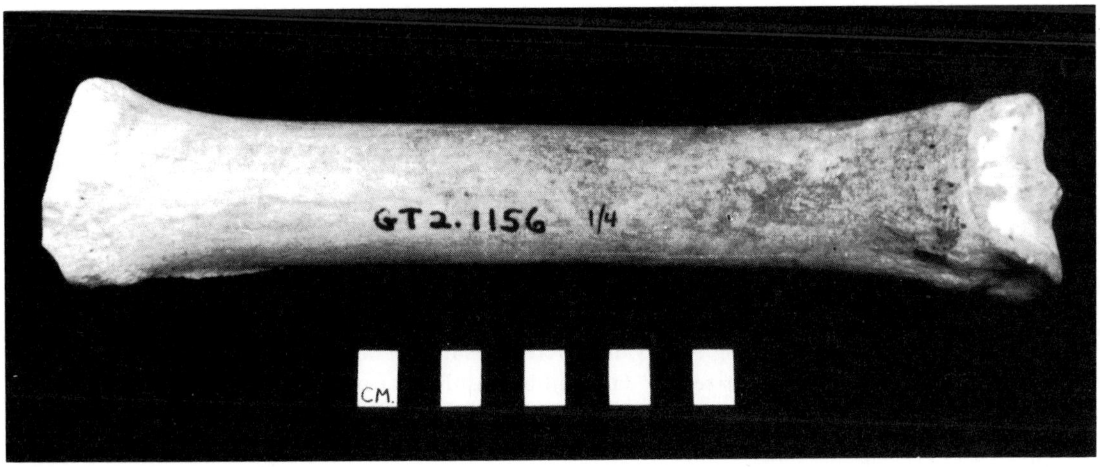

Plate 5: Complete right metacarpal III from *Equus caballus* (?) of Godin IV/V, dorsal view.

along the sagittal plane, raising the possibility that marrow grease or compact bone in a particular shape might also have been procured (Plate 4). This breakage pattern, curiously, is not observed on any of the hemione-sized first phalanges, which are either preserved intact or broken transversally.

Two possible caballine elements come from lots assigned to mixed Godin IV/V levels. These lots are probably to be understood as largely Godin IV, as their findspot contexts (including deposit codes #23, 28, and 36 in Table 2) are characteristic of uncontaminated Godin IV lots. For neither specimen, a complete right metacarpal III (Plate 5) and a right mandibular cheektooth (Figure 2d), is identification as *Equus caballus* certain.

The metacarpal has the following dimensions (in mm): Bp = 48.5, Dp = 33.7, SD = 32.6, Bd = 44.7, Dd = 33.4, GL = 237.2 with slenderness indices of 18.84 (100*Bp/GL: Bökönyi's index) and 13.74 (100*SD/GL: Hilzheimer's index). The latter index is just greater than those considered by Hilzheimer (1941: 11–14) to mark the lowest index representative of caballines (13.6) and the highest index representative of hemiones (13.5). Bökönyi (1972: 16), however, anticipates great variability in this measure of slenderness and relies instead upon Hilzheimer's use of distal epiphyseal form, which provides two applicable diagnostic features. First, Hilzheimer (1941: 9–10) recognized the existence of protuberances on the dorsal side of hemione metacarpals proximal to the epiphyseal line. Such protuberances do not occur on either the medial or lateral sides, which remain quite slender (see also Bökönyi 1972: 15). Horses, in contrast, possess smaller dorsal protuberances but larger medial and lateral bulges in connection with their flaring epiphyses. The Godin IV/V metacarpal shows definite mediolateral eminences and rather weak dorsal relief, thus suggesting it came from *Equus caballus.* Second, Hilzheimer noted that distal breadth is relatively greater in horses than in hemiones, there being a flaring of the epiphysis well beyond the breadth of the diaphysis. Hemiones show less epiphyseal spread, with distal breadth narrowed closer to the dimensions of the diaphysis. The metacarpal from Godin IV/V possesses a moderately broad distal epiphysis which, together with its greatest length, places it somewhere between hemione and the caballine metacarpals as plotted by Bökönyi (1972: 21).

Comparing the Godin metacarpal III to specimens of recent equids in the American Museum of Natural History yielded the data of Table 9. Using Hilzheimer's slenderness index, the Godin metacarpal compares favorably to the museum's *Equus caballus przewalskii* and hinny (*E.caballus* male X *E.asinus* female). *Equus hemionus* metarcarpals range substantially lower. Using Bökönyi's slenderness index, the Godin metacarpal ranks just above the uppermost hemione index and below the successively higher hinny and Przewalski metacarpals.[6]

In sum, the Godin metacarpal III seems horse-like based upon its metaphyseal protuberances and the flaring of its distal epiphysis, but its robustness as measured by two indices of slenderness suggests that it falls generally between the ranges of recent hemiones and Przewalski horses. Bökönyi (1972: 16) has remarked that European Iron Age horses (Bökönyi 1968) and

[6] Inspection of the accession numbers listed for the museum specimens used for comparison will reveal that both metacarpals from a single individual were measured in many cases. As paired elements will likely bear some correlation in size, the data will be unsuitable for any statistical exercise requiring independent cases. Although using such paired elements may have decreased the variance to be expected for a sample of metacarpals from different skeletons, the goal here was to illustrate general size relationships. To this end, the comparison appears justified, and the juxtaposition of values from both forelimbs gives some idea of biological variability within the animal.

TABLE 9

Equid remains from Godin Tepe (as of 1985):
metacarpal dimensions and indices of specimen from Godin IV/V compared with examples from the American Museum of Natural History collection, New York.

Source	Species	Side	SD	Bd	GL	HILZHEIMER'S Slenderness Index	BÖKÖNYI'S Slenderness Index
Godin IV/V	*caballus?*	R	32.6	44.7	237.2	13.74	18.84
AMNH 75334	hinny	L	28.9	41.7	212.7	13.59	19.61
AMNH 75334	"	R	28.9	41.9	211.7	13.65	19.79
AMNH 204139	"	R	27.0	36.8	190.0	14.21	19.37
(40082)	x̄		28.3	40.1	204.8	13.82	19.58
	s.d.		1.1	2.9	12.8	.34	.21
AMNH 202718	*caballus przewalski*	R	28.9	43.9	218.2	13.24	20.12
AMNH 204071	" "	R	31.5	46.3	222.5	14.16	20.81
AMNH 204071	" "	L	31.1	45.8	223.0	13.95	20.54
AMNH 35854	" "	R	30.0	44.6	215.7	13.91	20.68
AMNH 35854	" "	L	30.0	44.6	217.8	13.77	20.48
AMNH 80062	" "	R	29.5	43.1	217.5	13.56	19.82
AMNH 80062	" "	L	29.5	43.0	219.4	13.45	19.60
	x̄		30.1	44.5	219.2	13.72	20.29
	s.d.		.9	1.3	2.7	.32	.46
AMNH 57201	*hemionus*	R	28.1	40.3	234.5	11.98	17.19
AMNH 57201	"	L	28.2	40.1	235.0	12.00	17.06
AMNH 57208	"	R	27.1	40.5	239.2	11.33	16.93
AMNH 57208	"	L	26.7	40.3	241.0	11.08	16.72
AMNH 57209	"	L	29.1	42.5	231.5	12.57	18.36
AMNH 57212	"	R	28.2	40.2	231.2	12.20	17.39
AMNH 57212	"	L	28.3	40.1	231.6	12.22	17.31
AMNH 57214	"	R	28.3	42.5	234.7	12.06	18.11
AMNH 57214	"	L	27.9	44.0	234.7	11.89	18.75
AMNH 149409	"	R	28.3	38.9	239.6	11.81	16.24
AMNH 149409	"	L	28.1	38.6	239.3	11.74	16.13
AMNH 204201	*hemionus kiang*	L	28.3	41.1	241.8	11.70	17.00
	x̄		28.1	40.8	236.2	11.88	17.27
	s.d.		.6	1.6	3.8	.40	.80

the Caspian miniature horse (FIROUZ 1972) both have yielded indices below HILZHEIMER'S 13.5 cutoff point, thus indicating an overlap. In addition to this possibility of caballine indices dipping into the hemione range, hinny (and mule) metacarpals from a horse-hemione hybrid might also give similar indices. Thus, the possibility of a cross can not be ignored.

The other Godin IV/V specimen, the mandibular cheektooth (Figure 2d), bears traits generally acknowledged to be caballine, namely, a U-shaped *sinus lingualis,* a deeply penetrating *sulcus externus,* a moderately large hypoconulid, flattened boundaries of the metaconid and meta-

TABLE 10

Equid remains from Godin Tepe (as of 1985):
dimensions and rough occlusal areas of cheekteeth.

Possible Species?	Period	L (mm.)	B (mm.)	Approximate Occlusal Area (mm²)
E.caballus	IV/V	29.5	16.4	483.8
E.hemionus	III: 2	23.9	17.1	408.7
E.hemionus	III: 2	31.2(28.7)	14.6	455.5(419.0)
E.hemionus	III: 2	23.8	15.5	368.9

stylid folds, and a slight asymmetry in the double knot. The slightly larger size of the tooth compared to other recovered specimens also suggests a caballine. The size difference may best be seen by computing a rough occlusal area based upon length and breadth dimensions, as in Table 10.[7]

Even though it seems appropriate to identify this tooth as caballine, isolated teeth generally should not be relied upon for defining the presence of equid species because of potential variability in tooth characteristics even within a single jaw. Furthermore, the "blending" of diagnostic traits seen in other lower cheekteeth from Godin Tepe may suggest the presence of hybridization. The effects of cross-breeding on tooth cusp patterns are not known, but it is possible that hybridization leads to the presence of teeth with a combination or blurring of traits characteristic of the parent species. As there is much uncertainty among researchers about what dental (and skeletal) traits can be used reliably to discriminate specimens of known species, one is left with hardly any foothold on the problem of identifying hybrids.

A final problem with this particular tooth is the fact that, based on its low crown height, the specimen clearly comes from an old animal (if it were a modern horse, perhaps about 15 years). Thus, unusual enamel-fold characteristics may derive from the fact of advanced crown reduction.

In sum, on the basis of the metacarpal III and the lower cheektooth, the presence of *Equus caballus* is suggested at least as early as Godin IV. However, absence of bones showing definite caballine size and robustness well beyond the range of *Equus hemionus* together with the peculiar "blending" of caballine and hemione traits in some specimens, raises the possibility of horse-onager crosses. If true, Godin Tepe would preserve an osteological record of some of the earliest known interspecific animal breeds.

At this point, four possible ass bones from Godin Tepe must be considered. Two can be identified only tentatively on the basis of their small size, being well below that anticipated for a mature hemione. One proximal metacarpal III from Godin IV/V has a Dp dimension of 26.3 mm and a phalanx 3 from Godin III yielded a BF measurement of 31.7 mm (measurement

[7] The second hemione tooth listed in Table 10 (also Figure 2b) has an exaggerated tooth length due to the presence of a hypoconulid that is not worn by the immediately aboral tooth. The length would probably be about 28.7 mm if the projecting enamel fold were worn down to expected flatness. Such a shortening of the tooth would yield an approximate occlusal area of 419.0 sq. mm, a dimension comparable to those of the other hemione teeth.

definitions from VON DEN DRIESCH 1976). That these specimens may have come from unusually small or immature hemiones, however, cannot be ruled out.

Two other equid bones at the lower end of the hemione size range exhibit traits that have been described as being more characteristic of asses than of hemiones (BOESSNECK/KOKABI 1981, MEADOW 1986). One distal humerus and one proximal metacarpus III, both from Godin III: 4 (c. 2100–1900 BC), provide the following dimensions (in mm): humerus – Bd = 63.2, BT = 58.4, SD = 28.4; metacarpal – Bp = 44.0, Dp = 31.7. On the humerus, a thin ridge running down the anterior face of the shaft terminates in a small, irregular roughness just above the trochlea which possesses a sloping proximo-lateral border that continues smoothly back toward the lateral epicondyle. Hemiones show, in contrast, no muscular line on the shaft, a dimple in place of a roughness, a more horizontal trochlear border, and a padlike projection at the transition to the lateral epicondyle (MEADOW 1986: 275). The metacarpal reveals a fairly wavy volar edge on its proximal articular surface. This trait is said to be asinine, to be distinguished from the straighter edge characteristic of the hemiones (MEADOW 1986: 276).

Most of the well-preserved Godin equid bones bear traits characteristic of hemiones. A number of them (expecially among innominates and calcanea) were difficult to characterize on the basis of published characteristics, but the few bones showing any asinine traits come largely from Godin III. The rare presence of ass bones (if they are correctly identified) and their probable concentration in Godin III suggest that hybridization in the later Bronze Age might have involved horse-donkey crosses.

If the hybrid status of individuals represented by recovered bones could be established with certainty, there would then be no question about their status as domestic animals. If, however, the majority of the bones represent true breeding hemione stock, one must then address the question of cultural control. There is presently no unequivocal evidence for onager breeding at Godin Tepe or elsewhere in the Middle East, but an attempt at gauging the degree of human control over the animals was begun in 1984 in collaboration with BRIAN HESSE. With support from the University of Alabama at Birmingham, thin sections were cut from weight-bearing articular surfaces of bones from western Iran that have been identified as coming from hemiones. Four specimens from Godin Tepe were examined: one from Godin III: 4, one from Godin IV, and two from mixed Godin IV/V contexts. These bones represent either hunted individuals or animals held under some regime of human control perhaps for the purpose of serving principally as beasts of burden. Also sectioned were six bones from Tepe Ali Kosh (Deh Luran plain, Khuzestan, Iran) coming from the Bus Mordeh and Ali Kosh period levels (HOLE/FLANNERY/NEELY 1969: 295–298). Almost certainly these bones which date to the early Neolithic (c. 7500–6000 BC) come from wild onagers, and thus their microstructure can appropriately be compared with that from the Godin specimens in order to identify differences that might relate to state of cultural control. Changes in trabecular architecture and subchondral plate structure have been noted in other early domestic animals (DREW/PERKINS/DALY 1971), and although challenges have been leveled at the analytical techniques used and interpretations offered (MCCONNELL/FOREMAN 1971, WATSON 1975, ZEDER 1978, ØSTERGÅRD 1980), there remain good biomechanical reasons to suspect that changes would take place in early domesticates in spite of the difficulties inherent in identifying behavioral patterns as the causal factors behind the changes.

Without detailed discussion of the study and its potential usefulness (this will appear elsewhere upon completion of the study), it is possible to identify three features that differ significantly between the Godin and Ali Kosh specimens. First, the thickness of the subchondral plate appears generally greater in the Ali Kosh material. Second, crystallite orientation within the subchondral plate is generally greater in the Godin material. Third, intertrabecular spaces within the spongy bone beneath the subchondral plate tend to be larger in the Godin material. These features would suggest a "wild" status for the Ali Kosh hemiones and a "domestic" status for the Godin hemiones according to the criteria established by DREW/PERKINS/DALY (1971). "Controlled," however, is a better word than "domestic" in this context, as it does not imply the full range of husbandry practices. On biomechanical grounds, one might tentatively conclude that the Ali Kosh specimens had been subject to strong locomotor stresses that encouraged sturdier, less oriented, and more compactly structured bone growth that would best serve a wild hemione by absorbing the impact shock of footfall at high speeds. By contast, the Godin specimens reveal greater economy of mass and more structural sensitivity to specific stress vectors, a condition more likely to be found in controlled animals with more sedentary or less vigorous cycles of activity and thus reduced levels of osteological strain. The suggestion of hemione utilization during Godin IV and III therefore appears to be supported by the thin section data. The existence of a measure of cultural control over the Godin equids, together with the presence of caballines, continues to let us suspect the possibility of cross-breeding.

Because they represent a unique find, the horse bones from Godin III: 4 have not yet been sectioned for comparison. Worn spots on the subchondral plate of some poorly preserved, weight-bearing articular surfaces expose enough spongiosa to identify a lightly webbed, open texture similar to the trabecular arrangement of the other sectioned Godin examples. Until microscopic analysis confirms these initial observations, however, controlled or "domestic" status for the lone caballine is proposed principally on the basis of zoogeographic considerations.

3. Interpretation

3.1. Hybrids

This contribution to the study of Old World equids has conveyed some information about archaeological fauna excavated in western Iran. Although parallel findings at other sites are not yet forthcoming, it is possible that materials here described are not merely reflections of unique events at Godin, but that they signal more widespread trends taking place within the Zagros-Taurus area. The following thoughts are offered as a first attempt to place the analytic results from Godin into wider cultural perspective.

Most of the analyzed equid bones recovered from the late Uruk through Bronze Age at Godin Tepe (c. 3200–1600 BC) are the remains of hemiones, possibly including an unknown proportion of hybrids closely resembling hemiones in osteological characters and size. Without necessarily implying full domestication, microscopic details of skeletal architecture suggest that some form of cultural control was exercised over at least some, but probably most, of the equids.

Several lines of evidence raise the possibility that cross breeding may have been practiced. These include the following:

1) The remains of *Equus caballus* are unquestionably present at Godin from as early as 2000 BC (Godin III) with two other specimens, possibly caballine, coming from mixed Godin IV/V

deposits. In addition, there are a limited number of potential ass bones that occur in Godin III. Thus, two (and perhaps even three) species of equid may have been available for hybridization as early as the first half of the third millennium BC and certainly by the end of that same millennium.

2) Individual bones and teeth recovered from both Godin III and IV bear traits that have been considered diagnostic of both hemiones and caballines. Although there may be other reasons for such variability (including instability of the traits in question) genetic mixing is one possible explanation.

3) Godin Tepe was for much of its existence an important trade route town, situated on the principal east-west passage through the central Zagros mountains. It is conceivable that goods in moderate bulk were carried by humans in some instances but rather more likely that animals shouldered the transport burdens. The use of equids in the Old Assyrian trade to highland Anatolia implies their use in intermontane commerce during contemporary Godin III. In addition, the fact that Godin V deposits contained a higher percentage of the total equid bone sample than of all fauna from the recovered collection (see Section 2.2) is a possible, albeit inconclusive, indication of the importance of equids 1400 years earlier at a time of active commerce and communication between the Mesopotamian lowlands and the Persian plateau. Use of hemiones in such trade does not imply hybridization, especially during Godin V when the presence of *Equus caballus* is yet undocumented, although at a later date, it is possible that hybrids were perceived as advantageous in this capacity. Such hybrids would be stronger and easier to handle than the highly territorial and often intractable onager and climatically better adapted to the Near Eastern heat and aridity than the horse (suggestion of Heinz Heck). One might thus anticipate a turn to equid crosses once horses had been introduced into the area.

4) One might also expect initial cross-breeding of equids within the region in question to favor the production of hinnies, i.e., offspring sired by an imported horse and carried through gestation by a dam of the indigenous species. Such a practice would relieve the newly introduced caballine from the stresses of foaling in an unfamiliar and uncomfortable environment (suggestion of Mary Littauer). Although there is much size variation in modern hybrids, hemione-sized offspring appearing at the dawn of experimentation might, with greater probability, be identified as hinnies. If hybrids are, in fact, present in the Godin material, they are hemione-like rather than horse-like in size and thus provide a good fit to expectation on this point.

5) The acknowledged difficulties of producing *Equus caballus* x *Equus asinus* hinnies (e.g., Gray 1954: 46) have led others to suggest that, conversely, mules would have been the more common early hybrid (e.g., Zarins 1976: 366). It is unclear, however, to what degree the same behavioral and mechanical impediments to mating occur between horses and onagers or whether such difficulties as might have arisen during the breeding process were perceived as real obstacles by the early handlers. Onagers tend to be larger than asses, which alleviates somewhat the inherent size incompatibilities. The few modern reports on horse-onager crosses describe a hybrid with generally favorable characteristics. Gray (1954: 53) notes for onager-horse crosses:

> The hybrids are larger and of better conformation than mules or donkeys. They are easily tamed but probably could not be domesticated without prolonged selective breeding [of the parent stock.] ... Reciprocal crosses are possible.

Antonius (1944: 10–14) reviewed horse-onager crosses known to him. An onager (male) mated with a somewhat uncooperative Tarbes pony (female) produced a female hybrid in the

Jardin des Plantes in Paris in 1864. As reported by MILNE-EDWARDS (1869), it was "parfaitement conformé et très vigoureux," "plus grand que les hémiones du même âge," but resembling them "par la disposition générale de ses couleurs." The head was small in proportion to the body, and the ears were shorter than those of a hemione. After several months, the hybrid became squatter and stockier, darker brown in the coat, and blackened in the feet which revealed faint zebra-striping over the area of the proximal metapodials. A later article by EWART (1903) described the author's crossing of an Exmoor pony (female) with a Tibetan kiang. (ANTONIUS (1944: 12) reported that the father was, in fact, a northwest Persian onager.) The hybrid so resembled its father that EWART (1903: 463) remarked that it "might pass for a purebred wild ass." The mane and tail were mule-like, no bars or stripes appeared, and the first few days of the animal's life were marked by a tremendous spurt of activity. EWART (1903: 467) described the hybrid's sound as a barking reminiscent of the rasping call of the onager but not at all similar to a horse's neighing. Unfortunately, nothing was recorded of the hybrid's further development. The general resemblance of the hybrid to the hemione was found again in the case described by MOSSE (1935) in which an Indian onager (female) was crossed with a pony (male). The offspring took on more characteristics from the hemione (locally called *ghorkhar*) than from the horse, especially in its tail and body structure.

6) Ancient Mesopotamian texts make it fairly clear that hybridization was being practiced by at least the mid-third millennium BC. The commonly encountered term anše.BAR.AN is generally taken to mean mule (e.g., ZARINS 1976: 362–364; 1978: 11–15) with the cross presumed to have been mainly between *Equus caballus* (female) and *Equus asinus* (male).[8] The texts also suggest that hemiones may have been cross-bred, but a number of documents recommend against their use as tamed, service animals because of their difficult nature (ZARINS 1986). Antonius (1944: 14) noted that, in Roman times, hemiones were paired with horses to produce mules, and Zarins (1978: 11) furnishes additional references that trace cross-breeding of hemiones into the very recent past. In the third and fourth millennia, evidence for ass exploitation in North Africa, the Levant, and western Arabia is reported (Zarins 1986), so commercial acquisition and importation into Mesopotamia by as early as the Early Dynastic period would not be unlikely. East of Mesopotamia and farther from the area of ass domestication, however, onagers may have played the role of donkey. Leaving aside further speculation, it seems feasible that geographic differences in hybrid production may have existed. A more detailed account of the economic significance of each species, ancient perceptions about classifying animals according to distinctive traits and desirable qualities, and the breeding histories of each species remain to be written, however.

3.2. On the Matter of *Equus caballus*

Horse bones are rare at Godin Tepe. Excluding the 2 possible caballine elements from Godin IV/V, horses are represented by only the 21 bones from Godin III: 4 for the entire 1600 year long and 12 meter deep series of deposits in the eastern half of the deep sounding. Furthermore, the 21 bones may reflect but a single burial event. This dearth of specimens contrasts with the 313 equid bones and teeth that probably represent nearly as many instances of hemione, ass, or hybrid discard occurring throughout the whole sequence.

[8] But see POSTGATE (1986) who argues that anše.BAR.AN refers to hemione-donkey crosses.

The scarcity of horse bones at Godin could result from a rarity of horses in the economy of the site but could also be due to a tendency on the part of the ancient inhabitants not to bring the remains of caballines into the site itself. Both are plausible hypotheses for two reasons. First, caballines are likely to have been rare in the Near East at the time of their introduction into the region because of the climatic difficulties they experience. Indigenous hemiones (and asses) experience no such problems and would have been relatively easier to maintain in good condition.

Second, hemiones were familiar to Near Easterners as a hunted resource from as early as the Paleolithic and, as a food animal, may have been preferred over caballines which were introduced from outside of the region and possibly bore cultural meanings or functions inimical to their being used also as a protein source. A similar, although not identical, phenomenon may be seen with dogs. The remains of canids at Godin Tepe were limited, only 57 specimens having been identified to date. Many of the canid bones come from foxes, *Vulpes vulpes,* which were a traditional Zagros-area edible consumed in significant quantity during the Neolithic (Hesse 1978, Hole/Flannery/Neely 1969: 314). Judging from the abundance of gnaw marks on recovered faunal remains (nearly a quarter of the analyzed collection shows evidence of canine scavenging), dogs were quite common at the site. Dogs appear not to have been considered food animals in the pre-Islamic Near East, and as a result, their remains are generally only infrequently encountered within site sediments. Canophagy was more widely practiced in East Asia, the Pacific, and parts of the New World (Manwell/Baker 1984) where the presence of dog bones in the form of the remains of meals is also better attested in the archaeological record (e.g., Higham/Kijngam/Manly 1980, Titcomb 1969, Chang 1977: 29, 56). Following this line of reasoning, one might speculate that the newness of the horse, its foreign sources of supply, a possibly more restricted or specialized range of uses, or different symbolic connotations introduced along with the species or independently evolved over time may have intensified the archaeological effect produced by a simple rarity of the animals by limiting even more the entry of parts of dead animals into occupied areas.[9]

If both reasons for the dearth of horse bones are, in some measure, correct, then environmental factors probably restricted the regional distribution of horses within the Near East while taphonomic factors involving food choice may have further limited the burial of their remains in places where subsequent archaeological recovery would be possible. This situation might

[9] One can also compare the behavior of protohistoric and historic Late Village period Plains Indians of North America. Ethnohistoric sources generally acknowledge that most groups were reluctant to eat their horses, newly introduced from the Old World, except during time of duress when bison, deer, antelope, or other traditional food animals were unobtainable (e.g., Ewers 1955: 222, Matthews 1877: 24, Wilson 1924: 142–145). However Roe (1963: 276), citing others, remarks that horse-eating was variable and that groups reportedly consuming horses frequently were those of the Apache and Comanche, southern tribes wealthy in animals and probably among the first to get horses from the Spanish explorers. Archaeological remains also suggest this variability, with many sites yielding only the bones of hunted prey, predominantly bison or deer, and others producing occasional horse elements, thus attesting to some use of the equid carcass for its serviceable raw materials (hair, hide, etc.) if not also for its meat (e.g., Wedel 1959, Gunnerson 1960: 177, Lorrain 1967). No analogy may be extended to Bronze Age Zagros mountain peoples, but it is instructive for the present discussion to note that horse-using Native Americans have not always left a faithful archaeological record of animals kept but not used principally for food.

account for the relative "invisibility" of caballines in Bronze Age faunal assemblages. How rare they really were at the time of their southward diffusion and how misleading their possible under-representation in the archaeological record really has been are unanswerable questions at present.

Turning now to the matter of how and from where horses were brought into the Near East, others have noted that it is historically possible that they could have been introduced into Southwest Asia from the north by peoples of the Yanik culture (GIMBUTAS 1963, ZARINS 1976: 197–198). If domestication of caballines has been correctly inferred for Srednij Stog contexts of south Russia in the late fifth millennium BC (BIBIKOVA 1967, 1969; BÖKÖNYI 1974: 238; ANTHONY, this volume; calibrated radiocarbon dates based on uncalibrated dates reported by TELEGIN 1977), then the species might have become available to inhabitants of Transcaucasia prior to their southward push into the Near East in the early third millennium.

The Yanik culture represents an eastern prong of advance originating in or somewhat north of Transcaucasia. The best documentation of its presence comes from sites located in the area between the Kura and Araxes rivers that may be dated at the earliest to the very end of the fourth millennium (BURNEY/LANG 1971: 46–47, CHUBINISHVILI 1964, DZHAPARIDZE 1964, KRUPNOV 1965). The movement of these groups southward as far as Hamadan may have been rapid, for Yanik occupation of the mound at Godin followed closely upon the abandonment of the period V compound, walls of which appear to have still been standing at the beginning of Godin IV (WEISS/YOUNG 1975: 2).

One problem in identifying Transcaucasia and its eneolithic peoples as the distribution source for Near Eastern horses is the absence so far within the region of caballine remains. Art historical data shed a faint light on the issue but raise additional problems of their own. One of the necked, globular silver jars recovered from the kurgan burial at Maikop (FARMAKOVSKII 1914, ROSTOVZEV 1920, HANČAR 1937, PHILLIPS 1965, GIBELLINO KRASCENINNICOWA 1957) bears animal decoration chased in three registers, one above the other, on the vessel's surface (Figure 4). The jar's ornamentation depicts a lion attacking two bovids and a caballine equid in a band around the shoulder and a lioness chasing a sheep, gazelle, and boar encircled about the rounded bottom. The predators may have been lying in wait at a stream or lake represented by two rope-like channels flowing into a larger body of water at the base of the jar. The neck displays a backdrop of high mountains (from which the two streams emerge), a bear, and two trees. The find belonged to a barrow interment of a chieftain or other politically important person of the late third millennium, but the site's location on the Byelaya, a tributary of the Kuban river, places it beyond the Caucasus mountains about 700 km northwest of the Kura-Araxes drainage.

Whether one reads any symbolism or mythical content from the representations on this vessel, or simply accounts for the composition through Mesopotamian influence, the style is poorly documented and not connected to any particular artistic tradition (K. RUBINSON, personal communication). Further, it is difficult to tell whether the intention was to depict only wild animals or both wild and domestic animals among the prey species. Perhaps allegorical, perhaps naturalistic, the decoration in the main frieze, however, does associate horses and cattle, both being attacked by a lion. As cattle were the most important early livestock in the region from south Russia to Transcaucasia according to the faunal evidence (HANČAR 1937, TRINGHAM 1971, DZHAPARIDZE 1964: 227), the linkage of cattle with horses might signify that both were preeminent economic species and that, despite their rarity in excavated bone collections, horses were also important domesticates north of Transcaucasia. It would be rash to insist upon the

Fig. 4: Representation on a silver jar from Maikop; taken from FRANKFORT (1956: fig. 45) after the original in FARMAKOVSKII (1914: plate 27).

economic importance of the horse in the absence of supporting faunal or artifactual evidence, but the situation may parallel the pattern observed in the Near East. The importance of caballines may be masked by the paucity of their remains, which might be due in part to specialized taphonomic channels that restrict their burial to areas outside of settlements. Not a substitute for direct archaeological evidence, the Maikop jar is only suggestive of some special attachment to horses in the nearby Kuban culture of c. 2300–2100 BC. If horse bones continue to elude recovery in the Kura-Araxes cultural sphere, then perhaps the region may best be seen as having served a middleman function, conveying caballines into Southwest Asia from the more northerly Kurgan pastoralists.

The Yanik intrusion into Iran may have been echoed in the west by another movement of peoples or of cultural influences into Turkey and the Levant that has been traced through the ceramic affinities of the East Anatolian Early Bronze ware, the Red-Black Burnished ware of the Amuq in northwestern Syria, and the Khirbet Kerak ware of Palestine. The extent to which large-scale migration penetrated the southernmost Levant is unclear, and Khirbet Kerak ware may represent the spread of a ceramic technique with only minimal influx of persons (but see, most recently, AMIRAN 1986). Indications of population shifts are clear in the north, however, with cultural elements of Kura-Araxes origin spreading west and south into the region of Elaziğ and Malatya to sites such as Korucutepe and Arslantepe, ushering in as they arrive the East Anatolian Early Bronze Age.

The main prongs of Transcaucasian expansion into Iran and Anatolia remain within the more temperate regions of Southwest Asia. Whether by coincidence or not, the Yanik culture drove southeastward along the Zagros thrust zone to just beyond the latitude of Hamadan, which marks the approximate cessation of Spring convectional rainfall within the Iran-Iraq borderlands under present climatic conditions (GANJI 1955). Assuming a similar precipitation regime 5000 years ago, the combination of high elevation and later onset of summer drought would have afforded a cooler, more mesic environment that would have been the least debilitating of Near Eastern habitats for horses. The east Anatolian mountains would also have offered lower temperatures and higher moisture levels than the lowlands farther south. If caballines formed a significant domestic resource among incoming Transcaucasian groups, their climatic needs may have contributed to the initial spread and geographic pattern of Early Bronze Age settlement. It is pertinent to mention in this context the importance for horse breeding of the Median areas of Luristan and Iranian Kurdistan during the mid first millennium BC. These are the very areas overrun by the Yanik culture some 2500 years earlier.

Early Transcaucasian expansion affected the vast communications network, established by late Uruk times, that served ostensibly for acquisition and transfer of raw materials into southern Mesopotamia. Trade links across the central Zagros were broken by the Yanik advance, apparently causing an economic loss of considerable magnitude. With an entire arm of Uruk period commercial influence severed, the collapse of most major towns of Luristan that depended upon the flow of intermontane trade followed (e.g., GOFF 1971: 145). In response, passage between Mesopotamia and the Persian plateau shifted to routes farther south to circumvent the intrusion. Contact between Khuzistan and highland Luristan may have been re-established by the start of Godin III (c. 2600 BC), and presumably from that moment until the end of the Bronze Age, mountain products from the Luri and Kurdish hill peoples were available to lowland populations through Susa and other mountain fringe communities. In return, domestic asses from Mesopotamia could have reached the intermontane zone, possibly stimulating changes in equid breeding strategies.

Aside from Godin Tepe, osteological evidence of horses occurs in limited quantity at several other sites in the mountain areas of Turkey and Iran. Remains of *Equus caballus* from the Anatolian Early Bronze Age are reported from Pulur Höyük (DENIZ 1975: 291–292), Tepeçik, Norşuntepe, and Tulintepe (BOESSNECK/VON DEN DRIESCH 1976) in the Keban dam area, Yarıkkaya (BOESSNECK/WEIDEMANN 1977: 121) near Boğazköy, and Demircihüyük (RAUH 1981: 100ff.) farther west. At Pulur Höyük, Tepeçik, Tulintepe, and Norşuntepe, horse bones are reportedly stratified in late fourth millennium Chalcolithic contexts as well. (See also BÖKÖNYI, this volume.) BÖKÖNYI (1967: 311–312) has identified further caballine material from the site of Tal-i Iblis in the Kerman region of southern Iran in deposits of the fourth millennium. Early appearance of horses as far south as Iblis would imply their trade or spread quite soon after domestication, if indeed all relevant dates are reliable and accurately reflect the events associated with them. The occurrence of caballine (or hybrid) remains in early third millennium levels of Early Bronze Age Arad in southern Israel (DAVIS 1976: 158–162) perhaps underscores this movement, as the Arad specimens arrived well ahead of the Khirbet Kerak pottery tradition.

Most researchers have identified these fourth and third millennium horse bones as representing wild animals, hunted perhaps from scattered remnants of Pleistocene herds (e.g., BOESS-

NECK/VON DEN DRIESCH 1976). Only at Pulur Höyük has a claim for domestic status been proposed for this early date preceding the more traditionally defined watershed of c. 2000 BC when size reduction evident in the recovered faunal material appears to signal the onset of selective breeding practices. This writer suspects, however, that these early caballines may eventually be found to represent culturally-controlled imports. The occasional entry of horses into Southwest Asia from southern Russia during the fourth millennium seems not to have led to any permanent or large-scale change in Near Eastern equid exploitation until the early third millennium when Transcaucasian migrants brought the possibility for closer contact with potentially horse-using and horse-breeding peoples.

4. Acknowledgements

The writer would like to thank MARY A. LITTAUER for advice and encouragement over several years of inquiry into this subject. T. CUYLER YOUNG, JR., ROBERT C. HENRICKSON, and VIRGINIA R. BADLER provided the study with its stratigraphic backbone without which the paper would have proceeded, much impeded, with the prosthetic use of provisional dating. SIDNEY ANDERSON of the American Museum of Natural History kindly provided access to comparative osteological material. Precise rendering of the equid tooth cusp patterns of Figure 2 is due to the diligence and craft of PADDY BAREHAM.

5. Literature

AMIRAN, R., 1986, The fall of the Early Bronze Age II city of Arad. Israel Exploration Journal **36,** 74–76.

ANTONIUS, O., 1944, Beobachtung an Einhufern in Schönbrunn. Der Zoologische Garten (N.F.) **16**(1–2), 1–14.

BADLER, V.R., in preparation, Godin V and the Chronology of the Late Uruk Period; Dissertation, University of Toronto, Toronto.

BAGTACHE, B./HADJOUIS, D./EISENMANN, V., 1984, Présence d'un *Equus* caballin (*E. algericus* n.sp.) et d'une autre espèce nouvelle d'*Equus* (*E.melkiensis* n.sp.) dans l'Atérian des Allobroges, Algérie. Comptes Rendus de l'Académie des Sciénces, Paris (Série II) **298**(14), 609–612.

BIBIKOVA, V.I., 1967, K izucheniyu drevneishikh domashnikh loshadei Vostochnoi Evropy (Studies of ancient domestic horses in Eastern Europe). Bulleten Moskovskogo Obshchestva Ispitatelei Prirody, Otdel Biologicheskii **72,** 106–118.

BIBIKOVA, V.I., 1969, Do istorii domestikacii konia na Pivdennom Skhode Evropy. Arkheologiia **22,** 55–67.

BOESSNECK, J./DRIESCH, A.V.D., 1976, Pferde im 4./3. Jahrtausend v. Chr. in Ostanatolien. Säugetierkundliche Mitteilungen **24**(2), 81–87.

BOESSNECK, J./KOKABI, M., 1981, Tierknochenfunde. Saarbrücker Beiträge zur Altertumskunde **31**(Halawa 1977–1979), 89–104.

BOESSNECK, J./WIEDEMANN, U., 1977, Tierknochen aus Yarıkkaya bei Boğazköy, Anatolien. Archäologie und Naturwissenschaften **1,** 106–128.

BÖKÖNYI, S., 1967, The prehistoric vertebrate fauna of Tal-i Iblis. In: CALDWELL, J.R., ed., Investigations at Tal-i-Iblis; (Illinois State Museum Preliminary Reports **9**); Springfield (IL): Illinois State Museum, 309–317.

Bökönyi, S., 1968, Data on Iron Age horses of Central and Eastern Europe. (Bulletin of the American School of Prehistoric Research **25**); Cambridge (MA): Peabody Museum, Harvard University, 1–71.

Bökönyi, S., 1972, Once more on the osteological differences of the horse, the half-ass, and the ass. In: Firouz, L., comp., The Caspian Miniature Horse of Iran; Miami: Field Research Projects, 12–23.

Bökönyi, S., 1974, History of Domestic Animals in Central and Eastern Europe; Budapest: Akadémiai Kiadó.

Burney, C.A./Lang, D.M., 1971, The Peoples of the Hills: Ancient Ararat and Caucasus; New York: Praeger.

Chang, K.-C. (ed.), 1977, Food in Chinese Culture: Anthropological and Historical Perspectives; New Haven: Yale University Press.

Chubinishvili, T.N., 1963–64, The historical significance of the southern Caucasus in the third millennium B.C. Soviet Anthropology and Archeology **2**(3), 3–16.

Davis, S.J., 1976, Mammal bones from the Early Bronze Age city of Arad, northern Negev, Israel: some implications concerning human exploitation. Journal of Archaeological Science **3**, 153–164.

Davis, S.J., 1980, Late Pleistocene and Holocene equid remains from Israel. Zoological Journal of the Linnean Society **70**, 289–312.

Deniz, E., 1975, Neolithic, Chalcolithic, and Early Bronze Age faunal remains from Pulur Höyük (Keban Dam Region), Turkey. In: Clason, A.T., ed., Archaeozoological Studies; Amsterdam/New York: North Holland/American Elsevier, 284–294.

Drew, I.M./Perkins Jr., D./Daly, P., 1971, Prehistoric domestication of animals: effects on bone structure. Science **171**, 280–282.

Driesch, A. von den, 1976, A Guide to the Measurement of Animal Bones from Archaeological Sites; (Peabody Museum Bulletin **1**); Cambridge (MA): Peabody Museum, Harvard University.

Ducos, P., 1968, L'Origine des Animaux domestiques en Palestine; (Publications de l'Institut de Préhistoire, Université de Bordeaux **6**); Bordeaux: Institut de Préhistoire.

Dyson, R.H., 1973, Archaeological evidence of the second millennium B.C. on the Persian plateau. In: Cambridge Ancient History II/1 (3rd edition); Cambridge: Cambridge University Press, 686–715.

Dzhaparidze, O.M., 1964, The culture of early agricultural tribes in the territory of Georgia. In: Proceedings of the 7th International Congress of Anthropological and Ethnological Sciences: **5**; Moscow, 225–230.

Ewart, J.C., 1903, The wild horse (*Equus prjevalskii,* Poliakoff). Proceedings, Royal Society of Edinburgh **24**, 460–468.

Ewers, J.C., 1955, The Horse in Blackfoot Indian Culture; (Bureau of American Ethnology, Bulletin **159**); Washington, D.C.: Smithsonian Institution.

Farmakovskii, B.V., 1914, Arkhaicheskii Period v Rossii; Petrograd: Imperial'skaia Arkheologicheskaia Kommissiia.

Firouz, L., 1972, The Caspian Miniature Horse of Iran; Miami: Field Research Reports.

Frankfort, H., 1956, The Art and Architecture of the Ancient Orient; Baltimore: Penguin Books.

Ganji, M.H., 1955, The climates of Iran. Bulletin, Société Géographique d'Égypte **28**, 195–299.

Gibellino Krasceninnicowa, M., 1957, Il Caucaso; Roma: Fratelli Palombi.

Gilbert, A.S., 1979, Urban Taphonomy of Mammalian Remains from the Bronze Age of Godin Tepe, Western Iran; Dissertation, Columbia University, New York.

Gilbert, A.S./Singer, B.H., 1982, Reassessing zooarchaeological quantification. World Archaeology **14**, 21–40.

Gilbert, A.S./Steinfeld, P., 1977, Faunal remains from Dinkha Tepe, northwestern Iran. Journal of Field Archaeology **4**, 329–351.

Gimbutas, M., 1963, The Indo-Europeans: archeological problems. American Anthropologist **65**, 815–836.

Goff, C.L., 1971, Luristan before the Iron Age. Iran **9**, 131–152.

GRAY, A.P., 1954, Mammalian Hybrids: a Check-list with Bibliography; Farnham Royal, Bucks: Commonwealth Agricultural Bureau.

GUNNERSON, J.H., 1960, An Introduction to Plains Apache Archeology - the Dismal River Aspect. Bureau of American Ethnology, Smithsonian Institution, Anthropological Paper **58**(Bulletin 173), 131–260.

HANČAR, F., 1937, Urgeschichte Kaukasiens; Wien: A. Schroll.

HENRICKSON, R.C., 1984, Godin Tepe, Godin III, and Central Western Iran ca. 2600–1500 B.C.; Dissertation, University of Toronto, Toronto.

HENRICKSON, R.C., 1985, The chronology of central western Iran 2600–1400 B.C. American Journal of Archaeology **89,** 569–581.

HENRICKSON, R.C., 1986, A regional perspective on Godin III cultural development in central western Iran. Iran **24,** 1–55.

HESSE, B.C., 1978, Evidence for Husbandry from the Early Neolithic Site of Ganj Dareh in Western Iran; Dissertation, Columbia University, New York.

HIGHAM, C.F.W./KIJNGAM, A./MANLY, B.F.J., 1980, An analysis of prehistoric canid remains from Thailand. Journal of Archaeological Science **7,** 149–165.

HILZHEIMER, M., 1941, Animal Remains from Tell Asmar; (Studies in Ancient Oriental Civilization **20**); Chicago: The University of Chicago Press.

HOLE, F./FLANNERY, K.V./NEELY, J.A., 1969, Prehistory and Human Ecology of the Deh Luran Plain; (Museum of Anthropology, Memoir **1**); Ann Arbor: University of Michigan.

KRUPNOV, E.I., 1964–65, The most archaic culture of the Caucasus and the Caucasian ethnic community. Soviet Anthropology and Archeology **3**(3), 31–45.

LORRAIN, D., 1967, Animal remains. (The Gilbert Site; a Norteño Focus Site in Northeastern Texas.) Bulletin, Texas Archeological Society **37,** 225–243.

MANWELL, C./BAKER, C.M.A., 1984, Domestication of the dog: hunter, food, bed-warmer, or emotional object? Zeitschrift für Tierzüchtung und Züchtungsbiologie **101,** 241–256.

MATTHEWS, W., 1877, Ethnography and Philology of the Hidatsa Indians; (U.S. Geological and Geographical Survey, Miscellaneous Publication 7); Washington, D.C.: Department of the Interior.

MCCONNELL, D./FOREMAN, D., 1971, Texture and composition of bone. Science **172,** 972.

MEADOW, R.-H., 1986, Some equid remains from Çayönü, Southeastern Turkey. In: MEADOW, R.H./UERPMANN, H.P., eds., Equids in the Ancient World; (Beihefte zum Tübinger Atlas des Vorderen Orients, Reihe A **19/1**); Wiesbaden: Dr. Ludwig Reichert Verlag, 266–301.

MILNE-EDWARDS, A., 1869, Note sur un métis d'hémione et de jument sur l'hémippe ou hémione de Syrie et sur l'onagre d'Abyssinie. Bulletin de Nouvelles Archives du Muséum d'Histoire Naturelle de Paris **5,** 37–41.

MOSSE, A.H.E., 1935 (Dec. 7), India's wild ass: a species distinct in its coloration, with hybrid progeny. Illustrated London News (5042), 1058.

ØSTERGÅRD, M., 1980, X-ray diffractometer investigations of bones from domestic and wild animals. American Antiquity **45,** 59–63.

PHILLIPS, E.D., 1965, The Royal Hordes: Nomad Peoples of the Steppes; New York: McGraw-Hill.

POSTGATE, J.N., 1986, The equids of Sumer, again. In: MEADOW, R.H./UERPMANN, H.P., eds., Equids in the Ancient World; (Beihefte zum Tübinger Atlas des Vorderen Orients, Reihe A **19/1**); Wiesbaden: Dr. Ludwig Reichert Verlag, 194–206.

RAUH, H., 1981, Knochenfunde von Säugetieren aus dem Demircihüyük (Nordwestanatolien); Dissertation, Ludwig-Maximiliens-Universität, München.

ROE, F.G., 1963, The Indian and the Horse; Norman (OK): University of Oklahoma Press.

ROSTOVZEV, M., 1920, L'âge du cuivre dans le Caucase septentrionale et les civilisations de Soumer et de l'Égypte protodynastique. Révue archéologique **12** (5ème série), 1–37.

SILVER, I.A., 1969, The ageing of domestic animals. In: BROTHWELL, D./HIGGS, E.S., eds., Science in Archaeology (2nd edition); London: Thames and Hudson, 283–302.

SOLECKI, R.L./SOLECKI, R.S., 1983, Late Pleistocene-Early Holocene cultural traditions in the Zagros and the Levant. In: The Hilly Flanks and Beyond: Essays on the Prehistory of Southwestern Asia; (Studies in Ancient Oriental Civilization **36**); Chicago: The Oriental Institute of the University of Chicago, 123–137.

STEHLIN, H.G./GRAZIOSI, P., 1935, Richerche sugli Asinidi fossili d'Europa. Mémoires de la Société Paléontologique Suisse **56**(3), 1–73.

SUMNER, W.M., 1978, revised 1985, Instructions for Excavation and Recording: Malyan; Columbus (OH): distributed by the author from The Oriental Institute, University of Chicago.

TELEGIN, D.I., 1977, Ob absoliutnom vozraste iamnoi kul'tury i nekotorye voprosy khronologii ènеolita iuga Ukrainy. Sovetskaia Arkheologiia, 5–19.

TITCOMB, M., 1969, Dog and Man in the Ancient Pacific; (Special Publication **59**); Honolulu: Bernice P. Bishop Museum.

TRINGHAM, R., 1971, Hunters, Fishers, and Farmers of Eastern Europe: 6000–3000 B.C; London: Hutchinson.

TURNBULL, P.F./REED, C.A., 1974, The fauna from the terminal Pleistocene of Palegawra Cave, a Zarzian occupation site in northeastern Iraq. Fieldiana: Anthropology **63**(3), 81–146.

WATSON, J.P.N., 1975, Domestication and bone structure in sheep and goats. Journal of Archaeological Science **2**, 375–383.

WEDEL, W., 1959, An Introduction to Kansas Archaeology; (Bureau of American Ethnology, Bulletin **174**); Washington, D.C.: Smithsonian Institution.

WEISS, H. /YOUNG JR., T.C., 1975, The merchants of Susa: Godin V and plateau-lowland relations in the late fourth millennium B.C. Iran **13**, 1–17.

WILSON, G.L., 1924, The Horse and Dog in Hidatsa Culture; (Anthropological Papers **15**(2)); New York: American Museum of Natural History.

YOUNG JR., T.C., 1969, Excavations at Godin Tepe: First Progress Report; (Art and Archaeology Occasional Paper **17**); Toronto: Royal Ontario Museum.

YOUNG JR., T.C., 1985, Mesopotamia: perforce multicultural. Bulletin, Society for Mesopotamian Studies (May), 5–13.

YOUNG JR., T.C.,/LEVINE, L.D., 1974, Excavations of the Godin Project: Second Progress Report; (Art and Archaeology Occasional Paper **26**); Toronto: Royal Ontario Museum.

YOUNG JR., T.C./WEISS, H., 1974, The Godin project: Godin Tepe. Iran **12**, 207–211.

ZARINS, J., 1976, The Domestication of Equidae in Third Millennium B.C. Mesopotamia; Dissertation, University of Chicago, Chicago.

ZARINS, J., 1978, The domesticated Equidae of third millennium B.C. Mesopotamia. Journal of Cuneiform Studies **30**, 3–17.

ZARINS, J., 1986, Equids associated with human burials in third millennium B.C. Mesopotamia: two complementary facets. In: MEADOW, R.H./UERPMANN, H.-P., eds., Equids in the Ancient World; (Beihefte zum Tübinger Atlas des Vorderen Orients, Reihe A **19/1**); Wiesbaden: Dr. Ludwig Reichert Verlag, 164–193.

ZEDER, M.A., 1978, Differentiation between the bones of caprines from different ecosystems in Iran by the analysis of osteological microstructure and chemical composition. In: MEADOW, R.H./ZEDER, M.A., eds., Approaches to Faunal Analysis in the Middle East; (Peabody Museum Bulletin **2**); Cambridge (MA): Peabody Museum, Harvard University, 69–84.

Allan Gilbert, Department of Sociology and Anthropology, Fordham University, Bronx, NY, 10458, U.S.A.

APPENDIX 1

MEASUREMENTS OF EQUID TEETH FROM GODIN TEPE
all measurements in mm.; [mm.] indicates estimated measurement

Maxillary

tooth	P2	P2	P2	P3-M2	P3-M2	P3-M2	P3-M2
lot	AA2.225	A1.1138	A1.1156	A1.1156	A1.1156	AA2.150	AA2.154
period	III: 6	IV/V	IV/V	IV/V	IV/V	III: 2	III:2
L	38.1	33.8	34.5	26.9	25.8	27.8	24.4
B	21.5	24.5	25.5	25.6	25.8	25.9	26.3
			1 indiv.; 24: 85.4 (A1.1156, columns 3–5)				

tooth	P3-M2	P3-M2	P3-M2	P3-M2	P3-M2	P3-M2	M3
lot	A1.1159	A1.1139	A1.1139	A1.1138	A2.1192	AA2.302	A2.1196
period	IV	IV/V	IV/V	IV/V	V	?	IV/V
L	29.2	-	24.0	25.1	25.5	28.0	23.8
B	23.5	27.8	25.0	26.2	[27.0]	25.9	23.5
	just in wear						

Mandibular

tooth	P2	P2	P3-M2	P3-M2	P3-M2	P3-M2	P3-M2
lot	AA 270	A1.1158	AA2.150	AA2.8D-11	AA1.8D-11	A1.1138	A2.1169
period	III: 6	IV/V	III: 2	III: 2	III: 2	IV/V	IV/V
L	31.2	28.9	23.9	31.2	23.8	29.5	25.9
B	13.2	14.5	17.1	14.6	15.5	16.4	-
						E.cab.?	

tooth	M3	M3	M3	M3
lot	A2.146	AA2.181	A1.1133	A1.1156
period	III: 2	III: 4	IV/V	IV/V
L	28.1	34.0	27.0	27.5
B	14.5	14.0	13.0	14.5
		just in wear		

APPENDIX 2

MEASUREMENTS OF EQUID TEETH FROM ALI KOSH
all measurements in mm.; [mm.] indicates estimated measurement

Maxillary

tooth	P2	P2	P2	P3	P4	M1
locus	79/439	13/90-100	85/400	85/400	85/400	85/400
period*	AK	MJ	AK	AK	AK	AK
L	31.8	37.1	35.9	26.1	24.6	20.9
B	25.0	21.6	25.2	26.7	27.8	24.0

1 indiv.; 24: 87.2

tooth	P3-M2	P3-M2	P3-M2	P3-M2
locus	91/240-260	79/635	61/520-530	82/450-460
period*	AK	BM	AK	AK
L	25.5	28.2	26.5	21.0
B	26.0	23.2	26.3	25.8

Mandibular

tooth	P2	P2	P2	P3-M2	P3-M2	P3-M2
locus	18/130-150	99/250-270	69/220-240	79/502	83/420-430	99/250-270
period*	MJ	AK	AK	AK	AK	AK
L	30.1	29.5	28.0	23.1	22.3	25.0
B	13.0	12.1	13.8	15.8	13.4	18.1

tooth	P3-M2	P3-M2	P3-M2	M3	M3	M3
locus	63/180-190	1/150-160	45/170-180	12/120-130	98/260-270	45/170-180
period*	AK	AK?	AK?	MJ	AK	AK?
L	31.7	21.8	20.7	28.3	30.8	26.7
B	17.0	15.7	15.0	12.9	14.1	13.0

tooth	M3
locus	AK-B2
period*	?
L	29.0
B	14.0

* BM = Bus Morden (8200-7200 BC)
AK = Ali Kosh (7200-6400 BC)
MJ = Mohammed Jaffar (6400-6100 BC)

APPENDIX 3

Equus cf. asinus

Metacarpus III

Site:Unit	Layer	1	2	3	4	5	6	7	8
GODI:A01.34	IV/V	--	26.3	--	--	--	--	--	--
DINK:DI52	IVA	--	--	--	32.2	26.0	--	18.0	--
DINK:DI52	IVA	--	--	--	35.0	27.0	--	--	--

Columns 1–8: Measurements

2 - Dp (proximal depth)
4 - Bd (distal breadth)
5 - Dd (distal depth)
7 - DD (depth of diaphysis)

Astragalus

Site:Unit	Layer	1	2	3	4	5	6	7	8
DINK:DI48	III/II	44.9	44.6	38.1	--	45.4	36.0	--	--
DINK:DI15A	IVA	46.5	44.9	39.3	--	48.3	38.4	--	--
DINK:DI57A	III/II	46.6	46.4	39.7	--	47.4	37.1	--	--

Columns 1–8: Measurements

1 - GH (greatest height)
2 - LmT (medial length of trochlea)
3 - Dm (medial depth)
5 - GB (greatest breadth)
6 - BFd (breadth of dist. facies art.)

Phalanx 3 anterior or posterior

Site:Unit	Layer	1	2	3	4	5	6	7	8
GODI:A2.142	III:2	--	--	--	--	31.7	--	--	--

Columns 1–8: Measurements

5 - BF (breadth of facies art.)

Key to site abbreviations:

GODI : Godin Tepe
ALIK : Ali Kosh
DINK : Dinkha Tepe

Equus cf. caballus

Humerus

Site:Unit - Layer		Measurements 1	2	3	4	5	6	7	8
DINK:DI4	MIXED	75.3	70.2	--	--	--	--	--	--

1 - Bd (distal breadth)
2 - BT (breadth of trochlea)

Metacarpus III

Site:Unit - Layer		Measurements 1	2	3	4	5	6	7	8
GODI:AA2.185	III:4	--	37.5	--	--	--	--	--	--
GODI:A1.1156	IV/V	48.5	33.7	32.9	44.7	33.4	237.2	24.3	--
DINK:DI51	IVA	--	--	--	48.9	36.4	--	--	--
DINK:DI51	IVA	--	--	--	49.4	35.0	--	--	--

1 - Bp (proximal breadth)
2 - Dp (proximal depth)
3 - SD (smallest breadth of diaphysis)
4 - Bd (distal breadth)
5 - Dd (distal depth)
6 - GL (greatest length)
7 - DD (depth of diaphysis)

Femur

Site:Unit - Layer		Measurements 1	2	3	4	5	6	7	8
DINK:DI55A	IV/III	--	--	--	97.6	--	--	--	--

4 - Bd (distal breadth)

Tibia

Site:Unit - Layer		Measurements 1	2	3	4	5	6	7	8
GODI:AA2.193	III:4	--	--	72.5	49.5	--	--	--	--

3 - Bd (distal breadth)
4 - Dd (distal depth)

Astragalus

Site:Unit - Layer		Measurements 1	2	3	4	5	6	7	8
GODI:AA2.193	III:4	64.0	61.4	--	--	65.5	54.3	--	--

1 - GH (greatest height)
2 - LmT (medial length of trochlea)
5 - GB (greatest breadth)
6 - BFd (breadth of dist. facies art.)

Equus cf. caballus

Metatarsus III

Site:Unit	Layer	1	2	3	4	5	6	7	8
		Measurements							
GODI:AA2.185	III:4	56.5	--	33.0	54.5	42.5	263.3	271.2	31.6

1 - Bp (proximal breadth)
3 - SD (smallest breadth of diaphysis)
4 - Bd (distal breadth)
5 - Dd (distal depth)
6 - GLl (greatest lateral length)
7 - GL (greatest length)
8 - DD (depth of diaphysis)

Phalanx 1 anterior or posterior

Site:Unit	Layer	1	2	3	4	5	6	7	8
		Measurements							
GODI:A2.172	III:4	--	--	--	91.2	--	--	--	87.4
GODI:AA2.193	III:4	--	--	--	89.0	37.0	--	27.0	84.7

4 - GL (greatest length)
5 - Dp (proximal depth)
7 - Dd (distal depth)
8 - GL (greatest length)

Phalanx 2 anterior or posterior

Site:Unit	Layer	1	2	3	4	5	6	7	8
		Measurements							
GODI:A2.172	III:4	51.3	43.3	47.3	53.0	34.6	45.1	28.2	44.2

1 - Bp (proximal breadth)
2 - SD (smallest breadth of diaphysis)
3 - Bd (distal breadth)
4 - GL (greatest length)
5 - Dp (proximal depth)
6 - DS (depth of diaphysis)
7 - Dd (distal depth)
8 - GL (greatest length)

Phalanx 3 anterior or posterior

Site:Unit	Layer	1	2	3	4	5	6	7	8
		Measurements							
GODI:AA2.193	III:4	58.0	51.1	37.0	--	--	--	--	--

1 - GL (greatest length)
2 - Ld (dorsal length)
3 - HP (heigth at proc. extensorius)

Equus cf. hemionus

Scapula

Site:Unit - Layer		Measurements 1	2	3	4	5	6	7	8
GODI:A1.1119	IV	--	--	--	43.7	--	--	--	--
GODI:A2.1166	IV	--	80.4	53.5	--	--	--	--	--
GODI:A1.1143	IV/V	--	78.0	51.0	40.8	--	--	--	--
GODI:A1.1138	IV/V	55.2	82.1	50.8	41.7	--	--	--	--
GODI:A1.1138	IV/V	58.0	82.7	--	--	--	--	--	--
GODI:AO1.34	IV/V	54.5	79.2	49.8	42.1	--	--	--	--
GODI:A1.1133	IV/V	--	78.5	48.0	42.5	--	--	--	--
GODI:A1.1156	IV/V	--	83.0	53.4	45.8	--	--	--	--
GODI:A2.1190	IV/V	--	80.6	50.0	42.8	--	--	--	--
ALIK:4/150-160		--	74.7	48.0	--	--	--	--	--
ALIK:88/90-100		--	69.4	--	35.0	--	--	--	--
ALIK:91/540-550		--	--	44.8	38.3	--	--	--	--
ALIK:62/420-450		--	73.5	47.2	37.5	--	--	--	--
ALIK:10		--	--	49.1	38.1	--	--	--	--

1 - SLC (smallest length of collum)
2 - GLP (greatest length of proc. art.)
3 - LG (length of facies art.)
4 - BG (breadth of facies art.)

Humerus

Site:Unit - Layer		Measurements 1	2	3	4	5	6	7	8
GODI:AA2.193	III:4	63.2	58.4	28.4	--	--	--	--	--
GODI:AA2.8D11	III:2	72.6	67.7	--	--	--	--	--	--
GODI:AA2.235	III:5	--	--	28.0	--	--	--	--	--
GODI:A1.1136	IV	--	61.7	--	--	--	--	--	--
GODI:A1.1159	IV	71.2	66.6	--	--	--	--	--	--
GODI:A1.1138	IV/V	68.5	67.2	--	--	--	--	--	--
GODI:A1.1156	IV/V	69.2	64.0	--	--	--	--	--	--
GODI:A2.1184	IV/V	--	67.2	--	--	--	--	--	--
GODI:AA.304	?	--	68.0	--	--	--	--	--	--
ALIK:73/500-520		64.2	63.2	--	--	--	--	--	--
ALIK:92/430-440		67.3	64.0	--	--	--	--	--	--
ALIK:83/440-450		--	69.6	--	--	--	--	--	--
ALIK:72/420-440		66.5	63.0	--	--	--	--	--	--
ALIK:94/380-400		64.7	--	--	--	--	--	--	--
ALIK:83/440-450		64.0	--	--	--	--	--	--	--
ALIK:95/420-440		62.4	61.4	--	--	--	--	--	--

1 - Bd (distal breadth)
2 - BT (breadth of trochlea)
3 - SD (smallest breadth of diaphysis)

Equus cf. hemionus

Radius

Site:Unit -	Layer	1	2	3	4	5	6	7	8
				Measurements					
GODI:A2.179	III:4	--	--	59.8	--	--	--	--	48.0
GODI:A1.1105	III:6	--	--	62.4	--	--	--	--	54.3
GODI:A1.1112	IV	--	--	66.8	--	--	--	--	54.8
GODI:A1.1119	IV	--	--	62.3	--	--	--	--	53.9
GODI:A1.1159	IV	--	--	60.6	--	--	--	--	49.7
GODI:A101.36	IV/V	--	--	67.9	--	--	--	--	52.5
GODI:A1.1138	IV/V	--	--	66.9	--	--	--	--	55.1
GODI:A1.1138	IV/V	--	--	65.5	--	--	--	--	51.3
GODI:A1.1156	IV/V	--	--	65.3	--	--	--	--	54.4
GODI:A2.1169	IV/V	--	--	65.2	--	--	--	--	53.5
GODI:A2.1190	IV/V	--	--	66.4	--	--	--	--	55.8
GODI:A2.146	III:2	66.9	--	--	--	--	--	60.6	--
GODI:AA2.8D11	III:2	--	--	--	--	--	--	60.2	--
GODI:A2.181	III:4	63.0	--	--	--	--	--	58.5	--
GODI:A1.1105	III:6	71.0	--	--	--	--	--	65.5	--
GODI:A1.1104	III/IV	69.0	--	--	--	--	--	64.9	--
GODI:A1.1138	IV/V	--	--	--	--	--	--	66.4	--
GODI:A1.1138	IV/V	77.6	--	--	--	--	--	70.2	--
GODI:A1.1156	IV/V	73.5	--	--	--	--	--	66.2	--
ALIK:30		--	--	58.6	--	--	--	--	51.9
ALIK:72/500-250		--	--	56.8	--	--	--	--	--
ALIK:39/80-100		--	--	--	--	--	--	--	52.3
ALIK:93/410-420		--	--	55.3	--	--	--	--	--
ALIK:93/450-460		--	--	61.3	--	--	--	--	--

1 - Bp (proximal breadth) 3 - Bd (distal breadth)
7 - BFp (breadth of prox. facies art.) 8 - BFd (breadth of dist. facies art.)

Ulna

Site:Unit -	Layer	1	2	3	4	5	6	7	8
				Measurements					
GODI:A1.1105	III:6	41.6	55.2	--	--	--	--	--	--
GODI:A1.1156	IV/V	37.5	60.7	--	--	--	--	--	--
GODI:A1.1138	IV/V	39.4	--	--	--	--	--	--	--
GODI:A1.1138	IV/V	39.8	--	--	--	--	--	--	--
GODI:A2.146	III:6	--	48.2	35.7	--	--	--	--	--
ALIK:14/50-60		--	--	41.1	--	--	--	--	--

1 - BPC (breadth across coronoid proc.) 2 - DPA (depth across proc. anconaeus)
3 - SDO (smallest depth of olecranon)

Equus cf. hemionus

Metacarpus III

Site:Unit - Layer		1	2	3	4	5	6	7	8
GODI:A2.168	III:3	--	33.0	--	--	--	--	--	--
GODI:A2.179	III:4	44.0	31.7	--	--	--	--	--	--
GODI:A1.369	III:6	41.2	28.0	--	--	--	--	--	--
GODI:A1.1072	III:6	--	29.7	--	--	--	--	--	--
GODI:A1.1139	IV/V	46.7	32.0	29.5	--	--	--	--	--
GODI:A1.1156	IV/V	46.7	32.5	30.0	--	--	--	--	--
GODI:A1.1156	IV/V	--	--	--	40.5	31.6	--	--	--
ALIK:92/460-470		43.0	29.0	--	--	--	--	--	--

Columns 1–8: Measurements

1 - Bp (proximal breadth)
2 - Dp (proximal depth)
3 - SD (smallest breadth of diaphysis)
4 - Bd (distal breadth)
5 - Dd (distal depth)

Pelvis (Acetabula)

Site:Unit - Layer		1	2	3	4	5	6	7	8
GODI:A1.1159	IV	55.4	--	--	--	--	--	--	--
GODI:A2.1115	IV/V	68.5	--	--	--	--	--	--	--
GODI:A1.1156	IV/V	57.8	--	--	--	--	--	--	--
GODI:A1.1156	IV/V	59.0	--	--	--	--	--	--	--
GODI:A1.1156	IV/V	59.2	--	--	--	--	--	--	--
ALIK:79/300-320		56.0	--	--	--	--	--	--	--
ALIK:8/110-120		59.0	--	--	--	--	--	--	--
ALIK:72/150-170		51.4	--	--	--	--	--	--	--

Columns 1–8: Measurements

1 - LA (length of the acetabulum)

Femur

Site:Unit - Layer		1	2	3	4	5	6	7	8
GODI:A2.233	III:6	--	--	--	77.0	--	--	--	--
GODI:A2.1169	IV/V	--	--	--	83.1	--	--	--	--
ALIK:84/330-350		--	--	--	70.1	--	--	--	--

Columns 1–8: Measurements

4 - Bd (distal breadth)

Equus cf. hemionus

Tibia

Site:Unit	Layer	1	2	3	4	5	6	7	8
				Measurements					
GODI:82/A2	?	--	--	60.5	41.5	--	--	--	--
GODI:AA2.8D11	III:2	--	--	63.1	44.2	--	--	--	--
GODI:A1.237	III:6	--	--	57.5	40.5	--	--	--	--
GODI:A1.1159	IV	--	--	63.9	44.0	--	--	--	--
GODI:A2.1166	IV	--	--	65.1	--	--	--	--	--
GODI:A1.1124	IV/V	--	--	--	42.0	--	--	--	--
GODI:A1.1126	IV/V	--	--	62.5	46.1	--	--	--	--
GODI:A1.1134	IV/V	--	--	--	46.4	--	--	--	--
GODI:A1.1138	IV/V	--	--	62.8	43.6	--	--	--	--
GODI:A1.1156	IV/V	--	--	63.8	44.8	--	--	--	--
GODI:A2.1169	IV/V	--	--	61.8	42.2	--	--	--	--
ALIK:73/130-150		--	--	61.3	--	--	--	--	--
ALIK:72/420-440		--	--	57.7	40.8	--	--	--	--

3 - Bd (distal breadth) 4 - Dd (distal depth)

Astragalus

Site:Unit	Layer	1	2	3	4	5	6	7	8
				Measurements					
GODI:A2.1059	III:6	55.2	56.9	46.9	--	53.0	47.7	--	--
GODI:A1.1138	IV/V	53.8	57.1	47.0	--	51.0	45.8	--	--
GODI:A1.1138	IV/V	51.9	55.0	43.9	--	53.5	45.9	--	--
GODI:A1.1138	IV/V	54.0	57.0	47.0	--	55.0	46.5	--	--
GODI:A1.1138	IV/V	57.0	57.0	46.2	--	50.0	43.6	--	--
GODI:A1.1139	IV/V	54.7	57.6	45.9	--	54.1	46.8	--	--
GODI:A1.1143	IV/V	49.0	--	--	--	--	--	--	--
GODI:A1.1156	IV/V	57.0	59.2	49.2	--	53.0	46.2	--	--
GODI:A2.1187	V	--	--	--	--	--	41.2	--	--
GODI:B2/A2	?	55.0	57.3	48.1	--	53.2	44.4	--	--
DINK:DI35A	IVA	51.0	51.8	41.2	--	49.6	44.4	--	--
ALIK:73/130-150		--	--	--	--	51.8	44.0	--	--
ALIK:78/477		55.2	55.1	44.9	--	55.3	44.7	--	--
ALIK:62/420-450		47.8	49.6	41.2	--	45.5	41.2	--	--
ALIK:87/250-260		46.4	48.2	41.0	--	46.9	39.4	--	--
ALIK:40/50-60		52.5	54.0	45.2	--	--	41.5	--	--
ALIK:95/420-440		48.7	50.5	--	--	51.0	43.7	--	--
ALIK:9/60-70		50.6	51.2	43.2	--	47.5	41.8	--	--

1 - GH (greatest height) 2 - LmT (medial length of trochlea)
3 - Dm (medial depth) 5 - GB (greatest breadth)
6 - BFd (breadth of dist. facies art.)

Equus cf. hemionus

Calcaneus

Site:Unit -	Layer	Measurements 1	2	3	4	5	6	7	8
GODI:A2.150	III:3	102.4	44.0	47.4	--	--	--	--	--
GODI:A2.1059	III:6	103.7	48.3	48.0	--	--	--	--	--
GODI:A1.1133	IV/V	101.6	43.5	45.5	--	--	--	--	--
GODI:A1.1156	IV/V	103.1	47.4	48.7	--	--	--	--	--
GODI:A2.1169	IV/V	112.3	47.5	--	--	--	--	--	--
GODI:A1.1133	IV/V	--	--	48.1	--	--	--	--	--
GODI:B2/A2	?	--	44.5	--	--	--	--	--	--
ALIK:73/130-150		99.0	45.4	44.2	--	--	--	--	--
ALIK:63/430-460		89.9	40.0	--	--	--	--	--	--
ALIK:62/420-450		89.1	39.4	39.1	--	--	--	--	--
ALIK:95/420-440		95.2	44.0	43.5	--	--	--	--	--

1 - GL (greatest length)
2 - GB (greatest breadth)
3 - GD (greatest depth)

Metatarsus III

Site:Unit -	Layer	Measurements 1	2	3	4	5	6	7	8
GODI:AA2.147	III:2	--	--	--	45.0	33.7	--	--	--
GODI:A2.160	III:2	--	--	--	40.9	31.6	--	--	24.4
GODI:A2.209	III:6	--	--	--	40.6	32.9	--	--	22.5
GODI:A1.1104	III/IV	42.9	38.0	--	--	--	--	--	--
GODI:A1.1138	IV/V	41.0	40.5	--	--	--	--	--	--
GODI:A1.1156	IV/V	42.8	39.5	32.0	42.7	32.0	--	281.0	23.8
GODI:A1.1156	IV/V	44.7	--	--	--	--	--	--	--
GODI:A1.1156	IV/V	44.0	--	--	--	--	--	--	--
ALIK:62/420-450		38.8	33.8	--	--	--	--	--	--
ALIK:93/360-370		--	33.5	--	--	--	--	--	--

1 - Bp (proximal breadth)
2 - Dp (proximal depth)
3 - SD (smallest breadth of diaphysis)
4 - Bd (distal breadth)
5 - Dd (distal depth)
7 - GL (greatest length)
8 - DD (depth of diaphysis)

Equus cf. hemionus

Phalanx 1 anterior or posterior

Site:Unit	Layer	1	2	3	4	5	6	7	8
		Measurements							
GODI:AA2.8D11	III:2	40.5	26.6	36.9	70.2	27.8	39.5	19.5	67.1
GODI:A2.170	III:3	39.2	--	--	--	28.9	38.0	--	--
GODI:AA.80	III:4	--	25.8	36.2	79.0	--	--	21.2	76.2
GODI:A1.99	III:4	36.2	24.2	33.1	71.5	30.0	35.0	19.4	67.3
GODI:A1.394	III:6	41.8	26.6	36.9	73.5	31.9	39.1	21.0	69.7
GODI:A1.1088	III:6	43.4	--	--	82.0	34.0	40.2	--	78.0
GODI:A2.1120	III/V	42.0	28.1	39.3	80.0	33.3	41.0	22.8	75.5
GODI:A1.1064	IV	40.8	27.3	38.1	81.9	31.5	40.2	--	--
GODI:A1.1159	IV	43.7	25.1	34.7	72.0	32.5	39.9	20.0	67.9
GODI:A1.1159	IV	45.3	27.5	38.1	81.1	33.2	42.1	22.2	77.9
GODI:A1.1159	IV	42.6	26.5	37.9	77.5	32.9	40.3	20.8	73.7
GODI:A1.1124	IV/V	43.8	27.2	38.4	79.8	32.7	42.2	21.3	76.5
GODI:A1.1138	IV/V	43.8	27.8	38.5	84.3	34.6	43.1	22.8	80.5
GODI:A1.1138	IV/V	43.4	26.8	37.0	79.1	34.2	40.8	22.0	75.4
GODI:A1.1156	IV/V	47.6	30.9	42.2	79.9	35.9	46.3	23.1	75.5
ALIK:1/190-200		39.5	23.7	33.7	75.3	31.2	37.9	20.1	71.0
ALIK:61/290-300		39.2	22.5	34.1	79.7	29.2	36.1	18.5	74.3
ALIK:29		38.2	--	--	70.3	30.5	36.0	19.4	65.6
ALIK:17		--	--	34.5	--	--	--	19.1	70.7
ALIK:5/100-110		40.4	--	--	--	29.1	37.7	--	--
ALIK:74/490-520		--	--	--	--	--	--	19.9	--
ALIK:63/430-460		--	--	36.6	--	--	--	--	--
ALIK:27/160-170		--	--	38.6	--	--	--	18.6	--
ALIK:68/568		--	--	34.3	--	--	--	18.8	--

1 - Bp (proximal breadth)
2 - SD (smallest breadth of diaphysis)
3 - Bd (distal breadth)
4 - GL (greatest length)
5 - Dp (proximal depth)
6 - DS (depth of diaphysis)
7 - Dd (distal depth)
8 - GL (greatest length)

Phalanx 2 anterior or posterior

Site:Unit	Layer	1	2	3	4	5	6	7	8
		Measurements							
GODI:A2.170	III:3	38.1	35.0	39.0	37.7	27.0	33.6	22.2	29.5
GODI:A1.69	III:4	--	--	38.4	--	--	--	--	--
GODI:A2.172	III:4	39.8	33.2	33.0	41.7	29.3	36.8	22.0	32.1
GODI:AA2.183	III:4	39.0	33.6	35.6	44.1	28.8	35.9	23.5	34.0
GODI:AA2.193	III:4	49.9	43.5	46.4	47.5	32.7	46.2	27.9	39.2
GODI:AA2.193	III:4	52.4	43.0	48.2	45.1	30.9	45.1	--	37.3
GODI:A1.1096	III:6	39.9	35.2	36.5	42.3	29.6	--	24.7	33.6
GODI:A01.33	IV/V	42.9	38.7	41.4	41.5	28.5	40.1	25.9	32.3
GODI:A1.1138	IV/V	42.4	36.5	38.5	43.5	28.8	37.5	24.2	33.4
GODI:A1.1156	IV/V	--	33.4	--	40.0	27.0	35.2	23.3	31.5

continued

1 - Bp (proximal breadth)
2 - SD (smallest breadth of diaphysis)
3 - Bd (distal breadth)
4 - GL (greatest length)
5 - Dp (proximal depth)
6 - DS (depth of diaphysis)
7 - Dd (distal depth)
8 - GL (greatest length)

Equus cf. hemionus

Phalanx 2 anterior or posterior

Site:Unit	Layer	1	2	3	4	5	6	7	8
		Measurements							
		continued							
GODI:A2.1169	IV/V	40.8	34.4	35.5	41.4	28.5	36.1	23.4	31.2
GODI:A1.1155	V	42.3	35.6	37.6	43.5	29.0	36.5	24.5	33.5
GODI:A1.1155	V	40.2	--	37.3	41.0	28.8	37.0	23.8	31.3
ALIK:91/270-280		38.7	34.5	36.2	37.5	25.5	34.5	21.8	29.2
ALIK:76/250-270		38.0	32.4	36.0	38.9	26.2	36.0	21.9	32.3
ALIK:88/580		36.9	32.8	34.5	36.5	26.4	33.4	23.4	29.1
ALIK:92/430-440		37.7	33.5	35.2	34.6	23.5	34.0	19.9	27.1
ALIK:71/150-170		39.9	33.7	35.0	39.2	--	34.9	22.7	30.7
ALIK:28/120-130		40.0	35.9	--	37.2	25.9	37.2	22.3	29.0
ALIK:88/120-140		38.5	33.5	--	--	27.1	34.4	22.7	--
ALIK:53/140-150		39.7	37.5	38.5	38.2	--	36.5	--	30.0
ALIK:54/90-110		38.0	34.0	36.6	35.9	24.0	34.7	21.3	30.7
ALIK:76/200-230		41.7	37.8	39.1	40.1	27.0	38.0	23.5	33.5
ALIK:99/420-440		--	32.7	33.3	39.2	--	--	22.4	29.7
ALIK:29		39.7	37.9	38.7	39.7	--	36.0	--	30.5

1 - Bp (proximal breadth)
2 - SD (smallest breadth of diaphysis)
3 - Bd (distal breadth)
4 - GL (greatest length)
5 - Dp (proximal depth)
6 - DS (dèpth of diaphysis)
7 - Dd (distal depth)
8 - GL (greatest length)

Phalanx 3 anterior or posterior

Site:Unit	Layer	1	2	3	4	5	6	7	8
		Measurements							
GODI:A2.170	III:3	--	44.1	34.0	--	33.6	--	--	--
GODI:A2.179	III:4	--	--	--	--	35.8	--	--	--
GODI:AA2.183	III:4	52.1	46.8	35.0	55.5	40.0	--	--	--
GODI:AA2.193	III:4	--	52.2	39.0	60.4	46.4	--	--	--
GODI:A2.302	III:6	49.3	44.4	34.0	54.2	36.1	--	--	--
GODI:A1.1159	IV	50.0	46.4	32.5	53.2	35.8	--	--	--
GODI:A1.1156	IV/V	--	--	37.5	--	42.5	--	--	--
GODI:A1.1156	IV/V	--	46.9	31.5	52.0	35.1	--	--	--
ALIK:84/546		48.2	45.4	32.5	47.5	32.6	--	--	--
ALIK:62/524		--	47.2	32.4	--	36.0	--	--	--
ALIK:88/580		54.3	48.3	33.7	53.5	33.7	--	--	--
ALIK:79/502		--	41.8	33.1	55.5	35.6	--	--	--
ALIK:71/120-140		47.0	46.4	--	--	32.8	--	--	--
ALIK:84/460-480		56.1	47.5	35.3	--	--	--	--	--
ALIK:74/613		--	43.1	--	--	--	--	--	--
ALIK:84/500-520		--	--	29.0	--	35.0	--	--	--
ALIK:93/520-540		--	43.5	32.1	--	--	--	--	--

1 - GL (greatest length)
2 - Ld (dorsal length)
3 - HP (heigth at proc. extensorius)
4 - GB (greatest breadth)
5 - BF (breadth of facies art.)

SÁNDOR BÖKÖNYI

Late Chalcolithic Horses in Anatolia

Contents

1. Introduction

The presence of true horses in the late chalcolithic period of Anatolia has recently become the subject of lively discussion. The question, however, is not one of identification. Their large size and the enamel patterns of the upper and lower cheekteeth (particularly the presence of a *pli caballin* on the upper ones) clearly show that these late fourth millennium B.C. equid remains came from horses and not half-asses or asses. Instead, the argument is about whether this material came from wild or domestic horses. Horses at this period are particularly interesting because they are the earliest known from Anatolia and because they occur there immediately after the first domestication of the horse in eastern Europe in the middle of the fourth millennium B.C. (BIBIKOVA 1967, 1970).

2. Materials

Late chalcolithic horse bones have been found at five sites in Anatolia (Table 1). All of the sites listed lie in a comparatively resticted area of eastern Anatolia in a narrow strip approximately seventy kilometers long between the towns of Elaziğ and Malatya on both banks of the Euphrates River. It is certainly true that one reason for this localization is the amount of archaeological activity that has been concentrated in this area particularly as the result of rescue excavations connected with the building of the Keban Dam. Nevertheless, the occurrence of horse bones must also be due to the geographical situation of the region lying as it does with ready access to the southern slopes of the Caucasus Mountains. Passes through these mountains formed natural bridges between Anatolia and the early horse-raising areas of eastern Europe and Central Asia. Assuming that the horses of Anatolia came from these areas, we would also expect to find horse remains at contemporary sites all along the routes from the Caucasus into eastern Anatolia.
BOESSNECK/VON DEN DRIESCH (1976: 86) have stated that they believe the horse bones and teeth from Norşun-tepe, Tülintepe, and Tepecik came from wild animals. They base this opin-

TABLE 1

Chalcolithic Horse Bones from Anatolia

	no. of all bones	no. of horse bones	reference
Norşun-tepe	9,000	c. 60	BOESSNECK/VON DEN DRIESCH 1976: 82
Tulintepe	?	few	*ibid.*: 83
Tepecik	c. 3,600	few	*ibid.*: 83
Arslantepe	3,880	16	BÖKÖNYI 1988
Değirmentepe	5,860	21	BÖKÖNYI, in preparation

ion on the stocky proportions and large size of the bones and on the fact that in the succeeding Early Bronze Age levels of Norşun-tepe, no horse bones were found. They believe that it is unlikely that horses had been kept on the Altinova Plain (where the sites are located) during the chalcolithic and then reintroduced a thousand years later having disappeared from the area in the meantime. Instead, they argue that wild horses lived on the Altinova Plain during the late chalcolithic period and were hunted by the local inhabitants. These wild horses became rare or extinct, and domestic horses were introduced into the area at the end of the third or beginning of the second millennium B.C.

I have never been particularly satisfied with this explanation. In fact, I have always believed that the possibility of these late chalcolithic horses being domestic schould not be excluded (BÖKÖNYI 1978: 55, 1980a: 923). Now I also have found late chalcolithic horse remains amongst the faunal remains from two sites close to the Altinova Plain. One of these sites, Arslantepe, lies a little north of Malatya and about twelve kilometers south of the Euphrates River (as the crow flies). The other site, Değirmentepe, is situated north of Arslantepe directly on the southern bank of the Euphrates. Arslantepe was excavated by a team from the University of Rome directed by Professors S. Puglisi and A. Palmieri, Değirmentepe by a team from the University of Istanbul directed by Professor Ufuk Esin.

Among the Arslantepe horse remains were upper cheekteeth (Plate 1) that make the identification certain. There were, however, no whole bones and only three specimens could be measured (distal metacarpal and metatarsal, anterior second phalanx; see Table 2). A well-arched, narrow anterior third phalanx (Plate 2) was also discovered at the site, this showing that the horses of Arslantepe had a steppic character (BÖKÖNYI 1988: 587). The measureable bones suggest heavily built horses similar to both the Norşun-tepe horses and to the earliest domestic horses of eastern and central Europe.

An important feature of Arslantepe that supports my contention that the chalcolithic horses there were domestic is the presence of horse bones in every phase of the succeeding Early Bronze Age at the site. Thus there was uninterrupted horse-keeping at Arslantepe since the late fourth millennium B.C.

The excavations of Değirmentepe yielded twenty-one horse bones. These included cheekteeth (although only upper and lower deciduous premolars, Plate 3) but, as at Arslantepe, no complete long bones. Once again, a third phalanx was found (Plate 4), but there was only a single measurable bone, a distal metacarpal (Plate 5, Table 2).

Plate 1: Upper cheekteeth of horses from Arslantepe (scale only for specimen at right)

TABLE 2

Measurements of Horse Bones (in millimeters)

	Arslantepe McIII	Değirmentepe McIII	Arslantepe MtIII	Arslantepe Ph. 2 ant.
Sagittal Length	-	-	-	43
Proximal Breadth	-	-	-	54.5
Smallest Breadth	-	-	-	48
Distal Breadth	(52)	52	(54)	50.5
Proximal Depth	-	-	-	34
Smallest Depth	-	-	28	25
Distal Depth	38.5	37	41	29

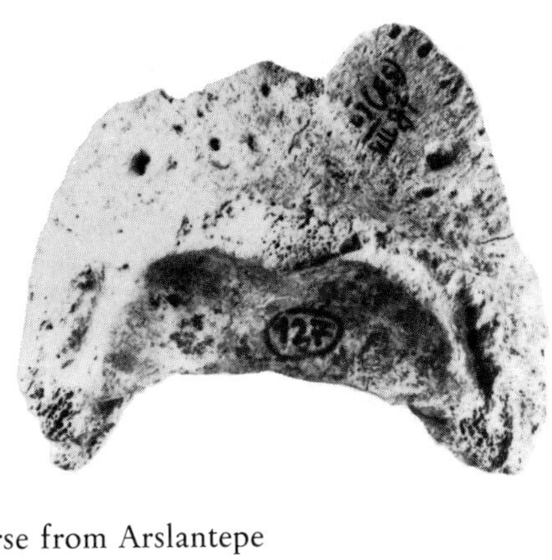

Plate 2: Third phalanx of a horse from Arslantepe

Plate 3: Upper deciduous cheekteeth of horses from Değirmentepe

Plate 4: Third phalanx of a horse from Değirmentepe

Plate 5: Distal metacarpal of a horse from Değirmentepe

When the measurements for the distal metacarpals from the Norşun-tepe, Arslantepe, and Değirmentepe horses are plotted on a scatter diagram (Figure 1), they lie in the middle of the range of variation for the earliest eastern European horses. (Since the distal depth of the Norşun-tepe metacarpus has not been published, its value was only approximately placed in Figure 1.)

Unfortunately, the Early Bronze Age level of Değirmentepe yielded very few bones (79 total). Although this small collection did not contain any horse bones, it clearly cannot be considered representative. Thus, the question of uninterrupted horse-keeping remains open for Değirmentepe.

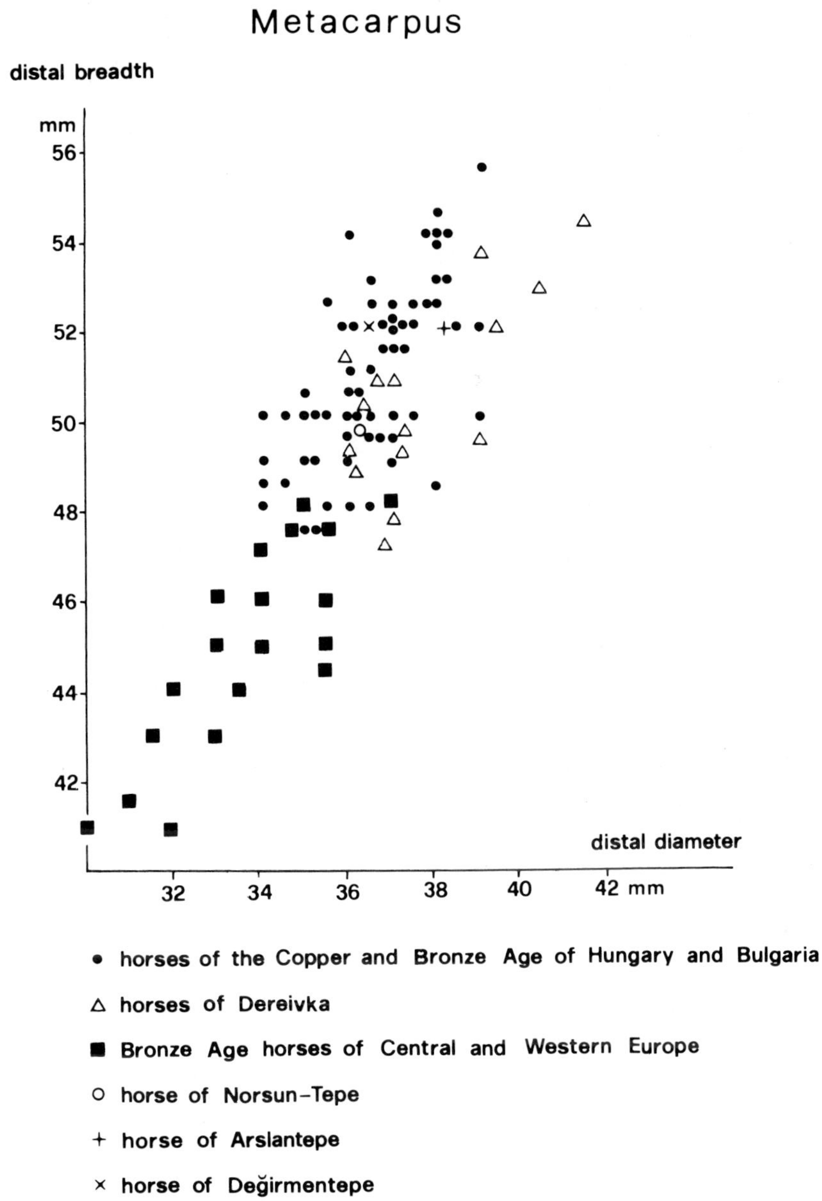

Figure 1: Metacarpus measurements of prehistoric horses from Europe and Anatolia

Değirmentepe did, however, yield some important finds from the late chalcolithic levels that can serve as evidence to support my contention that horses were kept. This material is part of a horse foetus (Plate 6). Only the diaphyses of six long bones could be saved, however; these are, with their greatest length measurements (in mm.): humerus (42), radius (58), metacarpus (46), femur (53), tibia (58), and metatarsus (59).
From the sizes of these bones, Dr. Gy. Feher, Professor of Anatomy at the Veterinary Faculty of Budapest, whose special field is the embryology of domestic animals, could determine that

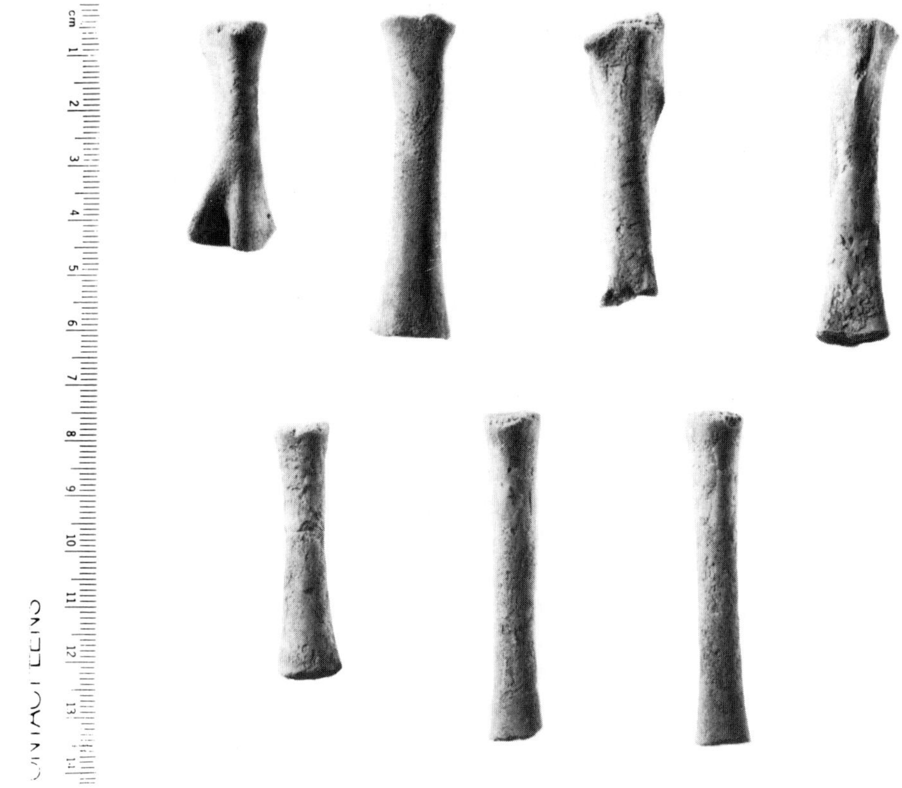

Plate 6: Fetal horse bones from Değirmentepe

they had come from an approximately eight-month old foetus. This means that since wild and primitive domestic horses foal some time in the Spring months, the miscarriage represented by these bones could have taken place in late winter.

As far as I am aware, developed hunters do not kill pregnant wild animals. In my more than thirty years as a practising archaeozoologist, I have never found remains of a foetus of a wild animal at a site although I have seen quite a few examples from different domestic species. Thus, I believe that this foetus from Değirmentepe provides further evidence for horses being kept at the site.

3. Discussion

Reviewing all the evidence we presently have for late chalcolithic horses from Anatolia, I cannot imagine that they were wild animals. I base my opinion on eight principal arguments.

First, there have not yet been found any early Holocene wild horse remains in Anatolia. While it is true that the amount of archaeozoological research carried out there has not been great, nevertheless it is noteworthy that not a single specimen from a horse has been found in neolithic context.[1]

Second, there are no data supporting the extension of Przewalsky horses into southern Anatolia. Just the opposite is the case: the range of distribution of the species shrank during the Holocene. Third, on the basis of the large collections of early domestic horse bones from sites in eastern Europe and the Carpathian Basin, it is clear that those horses were stocky and close to the wild species in proportion. This is not surprising since those horses were the first domestic horses in the region and had not had the time to undergo changes in their skeleton. In Hungary where a large number of prehistoric and early historic horse remains have been studied, I have documented that horses of the Copper and Bronze Ages were sturdier and resembled the wild form more than those of later periods (Bökönyi 1974: 242), and there has never been any doubt concerning the domestic status of these animals.

As a matter of fact, I am disturbed that the practice of the "good old times" seems to be coming back, that is, the tendency for every horse bone from before the beginning of the Bronze Age to the considered as having come from a wild animal and every horse bone from after as having come from a domestic animal. We should avoid this tendency since we fortunately have a large collection of eneolithic (chalcolithic) horse bones from Dereivka, the careful analysis of which has demonstrated that those horses were different from the wild horses of the region and that the differences that exist point toward the animals having been domesticated (Bibikova 1967: 106 and Nobis 1971: 60, 75, 79). Finally, from that same site come six fragmentary antler toggles or cheekpieces of bits (Telegin 1973: 324).

Fourth, measurements of the early Anatolian horses fit within the size variation defined for early domestic horses from centers of early horse domestication in eastern Europe. In fact, the Anatolian measurements form a small cluster in the middle of the range of variation for early eastern European horses and thus do not represent particularly stocky animals at all.

Fifth, the occurrence of horse bones in late chalcolithic levels of Anatolian sites falls within the time period for the first great diffusion of domestic horses from the center of their earliest domestication. Horses reached the Carpathian Basin, Moravia, Moldavia, eastern Roumania, the forest steppe and steppe belts between the Dnieper and Volga rivers, the northern foothills of the Caucasus, and the Caucasus mountains themselves during this period (Bökönyi 1978: 25), and they even appear in late chalcolithic sites of Transcaucasia (V. Titov, personal communication). Thus there seems little reason to suppose that horses were not brought into Anatolia by the end of the fourth millennium B.C. These earliest domestic horses of Southwest Asia need not have come from the southern Ukraine or southern Russia. Since Przewalsky

[1] In this respect, the case of the earliest neolithic "horses" of Çayönü, East Anatolia, should be mentioned. B. Lawrence (1980) noted the presence of twelve *Equus* bones from the site. (Apart from sheep and goats, she provided identifications to only the genus level because of the preliminary nature of her report.) When I visited the Çayönü excavations in 1979, I could see that these equid remains came from small half-asses. Then, to my great surprise, I saw the term "horse" ("Pferd") on the list of the hunted animals from the site in the German text of the Bittel Festschrift (Çambel/Braidwood 1983). I immediately contacted Professor Braidwood who informed me that the word "Equus" of the original English version had been translated as "Pferd" by Professor Halet Çambel who, as a prehistorian, was not aware of the important distinction between identification to the genus and species level. This is merely one example of how careful one must be with the publication of animal remains from a critical area.

horses lived in many areas east of the Urals, chalcolithic centers of domestication could have developed anywhere there where socio-economic developments reached a suitable level.
Sixth, the fact that the numbers of horse bones from Anatolian chalcolithic deposits are small is not surprising, given that in these cases we are dealing with a situation similar to that found at Copper Age sites in Central Europe. Horses there were generally nothing but status symbols and thus genuine horse breeding programs were not begun (Bökönyi 1978: 25, 1980 b: 16). They were owned by high status individuals in small numbers and thus their remains are more likely to occur at larger regional centers like Arslantepe than at smaller settlements like Değirmentepe where they are not found in the Bronze Age.
Seventh, the occurrence of the remains of a horse foetus strongly supports the domesticated status of at least the chalcolithic horses at Değirmentepe.
Eighth (and finally), the fact that domestic horses occurred from c. 3000 B.C. without local wild predecessors in several parts of Southwest Asia (Zarins 1976: 519) certainly suggests that in Anatolia, which lay on the route to Mesopotamia and the Zagros from early centers of horse domestication, domestic horses would be expected to occur somewhat earlier, hence in the chalcolithic period.
In conclusion, I am well aware of the fact that there are two kinds of direct evidence for chalcolithic domesticated horses that would be irrefutable, namely morphological changes in the skeleton and the presence of horse gear (e.g., cheekpieces or bits). Unfortunately, in horses morphological changes came only long after they are domesticated, and to find horse gear requires a lot of luck. Thus it is unlikely that this sort of evidence will be forthcoming, and we are forced to depend on more indirect lines of evidence and reasoning. The arguments presented above indicate that the chalcolithic horses of Anatolia were domesticated and should be evaluated as such when discussing Holocene horse evolution in the Middle East.

4. Literature

Bibikova, V.I., 1967, K izucheniu drevneishkh domashnikh loshadei Vostochnoi Evropy (Studies of ancient domestic horses in Eastern Europe). Bulleten Moskovskogo Obshchestva Ispitatelei Prirody, Otdel Biologicheskii **72**, 106–118.

Bibikova, V.I., 1970, K izucheniu drevneishkh domashnikh loshadei Vostochnoi Evropy (Studies of ancient domestic horses in Eastern Europe) 2. Bulleten Moskovskogo Obshchestva Ispitatelei Prirody, Otdel Biologicheskii **75**, 118–125.

Boessneck, J./Driesch, A.v.d., 1976, Pferde im 4./3. Jahrtausend v. Chr. in Ostanatolien. Säugetierkundliche Mitteilungen **24**(2), 81–87.

Bökönyi, S., 1974, History of Domestic Animals in Central and Eastern Europe; Budapest: Akadémiai Kiadó.

Bökönyi, S., 1978, The earliest waves of domestic horses in East Europe. Journal of Indo-European Studies **6**, 17–76.

Bökönyi, S., 1980 a, La doméstication du cheval. La Recherche **11**(114), 919–926.

Bökönyi, S., 1980 b, The importance of horse domestication in economy and transport. In: Sørbom, P., ed., Transport Technology and Social Change; (Tekniska Museet Symposia **2**); Stockholm, 15–21.

Bökönyi, S., 1988, Late chalcolithic animal remains from Arslantepe near Malatya, Anatolia: a preliminary report. Origini **12**(2): 581–598.

Çambel, H./Braidwood, R.J., 1983, Çayönü Tepesi: Schritte zu neuen Lebensweisen. In: Boehmer, R. M./Hauptmann, H., eds., Beiträge zur Altertumskunde Kleinasiens; (Festschrift für Kurt Bittel); Main: von Zabern, 155–166.

Lawrence, B., 1980, Evidences of animal domestication at Çayönü. In: Çambel, H./Braidwood, R.J., eds., The Joint Istanbul - Chicago Universities' Prehistoric Research in Southeastern Anatolia, I; Itanbul: Edebiyat Fakültesi Basimevi, 285–308.

Nobis, G., 1971, Vom Wildpferd zum Hauspferd; (Fundamenta, Reihe B, Band 6); Köln: Bohlau Verlag.

Telegin, D.I., 1973, Über einen der ältesten Pferdezuchtherde in Europe. In: Actes du VIIIe Congrès International des Sciences Préhistoriques et Protohistoriques 2; Beograd, 324–327.

Zarins, J., 1976, The Domestication of Equidae in Third Millennium B.C. Mesopotamia; Dissertation, University of Chicago, Chicago.

Sándor Bökönyi, Archaeological Institute of the Hungarian Academy of Sciences, Uri utca 49, 1250 Budapest 1, Hungary

SEBASTIAN PAYNE

Early Holocene Equids from Tall-i-Mushki (Iran) and Can Hasan III (Turkey)[1]

Contents

1. Introduction

The identification of equid remains from Holocene sites in the Near East presents considerable problems because of the number of species that might be represented. At least three wild species of *Equus* were present in the Near East during the Late Pleistocene: *E.caballus, E.hydruntinus,* and *E.hemionus* (GARROD/BATE 1937, DAVIS 1980), and DUCOS (1970) has suggested that wild *E.asinus* was present in north Syria in the early Holocene.[2] Further complication is added by the first appearance and spread of domestic horses, donkeys, and possibly, for a limited period, domestic onagers. As most equid species are interfertile [although the resulting hybrids are generally sterile (GRAY 1954, 1966)], the remains of hybrids such as the mule may also be found.

With so many closely-related forms, it is not surprising that identification is a matter of considerable difficulty and debate. Problems are compounded by the extinction or rarity of most of the wild species of *Equus* and the consequent scarcity in museum collections of reliable comparative material. *E.caballus* is extinct in the wild; *E.hydruntinus* is completely extinct; *E.hemionus* survives in Iran and farther east, but the Syrian onager (*E.hemionus hemippus*) is almost certainly now extinct; and wild asses are now confined to northeastern Africa [the wild asses reported from Oman are almost certainly feral donkeys (HARRISON 1972)]. Such wild

[1] I am most grateful to Gail BULL, Simon DAVIS, Richard MEADOW and John WATSON for reading and commenting on earlier drafts of this paper.

[2] BOHLKEN (1961) and GROVES (1971) have argued [a view followed by CORBET (1978)] that names based on domesticated forms should be ignored for the purposes of formal zoological nomenclature. *E.caballus* and *E.asinus* would thus be excluded in favour of *E.ferus* BODDAERT, 1785 and *E.africanus* FITZINGER, 1857. This paper retains the more traditional usage.

equid specimens as there are in museum collections are often poorly documented, often from zoos rather than from the wild, and often juvenile or senile individuals. Domestic horses and donkeys are well-represented in museum collections, but few of them are of Near Eastern breeds or types, and there are no modern domestic onagers. Equid hybrids are only poorly represented in museum collections. As many zoologists have only collected skulls, the situation is even worse when the postcranial skeleton is considered.

In keeping with normal zoological practice, the available modern reference material has, in the past, been used to describe the different taxa that we recognize and to establish their distinguishing characteristics. Data from the closest available samples are taken as substitutes when local samples are not available. For example, the characteristics of the extinct Syrian onager are reconstructed in part from the half-dozen skeletons that survive and in part from the larger Iranian onager. The scanty and patchy nature of the available modern reference material, however, makes such procedures hazardous and tends to create a simplistic picture of equid variation. Thus, local variation and the distinctive features of local populations may be largely ignored, and when we try to identify equid remains from archaeological sites, we are more likely to force them into straitjackets based on the limited modern reference material. Instead, we should make more use of fossil samples to provide us with a picture of past equid populations to add to and complement our picture of present equid taxa, and we should pay more attention to the study of morphological features that can be examined both in present and in past equid populations, such as dental and osteological characters. This would also make us more likely to recognize extinct and as-yet-unrecognized taxa, as REGALIA (1907; see also STEHLIN/GRAZIOSI 1935) did when he first described *Equus hydruntinus* from the Late Pleistocene of southern Italy. An explicit recognition of the usefulness of this approach is needed in order to encourage the fuller description and publication of fossil samples from what are, in palaeontological terms, very recent sites.

When we try to unravel the history of wild and domestic equids in the Near East, the early Holocene is a particularly crucial period, being a time at which essentially modern climatic conditions had become established but, as far as is currently known, domestic equids had not yet appeared on the scene. Fortunately equids are quite common from a wide range of sites of this date in the Near East. My purpose in this paper is to describe the equids from two such early Holocene sites: an equid from Can Hasan III in south-central Turkey that closely resembles *Equus hydruntinus* and an equid from Tall-i Mushki in western Iran that is probably *E.hemionus.* Emphasis is placed on objective description and, in particular, on measurements as providing a means of objective comparison between different equids and between samples from different sites. When applied to other sites, this approach should enable us to build up a detailed picture of the morphological and metrical characteristics of early Holocene wild equid populations in the Near East, thus allowing more reliable identification of fossil material and providing a more secure basis for the study of the more complex situation at sites where both wild and domestic equids may be present.

2. Methods

Most of the criteria used to identify recent fossil equids have relied on two kinds of character: differences in the pattern of enamel folding in the cheekteeth, and differences in the size and robustness of the postcranial bones.

More attention is given in this paper to cheekteeth than to postcranial bones. This mainly reflects the abundance of cheekteeth as compared with postcranial bones in the two samples that are described, a situation common in many samples of equids from archaeological sites. Dental characteristics may finally prove to be the best means of identifying fossil equid samples to the subgeneric and specific level, as size and robustness in the postcranial skeleton respond more to nutritional status and other environmental variables and probably respond more rapidly to selective pressures.

2.1. Cheekteeth

The literature concentrates most on differences in the pattern of enamel folding in the cheekteeth. While enamel patterns clearly vary significantly between different equid species, it is often hard to quantify these differences or to describe them objectively. The literature is full of subjective decriptions and page after page of drawings of enamel patterns, neither of which are easy to compare with specimens or data from other sites. For this reason particular attention is paid in the following descriptions to characters that can be measured or in some way reasonably objectively assessed and described. Also, as has been shown by serial sections through single teeth, the pattern of enamel folding at the occlusal surface changes considerably as the tooth is worn. It is important, therefore, to consider occlusal enamel measurements and morphology in relation to the varying extent to which different teeth are worn.

It has also long been recognized that the pattern of enamel folding differs in different teeth in the same tooth row (especially between molars and premolars). Most of the teeth in most archaeological samples are isolated teeth, and distinguishing isolated molars from premolars (apart from second premolars and third molars) is by no means as simple as many publications seem to suggest. Criteria that distinguish reliably between premolars and molars in some species and samples will not necessarily do so in others. For this reason the distinction between molars and premolars needs to be separately assessed for each sample.

2.1.1. Measurements

The approach used follows that of EISENMANN (1980: 80–81, 1981: 133–134), with some modifications and additions. The measurements and terms used are illustrated in Figure 1;[3] for the reasons given by EISENMANN, measurements were taken at and in the plane of the occlusal surface.

In both *upper and lower teeth*:

OL is the mesio-distal occlusal length, taken along the mesiodistal axis of the occlusal surface from the approximate center of its mesial side to the approximate centre of its distal side. External cement is included [EISENMANN's Lo excludes external cement (EISENMANN 1980: 81 and Fig. 43; 1981: 133–4 and Fig. 2A)].

[3] Figure 1 is partly based on drawings sent to me by Véra EISENMANN in 1975. I am most grateful to Dr. EISENMANN for her kindness in sending me details of her measurements some years before they were published.

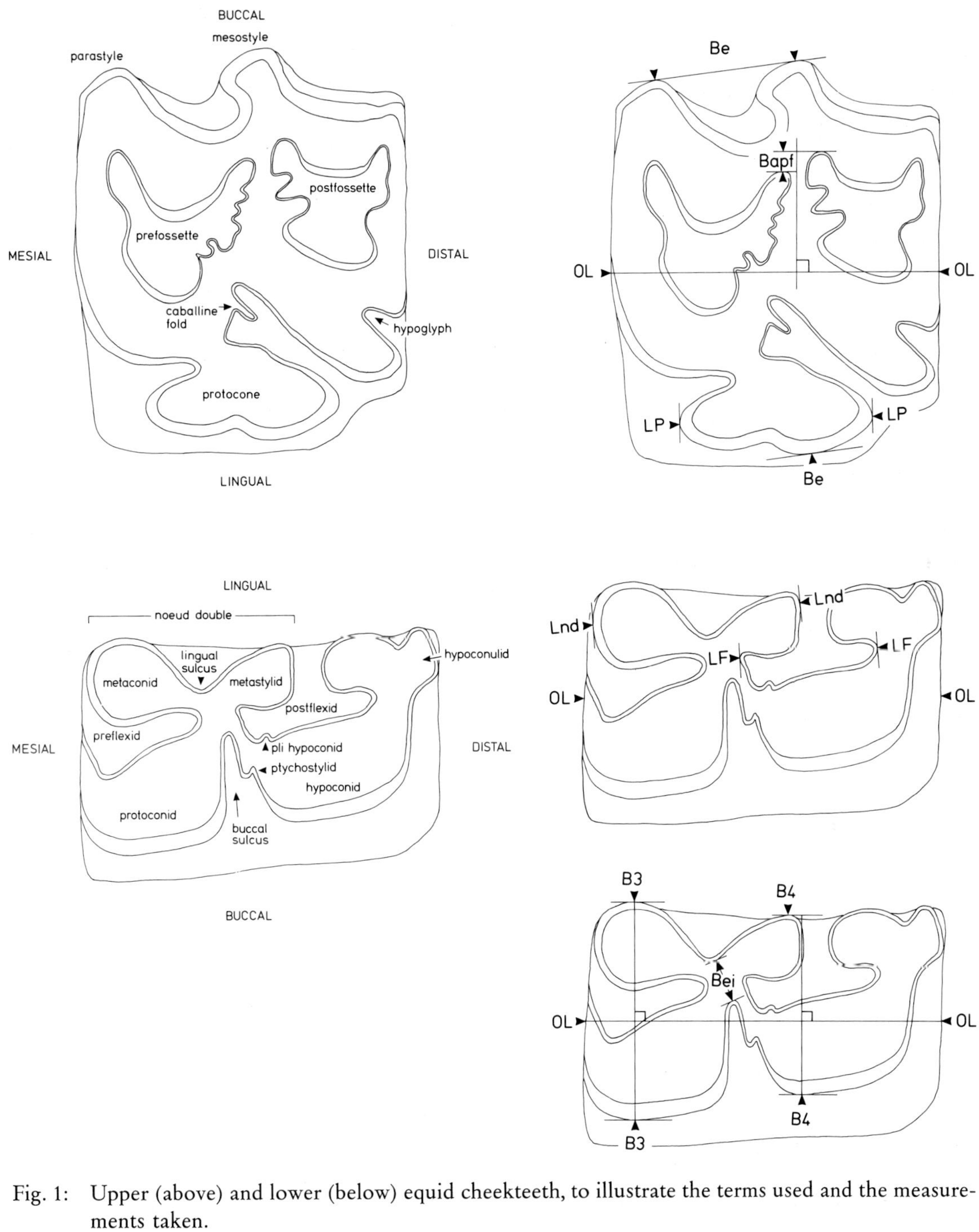

Fig. 1: Upper (above) and lower (below) equid cheekteeth, to illustrate the terms used and the measurements taken.

In *upper teeth*:

Be is the buccolingual occlusal length, taken with one jaw of the callipers in contact with the outer surface of the enamel of both the parastyle and the mesostyle, and the other jaw in contact with the outer surface of the enamel of the protocone; the external cement is not included. This is not the same as EISENMANN's lo, which is taken at right angles to Lo (EISENMANN 1980: 81 and Fig. 43).[4]
Bapf is the distance by which the mesial horn of the postfossette projects buccally beyond the distal horn of the prefossette, at right angles to the direction along which OL is taken. This was estimated to the nearest 0.2 mm from enlarged photographs of the occlusal surfaces.
LP (EISENMANN 1980) is the greatest length of the protocone, including the enamel, in whatever direction gives the greatest dimension (this is often a slightly different direction from that along which OL is taken).[5]

In *lower teeth*:

B3 is the width across the protoconid and metaconid, and B4 that across the hypoconid and metastylid, both at right angles to the direction along which OL is taken. Both B3 and B4 include the enamel but not the external cement.
Lnd is the greatest length of the "double knot" (metaconid + metastylid), and LF (EISENMANN 1981) is the greatest length of the postflexid. Like LP in the upper cheekteeth, both Lnd and LF include the enamel, and are taken in whatever direction gives the greatest dimension.[6]
Bei is the smallest distance between the internal enamel of the buccal sulcus and the internal enamel of the lingual sulcus.

2.1.2. Morphology

Selected morphological characters were scored as follows:

In *upper teeth*:

The development of the caballine fold was graded as follows (Figure 2):

O None (*e.g.*, 2MS 155-4).
tr Trace (*e.g.*, 2MS 163-2).
\+ Present (*e.g.*, 2MS 61-3).
\+ + Marked (*e.g.*, 2MS 1-5).

[4] Be was called B in notes sent in 1982 to Melinda ZEDER (ZEDER 1986); I have thought it preferable to rename it to avoid possible confusion with VON DEN DRIESCH's B which includes the external cement (VON DEN DRIESCH 1976).

[5] LP was called Lpc in notes sent in 1982 to Melinda ZEDER (ZEDER 1986); I have thought it preferable to rename it to follow EISENMANN (1980) which was not available to me at the time the notes were written.)

[6] LF was called Lent in notes sent in 1982 to Melinda ZEDER (ZEDER 1986); I have thought it preferable to rename it to follow EISENMANN (1981) which was not available to me at the time the notes were written.

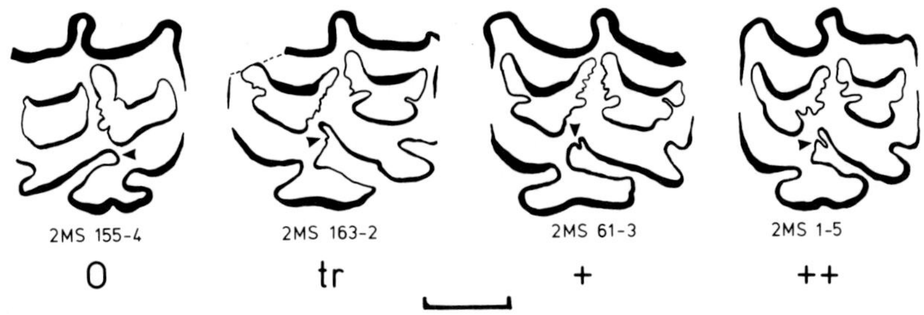

Fig. 2: Enamel patterns of selected upper cheekteeth from Tall-i-Mushki, to illustrate the classes used to record the development of the caballine fold (arrowed). The development of the pli hypoconid and the ptychostylid in the lower teeth were recorded in the same way.

In *lower teeth*:

The development of the pli hypoconid and ptychostylid were similarly graded:

O None.
tr Trace.
+ Present.
++ Marked.

The degree of penetration of the buccal sulcus was classified in four grades depending on the position of the tip of the sulcus in relation to the postflexid and the preflexid (Figure 3; see also EISENMANN 1981, where an essentially similar classification is used):

1 Tip does not reach the line joining the buccal-most parts of the enamel of the postflexid and preflexid.
2 Tip crosses that line, but does not reach the line joining the preflexid and postflexid at their nearest point of approach (as in the example in Figure 3).
3 Tip is across that line but is still more than 0.5 mm from the lingual sulcus.
4 Tip is within 0.5 mm of the lingual sulcus.

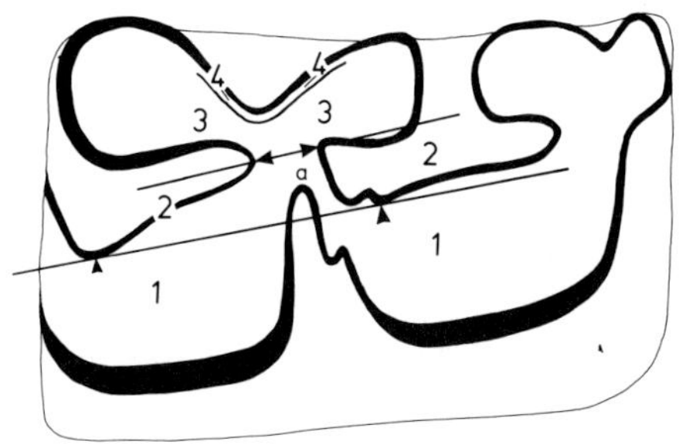

Fig. 3: Lower cheektooth, to illustrate the classes used to record the penetration of the buccal sulcus. In this example, the tip of the sulcus (a) crosses the line joining the buccal-most parts of the postflexid and the preflexid but does not reach the line joining the postflexid and the preflexid at their nearest point of approach. It would therefore be classed as grade 2.

2.1.3. Wear stages

In order to examine variation in relation to wear, teeth were classified in five wear stages[7]:

1 Early wear: some dentine exposed by wear, but exposed dentine only in limited areas which are not yet fully interconnected, and enamel pattern not yet fully established; fossettes not yet isolated in upper teeth.
2 Full wear; still in upper third of crown. In $M_{1/2}$, an early stage during which the hypoconulid projects distally (markedly overhanging the main part of the crown) was distinguished as 2*; $M^{1/2}$ pass through a similar stage, but this mainly happens during wear stage 1. The main reason for distinguishing stage 2* is that OL is considerably longer (sometimes by several mm) at this stage than during later wear stages (see Figure 7).
3 Full wear; in middle third of crown.
4 Full wear; in lower third of crown.
5 Crown worn close to roots; enamel pattern starting to simplify and break down, fossettes disappearing in upper teeth.
The boundaries between these stages are not sharply drawn, and the resulting classification is somewhat subjective. The results, however, clearly help us to understand some of the variation seen in the measurements and morphological data. It should also be noted that wear in stages 2, 3, and 4 cannot usually be distinguished if the teeth are still enclosed in the jaw, as the height of the crown cannot be seen.
No measurements or morphological data are given for teeth in wear stage 1. When a tooth is broken across the lower crown, morphological features can often be seen on the broken surface as they would appear at a later wear stage. Such data are given in Appendices 1 and 2 within square brackets: thus 2MS 4–18 (Appendix 1), an $M_{1/2}$ at wear stage 2* with well-marked pli hypoconid and ptychostylid, was broken across the lower part of the crown at a point equivalent to wear stage 4, where it could be seen that the pli hypoconid and ptychostylid had both disappeared.

2.2. Postcranial skeleton

Postcranial measurements were taken as described by von den Driesch (1976), with the following comments and additions:

Scapula, BGmin: as defined by von den Driesch, it is not clear whether BG should be taken as a minimum, or at right angles to GLP. Here, BGmin is taken as a minimum, whether or not at right angles to GLP.
Astragalus, LlT: taken in the same manner as LmT, but on the lateral trochlea.
Metapodia, Bda, Bdfp: as defined by von den Driesch, it is not clear whether Bd should be taken across the condyle or across the fusion plane – sometimes one is larger, sometimes the other. Bda is taken across the condyle, Bdfp across the fusion plane.

[7] This classification differs to some extent from that included in notes sent to Melinda Zeder (Zeder 1986): stage 1 is the same; stages 2, 3, and 4 here divide the crown into an upper, middle and lower third (previously the division was into an upper and a lower half); stage 5 here includes both 'worn almost to roots' and 'infundibula erased, uppers'.

Phalanx 1, GLcal: As defined by VON DEN DRIESCH, GL is taken in a measuring box. GLcal is taken with calipers, with the jaw at the proximal end in contact with the bone both medially and laterally.

A distinction is customarily made in archaeological samples between anterior and posterior phalanges. This distinction is probably not as simple and reliable as the literature suggests. As the present samples are small and the reliability of any distinction could not be usefully tested, no distinction is made here either between anterior and posterior phalanges or between distal metacarpals and distal metatarsals.

3. Samples

3.1. Tall-i-Mushki equids

Tall-i-Mushki is a ceramic Neolithic mound site near Shiraz, in southwestern Iran, excavated by a team from the University of Tokyo (FUKAI/HORIUCHI/MATSUTANI 1973).[8] The site is at an altitude of 1900 m, about 10 km south east of Persepolis. Its exact date is unknown, but in broad terms it almost certainly dates to some time in the fifth or the sixth millennia.
The animal bones collected during the second (1965) season of excavation are now housed in the University Museum of the University of Tokyo. Most of these are bones of *Equus, Bos, Capra* and *Gazella.* Occasional bones of *Sus* and carnivores were also noted.
The equid bones in this collection include about 150 cheekteeth, and about 50 postcranial bones complete enough to provide useful measurements.

3.1.1. Upper permanent cheekteeth (Plate 1)

Data for individual teeth (4 P^2, 49 $P^{3/4}/M^{1/2}$ and 10 M^3) are listed in Appendix 1.
The distinction between $P^{3/4}$ and $M^{1/2}$ gave problems. Figure 4 gives some separation: as in most equids, premolars are larger than molars, protocones are relatively slightly shorter in the premolars (Fig. 4 a), and there is greater asymmetry between the distal (posterior) horn of the prefossette and the mesial (anterior) horn of the prefossette in the premolars (Fig. 4 b, Bapf). There is considerable overlap in both scattergrams, however, and even when everything is taken into account, there are still some teeth that are difficult to identify and some that give inconsistent indications. Suggested identifications are given in Appendix 1, but these should be regarded with some caution, especially as considerably more teeth were identified as molars than as premolars. It is worth commenting that teeth in which the distal horn of the prefossette and the mesial horn of the postfossette are symmetrical, which is supposed to be characteristic of molars, are scarce (only two in this sample). There is greater asymmetry in the presumed premolars but some asymmetry in most of the presumed molars as well.

[8] I am indebted to Professor T. MATSUTANI of the University of Tokyo for permission to study the equid bones from Tall-i-Mushki and to the Japan Society for the Promotion of Science for making it possible for me to do this.

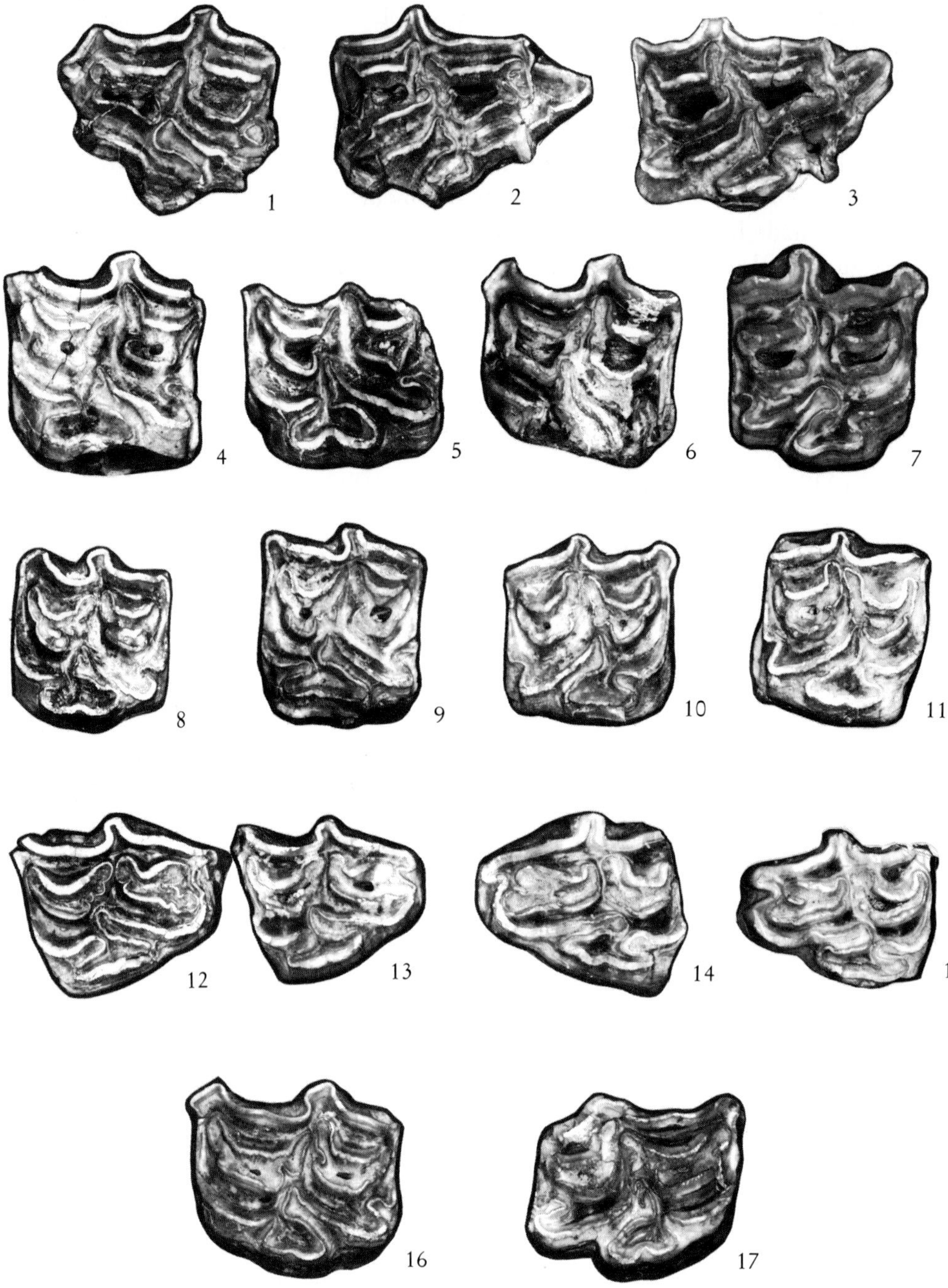

Plate 1: Tall-i-Mushki, upper cheek teeth 1–3 P^2, 4–7 $P^{3/4}$, 8–11 $M^{1/2}$, 12–15 M^3, 16–17 Pd

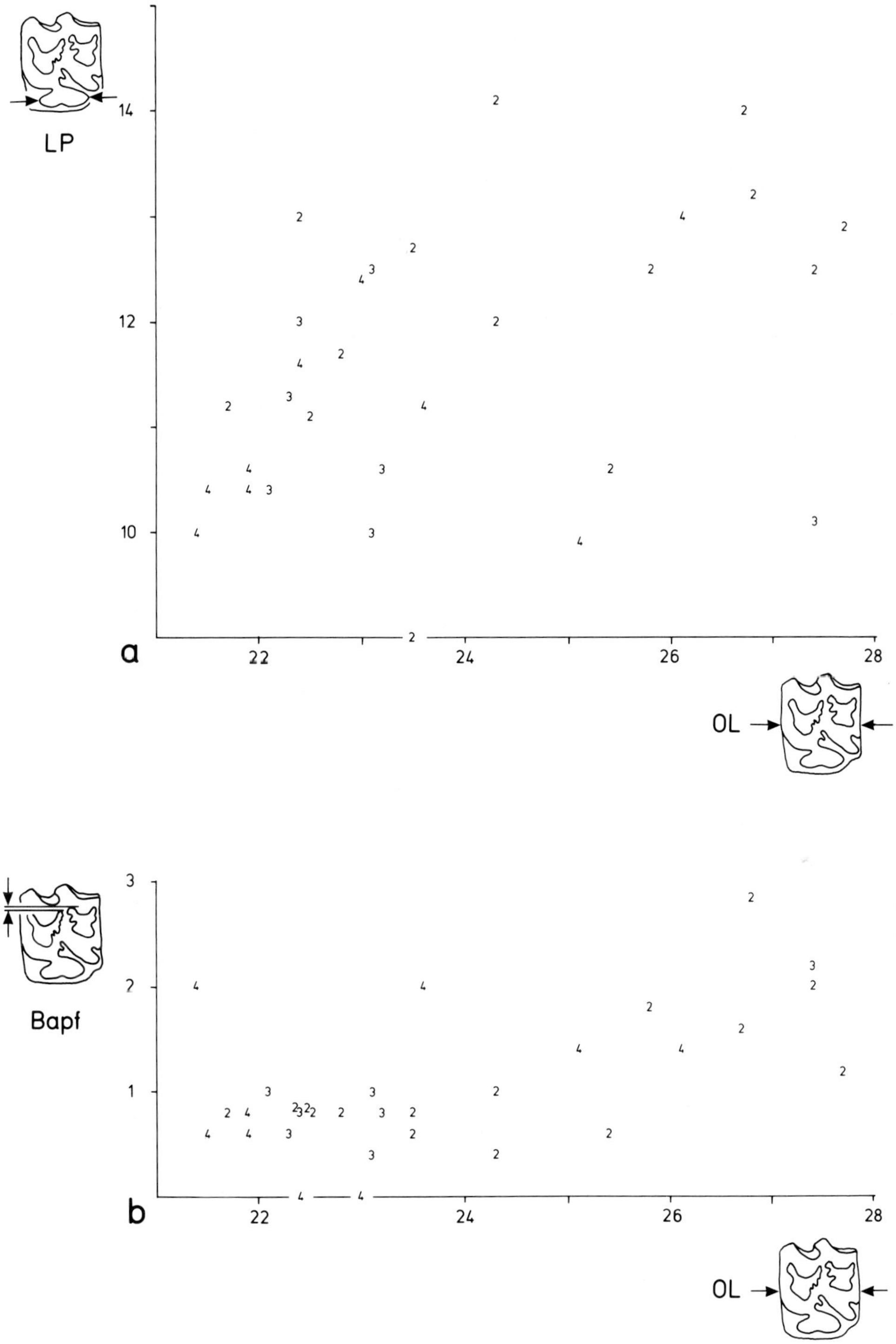

Fig. 4: Tall-i-Mushki, $P^{3/4}/M^{1/2}$. Both scattergrams show a partial separation between $P^{3/4}$(on the right) and $M^{1/2}$ (on the left). The numbers refer to the wear stage of the tooth concerned, as they also do in Figs. 6, 8 and 9.

3.1.1.1. Measurements and indices:

Average values are given in Table 1. Some caution is needed in calculating and comparing averages and indices because some of the measurements taken on equid teeth change (generally diminish) with increasing wear. This is illustrated in Figure 5, in which three measurements (OL, Bapf, and LP) and the protocone index (100 × LP/OL) are plotted against wear stage. Both OL and LP in $M^{1/2}$ appear to decrease with increasing wear. The protocone index, however, remains relatively unchanged. EISENMANN (1981) suggests that measurements taken on teeth that are relatively unworn or very worn should be excluded for comparative purposes. In keeping with this suggestion, averages are given in Table 1 both for teeth at wear stages 2 and 3 only and for teeth at wear stages 2, 3, and 4. Measurements on teeth at wear stages 2* and 5 are ignored. When teeth at wear stages 2 and 3 are considered, the protocone index averages 46.0 for $P^{3/4}$ (n = 4), and 49.7 for $M^{1/2}$ (n = 15). When teeth at wear stage 4 are added, the index is virtually unchanged at 45.6 for $P^{3/4}$ (n = 6) and 49.8 for $M^{1/2}$ (n = 20).

3.1.1.2. Morphology:

Protocones: Protocones are rather variable, sometimes short, sometimes relatively long. The lingual wall of the protocone is sometimes flat and sometimes markedly concave.
Caballine fold: As Table 2 shows, well-marked caballine folds are not common and are absent from teeth in later stages of wear.
Styles and interstylar walls: Parastyles are relatively strong and generally either square with rounded corners or more rounded. Sometimes they are more obliquely set or have a narrower neck. Mesostyles are less strong, and generally rounded with a narrow neck. A few teeth have stronger, squarer mesostyles. A few styles show some slight suggestion of channeling (*e.g.*, the mesostyle in 2MS 218-2, Plate 1.6), but none show any more strongly-developed channeling. Interstylar walls vary from slightly convex through flat and slightly concave to very concave, tending to become more concave with increasing wear.
Isolated hypoglyph: One P^2 (2MS 208-21, Plate 1.1) has an isolated hypoglyph.

TABLE 1

Tall-i-Mushki equids: average measurements and indices for upper permanent cheekteeth

	P^2		$P^{3/4}$		$M^{1/2}$		M^3	
	wear stages 2–3	wear stages 2–4	wear stages 2–3	wear stages 2–4	wear stages 2–3	wear stages 2–4	wear stages 2–3	wear stages 2–4
OL	35.8 (2)	35.8 (2)	27.1 (4)	26.6 (6)	23.1 (16)	22.8 (21)	24.9 (2)	25.2 (4)
Be			–	26.6 (2)	23.6 (15)	23.7 (18)	20.5 (1)	21.2 (2)
Bapf			2.0 (5)	1.9 (7)	0.7 (18)	0.7 (23)		
LP	10.1 (1)	9.5 (2)	12.4 (4)	12.1 (6)	11.5 (16)	11.4 (21)	12.1 (3)	12.5 (5)
100×LP/OL	28.1 (1)	28.1 (1)	46.0 (4)	45.6 (6)	49.7 (15)	49.8 (20)	47.0 (2)	49.1 (4)

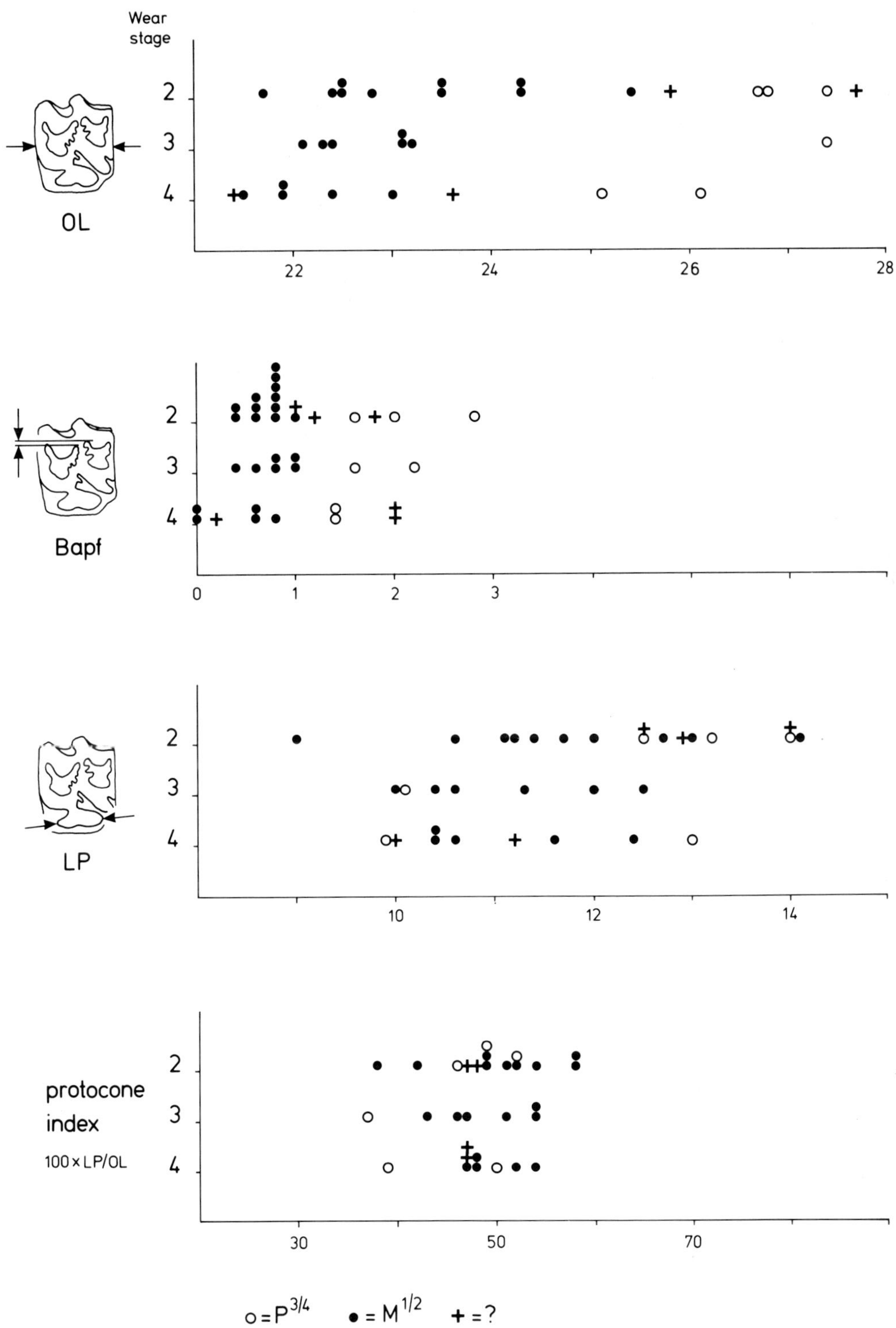

Fig. 5: Tall-i-Mushki, $P^{3/4}/M^{1/2}$. Measurements tend to diminish with increasing wear, but the protocone index remains stable.

TABLE 2

Tall-i-Mushki equids:
development of caballine fold in upper cheekteeth

1. By tooth, wear stages combined:

Class	P^2	$P^{3/4}$	$P^{3/4}/M^{1/2}$	$M^{1/2}$	M^3
++	.	.	1	1	.
+	.	2	.	2	.
tr	1	.	2	4	1
0	3	6	17	19	9

2. By wear stage, all teeth combined:

Class	Wear stage			
	2	2–3	3	4
++	2	.	.	.
+	2	.	2	.
tr	6	.	1	1
0	11	2	10	31

Note: includes data from broken crown-bases.

3.1.2. Lower permanent cheekteeth (Plate 2)

Data for individual teeth (8 P_2, 50 $P_{3/4}/M_{1/2}$ and 11 M_3) are listed in Appendix 1. Two scattergrams which appear to provide reasonably clear separation between $P_{3/4}$ and $M_{1/2}$ are shown in Figure 6 and have been used as the basis for the distinction of premolars and molars in this sample. As these scattergrams show, both B4 and B3 are fairly clearly bimodal (B4 more than B3), while Bei provides another separation especially once the reduction of Bei with increasing wear (Fig. 7) is taken into account. As in most equids (Eisenmann 1981), hypoconulids are generally narrower bucco-lingually and less distally compressed in the molars than in the premolars, although some premolars [*e.g.*, 2MS 50–3 and 88–1 (Plate 2.7,5)] have relatively narrow hypoconulids and might have been classified as molars if this had been used as a criterion.

3.1.2.1. Measurements and indices:

Average values are given in Table 3. Change with increasing wear is illustrated in Figure 7, in which four measurements (OL, Bei, Lnd, and LF) and two indices, the double knot index (100×Lnd/OL) and the postflexid index (100×LF/OL), are plotted against wear stage. OL, Bei, and LF all decrease with wear, and while Lnd remains fairly constant with increasing wear, the

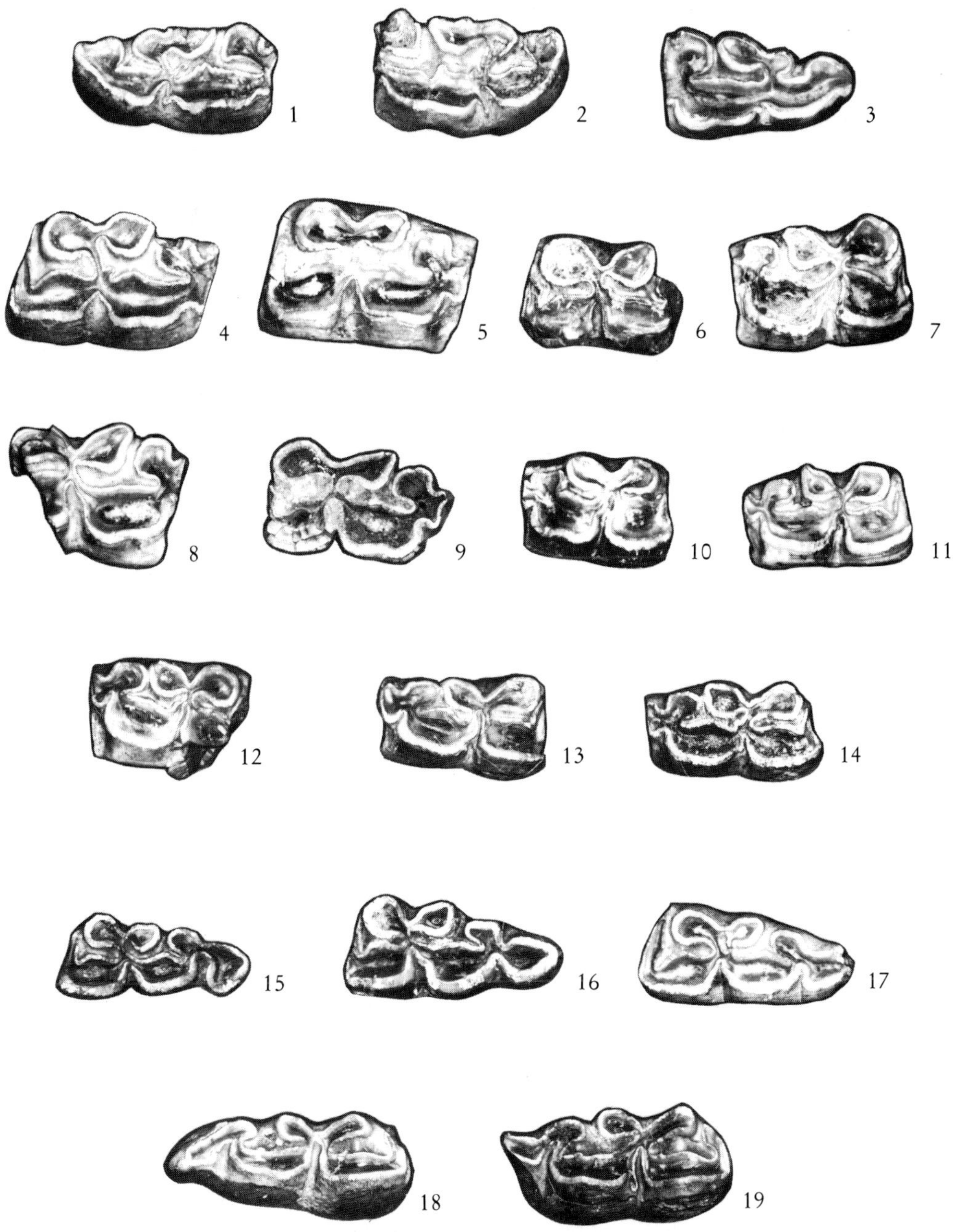

Plate 2: Tall-i-Mushki, lower cheek teeth 1–3 P_2, 4–9 $P_{3/4}$, 10–14 $M_{1/2}$, 15–17 M_3, 18–19 Pd

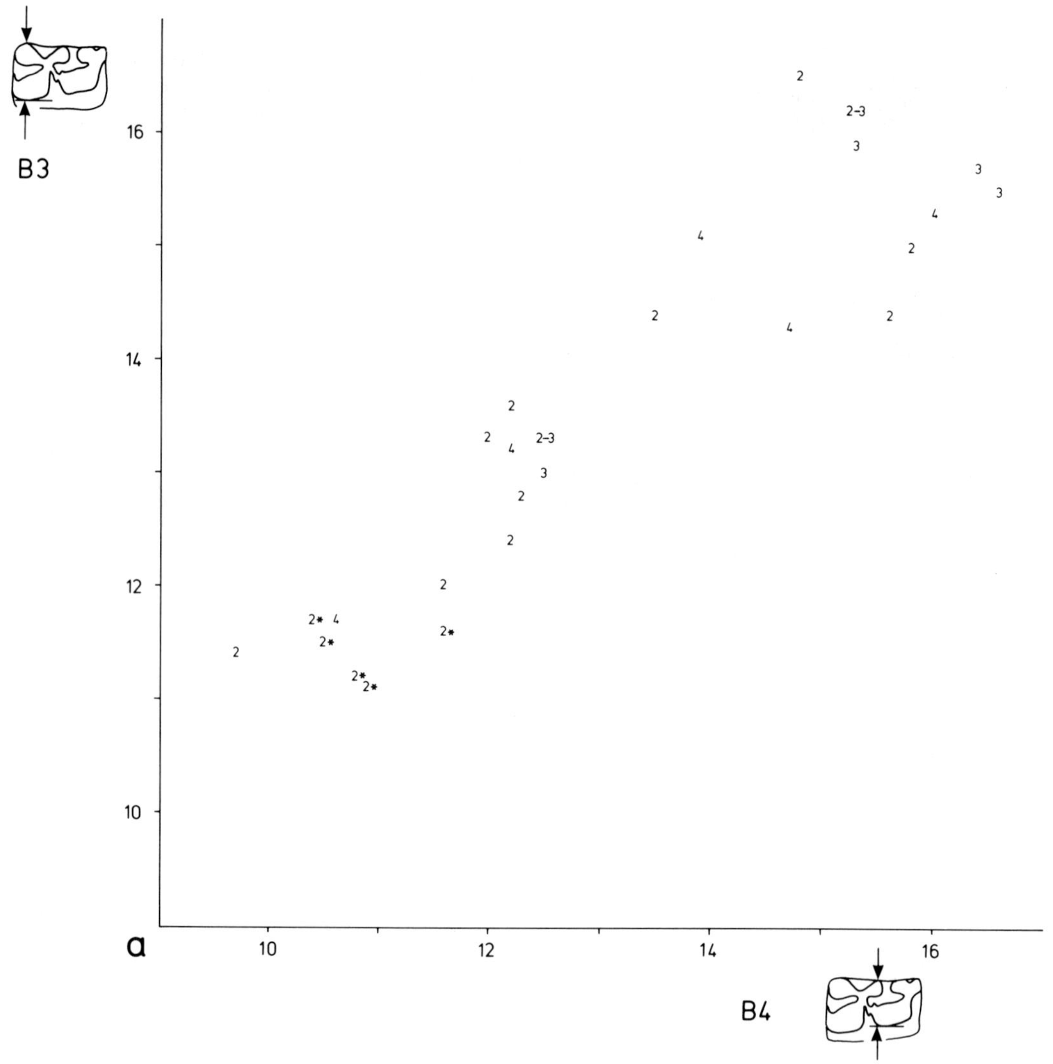

Fig. 6: Tall-i-Mushki, $P_{3/4}/M_{1/2}$. Both scattergrams give reasonable separation between $P_{3/4}$ (on the right) and $M_{1/2}$ (on the left.)

double knot index increases because OL diminishes. Molars in early wear (wear stage 2*) have particularly long occlusal surfaces because of the distal projection of the uppermost part of the hypoconulid, which provides a longer occlusal surface until the next molar erupts. As before, averages are given in Table 3 both for teeth at wear stages 2 and 3 only and for teeth at wear stages 2–4. When teeth at wear stages 2 and 3 are considered, the double knot index averages 60.4 for $P_{3/4}$ (n=9) and 57.7 for $M_{1/2}$ (n=10), and the postflexid index averages 48.6 for $P_{3/4}$ (n=9) and 43.5 for $M_{1/2}$ (n=11). When teeth at wear stage 4 are added, the double knot index becomes 61.6 for $P_{3/4}$ (n=12) and 58.5 for $M_{1/2}$ (n=12), and the postflexid index becomes 47.3 for $P_{3/4}$ (n=12) and 42.2 for $M_{1/2}$ (n=13).

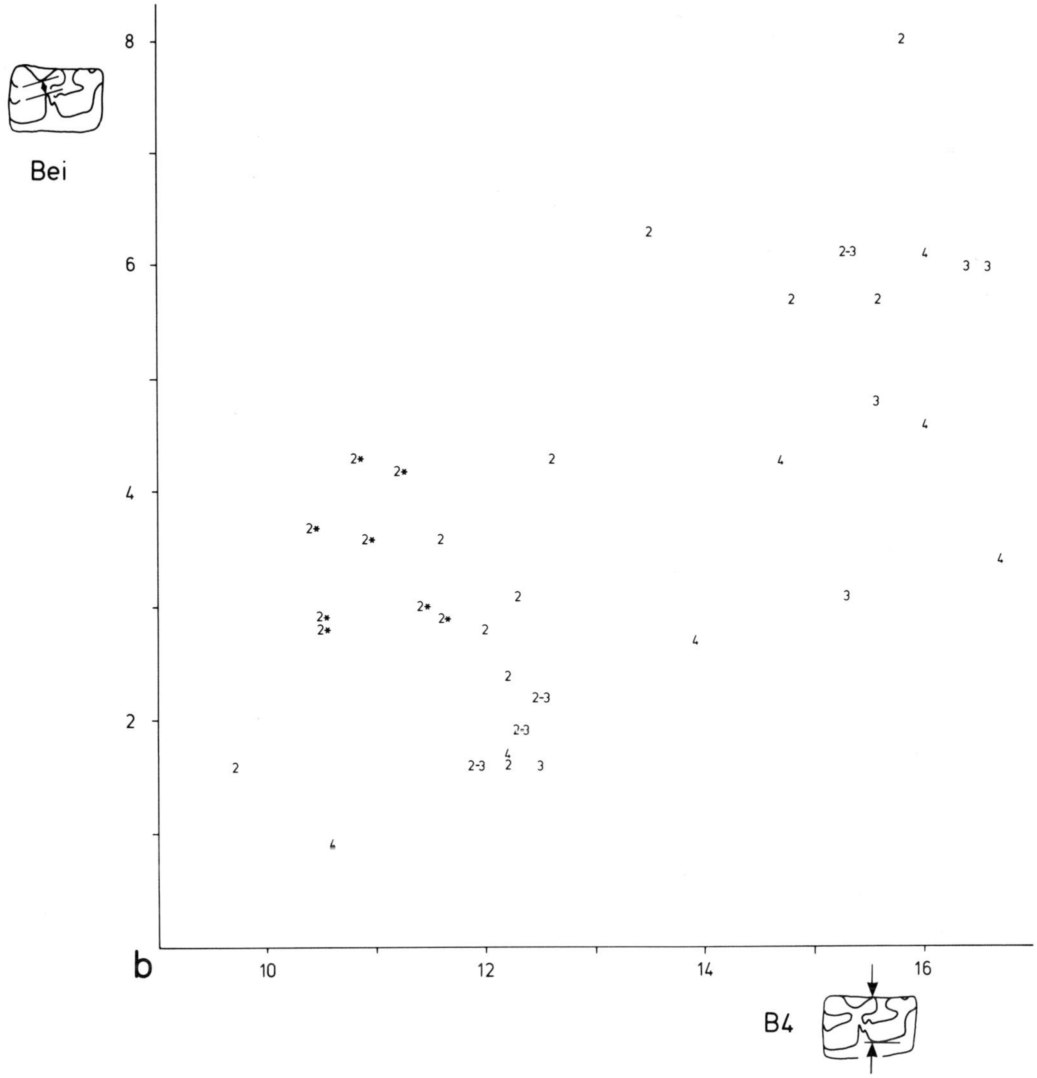

3.1.2.2. Morphology:

Penetration of the buccal sulcus: As Table 4 shows, this is again age-related, with penetration increasing as teeth are more worn [as is also implied by the decrease in Bei (Fig. 7)]. Penetration is generally shallow in premolars (increasing from grade 1 to grade 1–2) and deeper in molars (increasing from grade 1–2 to grade 2–3). The broken crown-base of one M_3 (2MS 38–9) has the buccal sulcus in contact with the lingual sulcus.

Shape of the lingual sulcus: This, as Plate 2 shows, varies widely. In P_2 it is usually poorly-marked. In $P_{3/4}$ it varies from a wide and rather shallow "U" [*e.g.*, 2MS 68–2 or 2MS 4–24 (broken base)] to a deeper wide "V" still somewhat rounded at the bottom (*e.g.*, 2MS47–2, 2MS 50–7). The range is similarly wide in the molars, including some young teeth in which the sulcus is still rather shallow (*e.g.*, 2MS 236–4). In older molars the sulcus is generally a deeper wide "V", usually rounded at the bottom, but sometimes more sharply "V"-shaped (*e.g.*, 2MS 128–5).

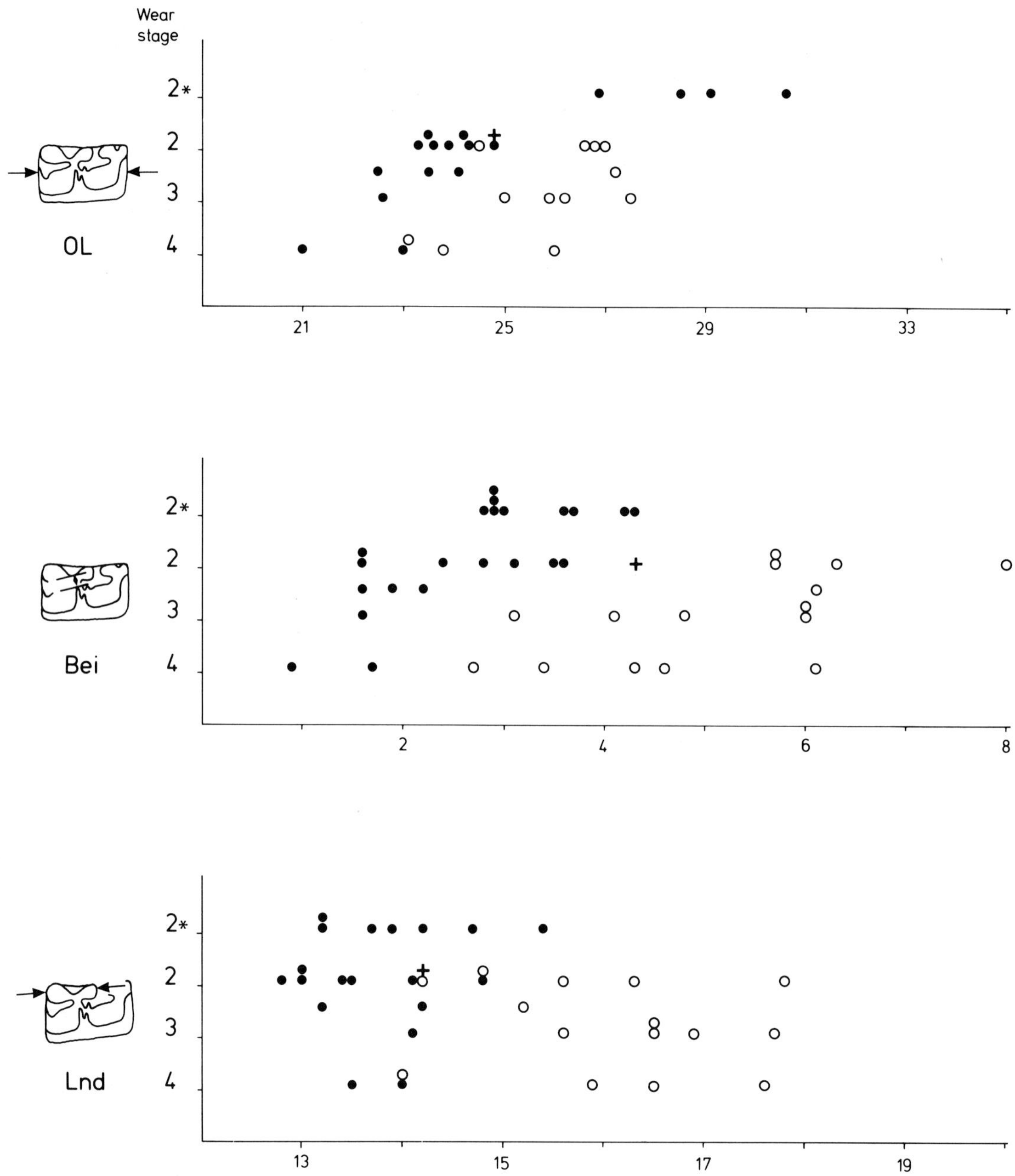

Fig. 7: Tall-i-Mushki, $P_{3/4}/M_{1/2}$. Some measurements diminish with increasing wear, but others (*e.g.* **Lnd**) remain stable. The double knot index increases while the postflexid index decreases.

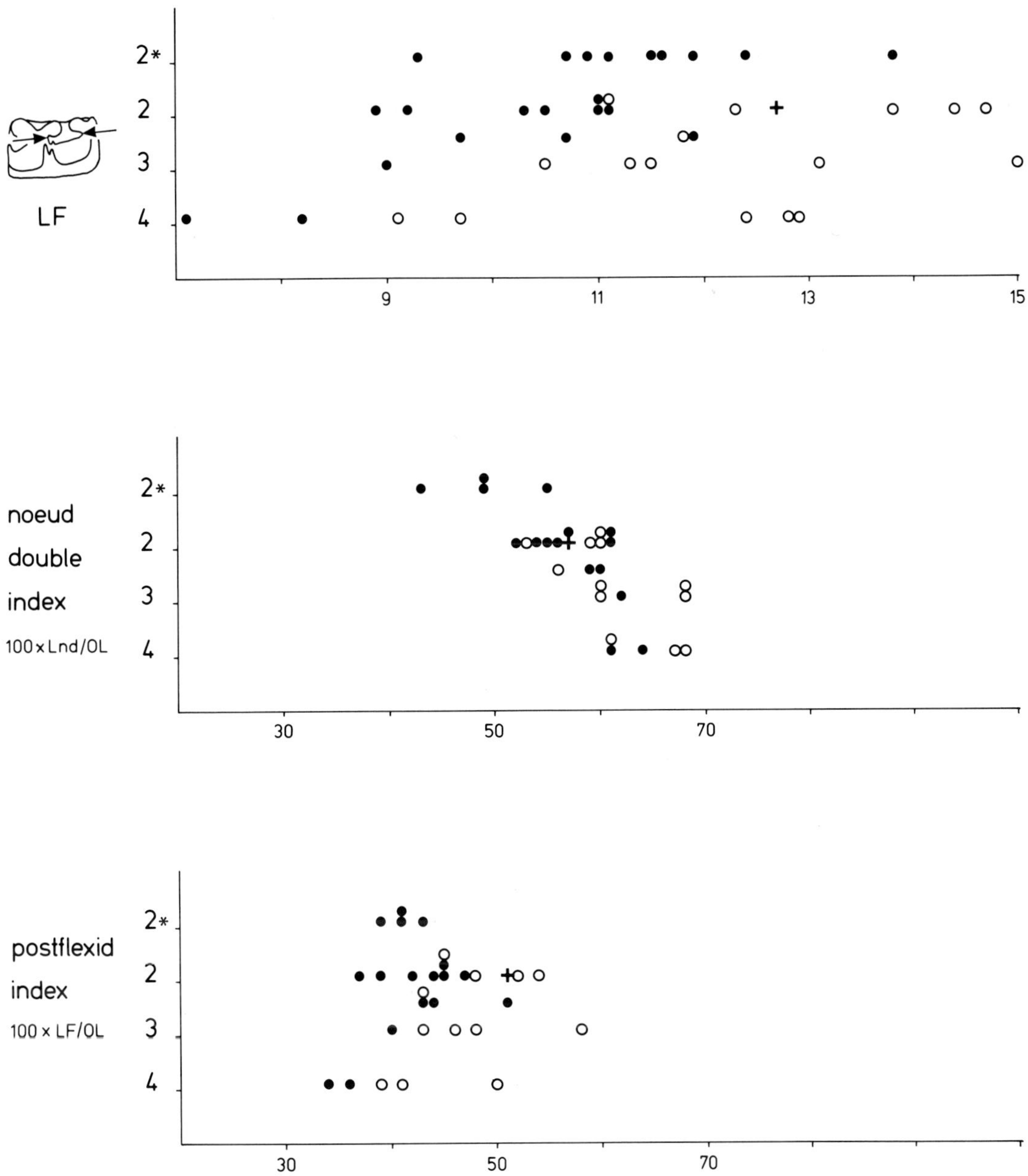
2*
2
3
4
LF
9
11
13
15
noeud
double
index
100×Lnd/OL
30
50
70
postflexid
index
100 × LF/OL

○ = $P_{3/4}$ ● = $M_{1/2}$ + = ?

TABLE 3

Tall-i-Mushki equids: average measurements and indices for lower permanent cheekteeth

	P_2		$P_{3/4}$		$M_{1/2}$		M_3	
	wear stages 2–3	wear stages 2–4	wear stages 2–3	wear stages 2–4	wear stages 2–3	wear stages 2–4	wear stages 2–3	wear stages 2–4
OL	28.4 (4)	28.1 (5)	26.3 (9)	25.8 (12)	23.7 (11)	23.4 (13)	28.5 (5)	28.5 (5)
B3	11.6 (4)	11.2 (5)	15.4 (8)	15.3 (11)	12.7 (8)	12.7 (10)	12.3 (9)	12.3 (9)
B4	14.0 (2)	13.7 (3)	15.4 (10)	15.5 (15)	11.9 (10)	11.8 (12)	10.9 (9)	10.9 (9)
Lnd	13.1 (4)	12.8 (5)	16.1 (11)	16.1 (15)	13.6 (10)	13.6 (12)	12.2 (9)	12.2 (9)
LF	15.1 (4)	14.4 (5)	12.7 (11)	12.3 (16)	10.3 (11)	9.9 (13)	9.9 (10)	9.9 (10)
Bei	7.4 (4)	7.1 (5)	5.6 (10)	5.1 (15)	2.4 (11)	2.2 (13)	3.1 (9)	3.1 (9)
100×Lnd/OL	46.0 (4)	45.4 (5)	60.4 (9)	61.6 (12)	57.7 (10)	58.5 (12)	44.1 (5)	44.1 (5)
100×LF/OL	53.3 (4)	51.1 (5)	48.6 (9)	47.3 (12)	43.5 (11)	42.2 (13)	34.9 (5)	34.9 (5)

TABLE 4

Tall-i-Mushki equids:
penetration of buccal sulcus in lower cheekteeth

1. By tooth, wear stages combined:

Penetration grade	P_2	$P_{3/4}$	$P_{3/4}/M_{1/2}$	$M_{1/2}$	M_3
1	10	14	2	5	1
2	1	7	5	20	8
3	.	.	2	8	1
4	.	.	.	.	1

2. By wear stage, all teeth combined:

Penetration grade	Wear stage				
	2*	2	2–3	3	4
1	4	13	2	7	6
2	5	10	10	4	12
3	.	1	1	1	8
4	.	.	1	.	.

Note: includes data from broken crown-bases.

Pli hypoconid: As Table 5 shows, the development of the pli hypoconid is related to wear stage, gradually disappearing in teeth that are more worn. It is often well-marked in young teeth, but absent in most well-worn teeth. It probably disappears relatively earlier in molars than in premolars, but the sample is not large enough to establish this definitely.
Ptychostylid: As Table 6 shows, the development of the ptychostylid is again related to wear stage, and it disappears in teeth that are more worn. Even in young teeth, however, it is only well-marked in a minority of cases and often absent.

3.1.3. Milk molars

Of nine upper milk molars, five are unworn or in early wear, and two are very battered. The remaining two are illustrated in Plate 1. Their protocones are relatively short, and one (2MS 255–20) has a well-marked caballine fold.
Two lower milk molars are illustrated in Plate 2. One (2MS 182–36) has a "V"-shaped lingual sulcus, moderate penetration of the buccal sulcus, and a well-marked ptychostylid. The other (2MS 182–29) has a "U"-shaped lingual sulcus, deeper penetration of the buccal sulcus, and no ptychostylid.

3.1.4. Postcranial skeleton

Measurements are listed in Appendix 1.

TABLE 5

Tall-i-Mushki equids:
development of pli hypoconid in lower cheekteeth

1. By tooth, wear stages combined:

Class	P_2	$P_{3/4}$	$P_{3/4}/M_{1/2}$	$M_{1/2}$	M_3
++	1	4	1	4	.
+	4	5	.	5	.
tr	2	4	1	5	2
0	4	8	7	20	9

2. By wear stage, all teeth combined:

Class	Wear stage					
	2*	2	2–3	3	4	5
++	4	6	.	.	.	.
+	3	5	2	3	1	.
tr	1	7	1	2	3	.
0	1	6	11	7	22	1

Note: includes data from broken crown-bases.

TABLE 6

Tall-i-Mushki equids:
development of ptychostylid in lower cheekteeth

1. By tooth, wear stages combined:

Class	P_2	$P_{3/4}$	$P_{3/4}/M_{1/2}$	$M_{1/2}$	M_3
++	3	.	1	2	.
+	3	.	.	2	.
tr	2	4	2	8	3
0	3	17	7	22	8

2. By wear stage, all teeth combined:

Class	Wear stage					
	2*	2	2–3	3	4	5
++	2	3	1	.	.	.
+	2	1	.	1	1	.
tr	2	7	4	3	3	.
0	3	13	9	8	22	2

Note: includes data from broken crown-bases.

3.2. Can Hasan III equids

Can Hasan III is an aceramic Neolithic site near Karaman, in south-central Turkey, excavated by a team from the British Institute of Archaeology at Ankara (FRENCH 1972).[9] The site lies at an altitude of 1000 m about 15 km northeast of Karaman. Its occupation dates from the late seventh and early sixth millennia.

The animal bones from the Can Hasan III excavations are currently housed in the British Institute of Archaeology at Ankara. Among the larger mammals, *Bos, Equus, Ovis, Capra,* and *Sus* are all fairly common. There is a wide range of smaller mammals among which *Lepus* and *Vulpes* are especially abundant.

The equid bones from Can Hasan III include about 80 cheekteeth and 27 postcranial bones complete enough to give useful measurements. As the Can Hasan III equids were studied some time before the Tall-i-Mushki equids, the following description lacks some data that were recorded for the Tall-i-Mushki sample.

[9] I am indebted to Dr. D. H. FRENCH, of the British Institute of Archaeology at Ankara, for inviting me to study the animal bones from Can Hasan III.

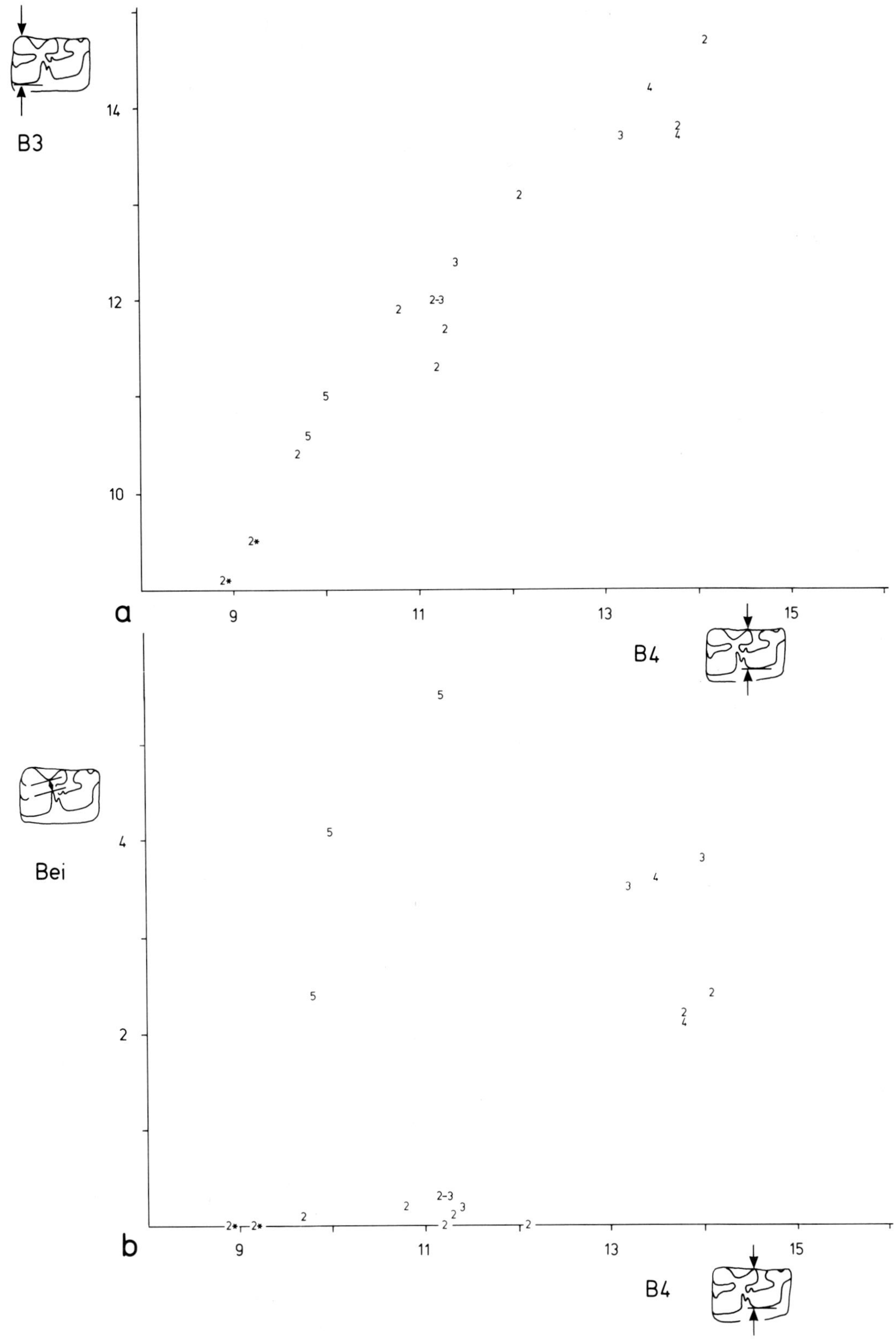

Fig. 8: Can Hasan III, $P_{3/4}/M_{1/2}$. a shows a partial separation and b a better separation between $P_{3/4}$ (upper right) and $M_{1/2}$ (lower left). As in Figure 8, teeth in wear stages 2* and 5 should be ignored.

3.2.1. Upper permanent cheekteeth (Plate 3)

Data for individual teeth (3 P^2, 14 $P^{3/4}$/$M^{1/2}$ and 4 M^3) are listed in Appendix 2.
Although numbers are small, Figure 9 suggests a separation between $P^{3/4}$ and $M^{1/2}$: the group of teeth that are probably premolars are larger, with relatively shorter protocones. If this is correct, asymmetry between the distal horn of the prefossette and the mesial horn of the postfossette is an unreliable discriminant in this sample: the molars have fairly symmetrical horns, as in most equids (EISENMANN 1980), but the premolars vary, some showing the expected asymmetry (*e.g.*, 49L 105.11 110T1.1), while others (*e.g.*, 49L 109.13 268T1.2) have horns placed as symmetrically as in the molars (Plate 3.2,3).

3.2.1.1. Measurements and indices:

Average values are given in Table 7. The protocone index averages 33.9/32.9 (n = 3,4) for $P^{3/4}$, and 43.0 for $M^{1/2}$ (n = 4).

3.2.1.2. Morphology:

Protocones: Protocones are relatively short, especially in the premolars, and asymmetrical. The lingual wall of the protocone is often markedly concave.
Caballine fold: Two $P^{3/4}$ in wear stage 2 had caballine folds. None of the molars and none of the premolars in later wear stages had more than a trace (Table 8).
Isolated hypoglyph: One M^3 (49F 929.1 32T1.1, Plate 3.13) had an isolated hypoglyph.

3.2.2. Lower permanent cheekteeth (Plate 4)

Data for individual teeth (5 P_2, 21 $P_{3/4}$/$M_{1/2}$ and 4 M_3) are listed in Appendix 2.
While numbers are rather small, what seems to be a very clear separation between the $P_{3/4}$ and $M_{1/2}$ from Can Hasan III is given by the bimodal distribution of Bei in teeth in wear stages 2-4: in the molars Bei is between 0.0 and 0.3 mm, while in the premolars it is above 2 mm (Figure 8). This has been used as the basis for the separation between molars and premolars in this

TABLE 7

Can Hasan III equids: average measurements and indices for upper permanent cheekteeth

	P^2		$P^{3/4}$		$M^{1/2}$		M^3	
	wear stages 2–3	wear stages 2–4	wear stages 2–3	wear stages 2–4	wear stages 2–3	wear stages 2–4	wear stages 2–3	wear stages 2–4
OL	31.3 (1)	31.3 (1)	23.9 (4)	23.9 (5)	21.1 (4)	21.1 (4)	–	–
Be			22.9 (3)	22.9 (4)	22.0 (2)	22.0 (2)	–	–
LP	5.0 (1)	5.0 (1)	8.0 (3)	7.8 (4)	9.1 (4)	9.1 (4)	10.4 (2)	10.1 (4)
100×LP/OL	16.0 (1)	16.0 (1)	33.9 (3)	32.9 (4)	43.0 (4)	43.0 (4)	–	–

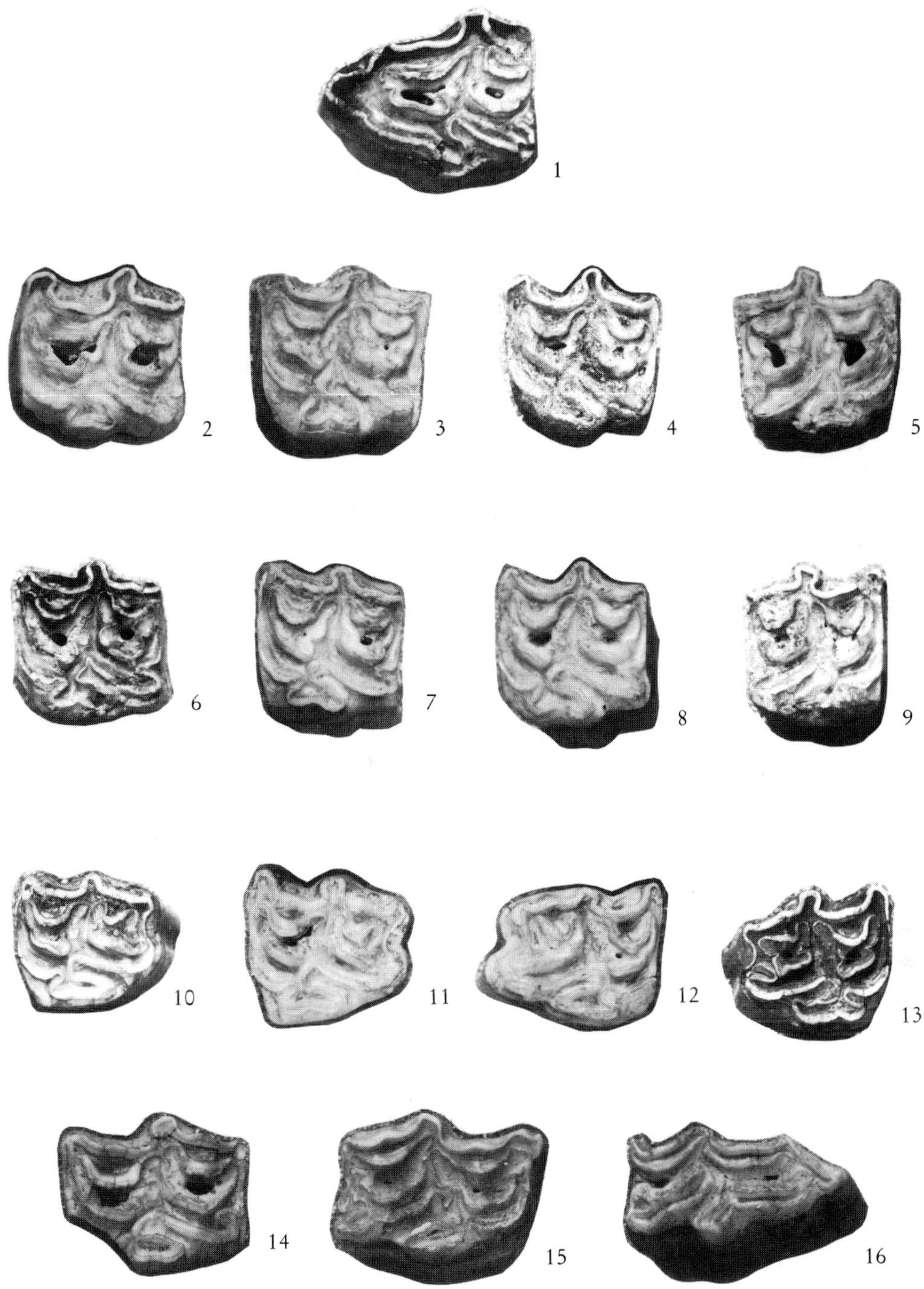

Plate 3: Can Hasan III, upper cheek teeth 1 P^2, 2–5 $P^{3/4}$, 6–9 $M^{1/2}$, 10–13 M^3, 14–16 Pd

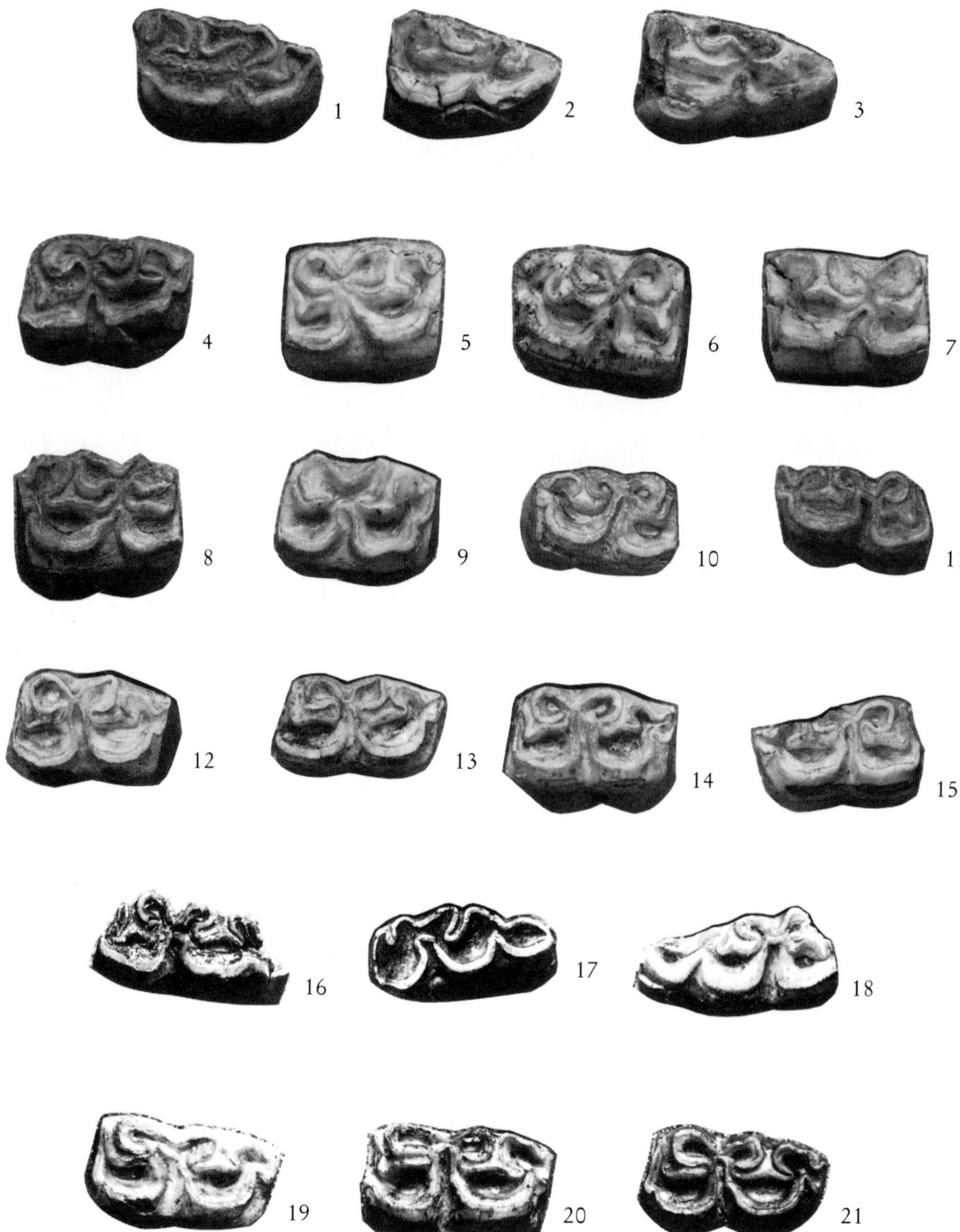

Plate 4: Can Hasan III, lower cheek teeth 1-3 P_2, 4-9 $P_{3/4}$, 10-15 $M_{1/2}$, 16-18 M_3, 19-21 Pd

TABLE 8

Can Hasan III equids:
development of caballine fold in upper cheekteeth

1. By tooth, wear stages combined:

Class	P^2	$P^{3/4}$	$P^{3/4}/M^{1/2}$	$M^{1/2}$	M^3
++	.	1	.	.	.
+	.	1	.	.	.
tr	.	.	2	.	1
0	3	3	3	4	3

2. By wear stage, all teeth combined:

Class	Wear stage				
	2*	2	3	4	5
++	.	1	.	.	.
+	.	1	.	.	.
tr	2	.	1	.	.
0	.	4	4	3	5

sample. As Bei increases once the enamel pattern starts to simplify, teeth in wear stage 5 have been left unclassified.

3.2.2.1. Measurements and indices:

Average values are given in Table 9. As for the Tall-i-Mushki sample, averages are given both for teeth at wear stages 2 and 3 only and for teeth at wear stage 2–4. The average postflexid index (100×LF/OL) is 46.5/43.4 for $P_{3/4}$ (n=5,7), and 32.7 for $M_{1/2}$ (n=8). The average double knot index (100×Lnd/OL) is 60.2/60.3 for $P_{3/4}$ (n=4,6), and 55.8 for $M_{1/2}$ (n=7).

3.2.2.2. Morphology:

Penetration of the buccal sulcus: Penetration is deep (usually grade 4) in molars, but relatively shallow (grade 2) in premolars (Table 10). Penetration is much reduced in extreme wear (wear stage 5) as the enamel pattern simplifies.
Shape of the lingual sulcus: Lingual sulcuses are mostly sharply "V"-shaped, although some are a more rounded "V" or a fairly deep "U".
Ptychostylid: With the exception of one P_2 at wear stage 2 with a well-developed ptychostylid, ptychostylids are either absent or present only as a slight trace.

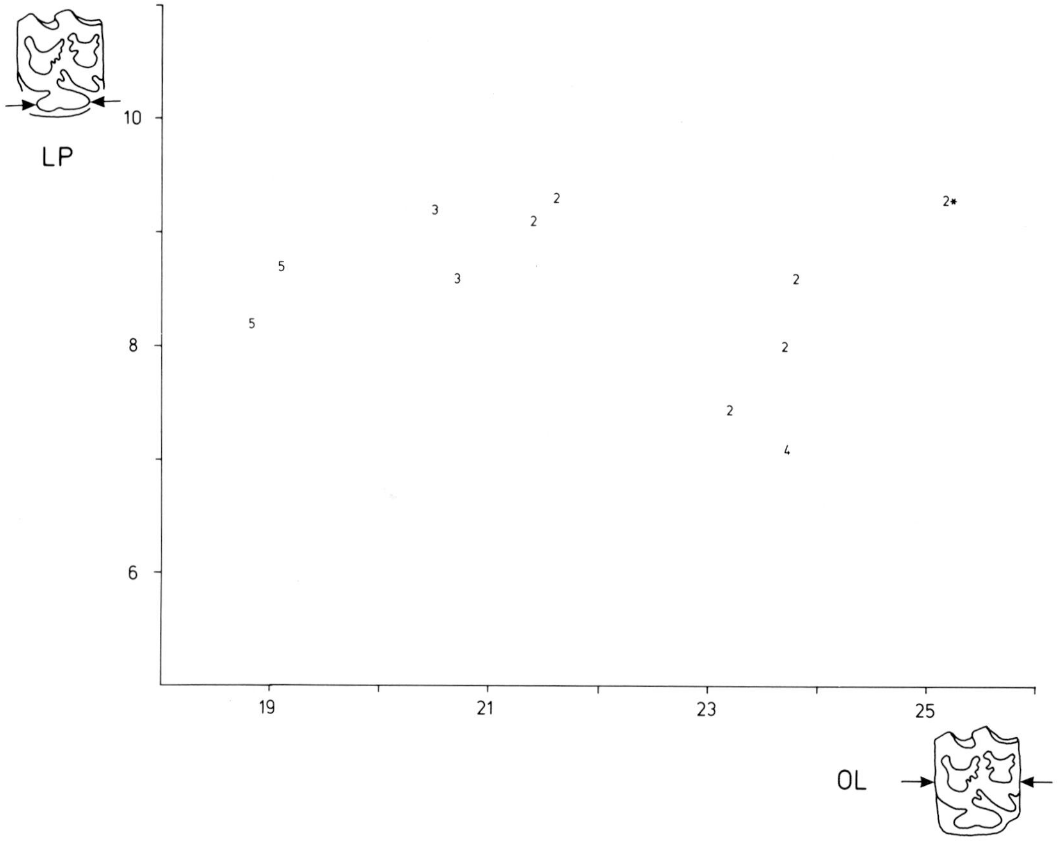

Fig. 9: Can Hasan III, $P^{3/4}/M^{1/2}$. The number of specimens is small, but the scattergram probably gives a separation between $P^{3/4}$ (on the right) and $M^{1/2}$ (on the left). Teeth in wear stages 2* and 5 should probably be ignored.

TABLE 9

Can Hasan III equids: average measurements and indices for lower permanent cheekteeth.

	P_2		$P_{3/4}$		$M_{1/2}$		M_3	
	wear stages 2–3	wear stages 2–4	wear stages 2–3	wear stages 2–4	wear stages 2–3	wear stages 2–4	wear stages 2–3	wear stages 2–4
OL	25.6 (1)	24.7 (2)	22.9 (5)	22.8 (7)	21.7 (8)	21.7 (8)	27.5 (1)	27.5 (1)
B3	10.8 (2)	10.8 (2)	13.7 (4)	13.8 (6)	11.7 (8)	11.7 (8)	10.7 (1)	11.0 (2)
B4	12.4 (1)	12.4 (2)	13.8 (4)	13.7 (6)	11.1 (7)	11.1 (7)	9.5 (1)	10.0 (2)
Lnd	12.3 (2)	12.4 (3)	13.9 (4)	13.8 (6)	12.0 (7)	12.0 (7)	11.4 (1)	11.2 (2)
LF	13.3 (2)	12.5 (3)	10.6 (5)	9.9 (7)	7.1 (8)	7.1 (8)	7.4 (2)	7.1 (3)
Bei	5.5 (2)	5.3 (3)	3.3 (5)	3.2 (7)	0.1 (8)	0.1 (8)	0.6 (2)	0.5 (3)
100×Lnd/OL	47.3 (1)	50.2 (2)	60.2 (4)	60.3 (6)	55.8 (7)	55.8 (7)	41.5 (1)	41.5 (1)
100×LF/OL	58.2 (1)	52.1 (2)	46.5 (5)	43.4 (7)	32.7 (8)	32.7 (8)	27.3 (1)	27.3 (1)

TABLE 10

Can Hasan III equids:
penetration of buccal sulcus in lower cheekteeth

1. By tooth, wear stages combined:

Penetration grade	P_2	$P_{3/4}$	$P_{3/4}/M_{1/2}$	$M_{1/2}$	M_3
1	5	.	3	.	.
2	.	7	.	.	.
3	.	.	.	.	2
4	.	.	.	10	2

2. By wear stage, all teeth combined:

Penetration grade	Wear stage					
	2*	2	2–3	3	4	5
1	.	2	.	.	1	5
2	.	3	.	2	2	.
3	.	1	.	.	.	1
4	2	7	1	1	1	.

3.2.3. Milk molars

Most of the eleven upper milk molars are either unworn or in early wear. Three that are in full wear are illustrated in Plate 3. They have short asymmetric protocones, and one $dP^{3/4}$ (49L 109.14 271T1.3) has a well-marked caballine fold.

Some of the seven lower milk molars are illustrated in Plate 4. Penetration of the buccal sulcus is shallow (grade 1 or 2) in dP_2, but deep (grade 3 or 4) in $dP_{3/4}$. Lingual sulcuses are shallow. One dP_2 (49B 900.11 11W1.3) has a well-marked ptychostylid.

3.2.4. Postcranial skeleton

Measurements are listed in Appendix 2.

4. Discussion

No reason was found to suspect the presence of more than one species of equid in either sample. The measurements and morphological data presented in each appendix are no more variable than would be expected from a single population. Equally important, if admittedly more subjective, is the strong feeling of homogeneity that each sample gave when examined, despite constant awareness of the possibility that more than a single species might be represented.

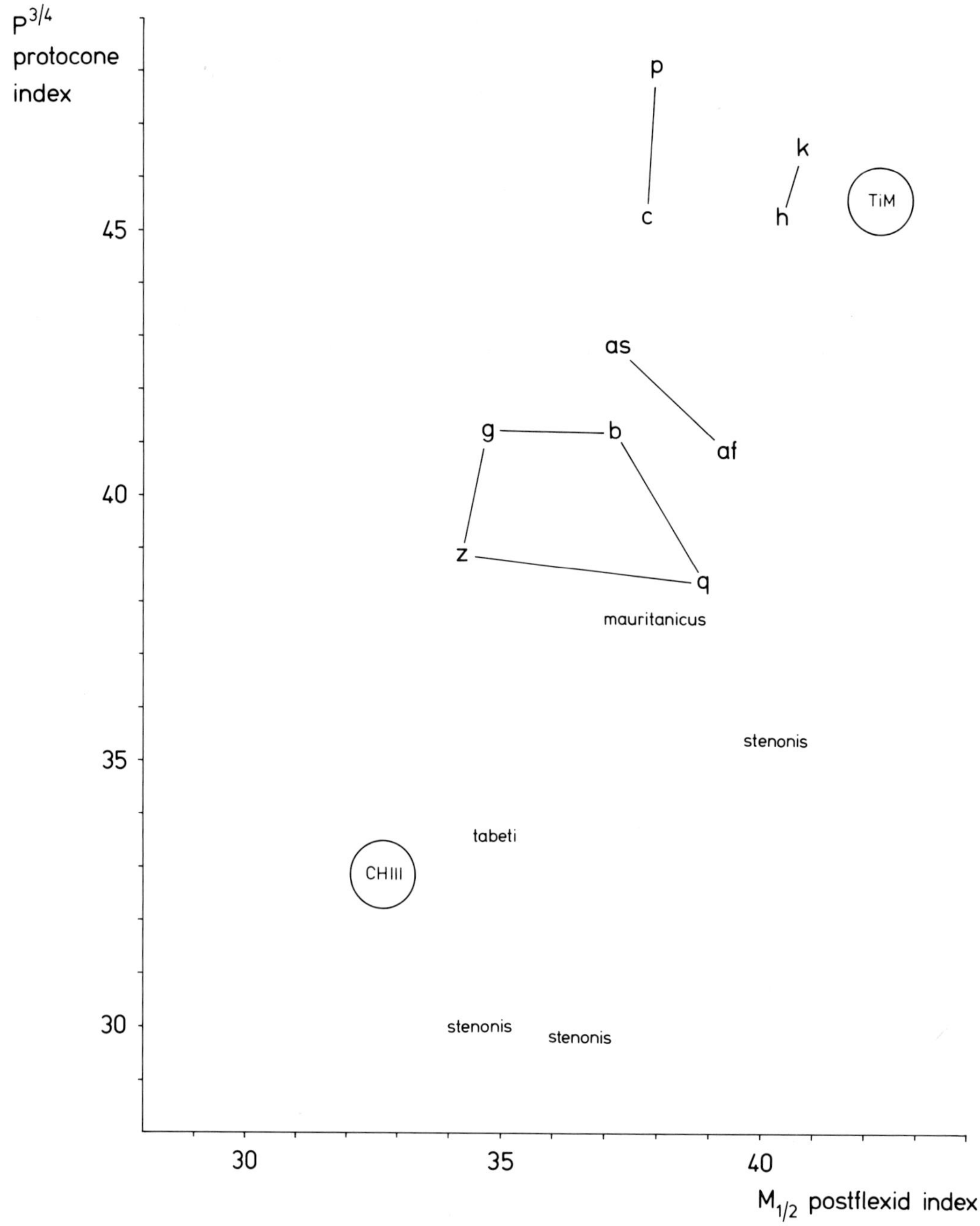

Fig. 10: The protocone index in $P^{3/4}$ and the postflexid index in $M_{1/2}$ vary considerably in different equid species. On this basis, the Tall-i-Mushki equid (TiM) falls close to *E.hemionus* (h) and *E.kiang* (k), while the Can Hasan III (CHIII) equid falls close to the extinct species *E.tabeti* and *E.stenonis.* Data for the different equid species are taken from Eisenmann (1980, 1981). Letters are used for living species of *Equus*: Zebras: *E.zebra* (z), *E.burchelli* (b), *E.grevyi* (g), and *E.quagga* (q); Asses: *E.asinus* (as) and *E.africanus* (af); Hemiones: *E.hemionus* (h) and *E.kiang* (k); Caballines: *E.caballus* (c) and *E.przewalskii* (p). *E.hydruntinus* could not be plotted because no data are available for the postflexid index (it is not included in Eisenmann 1980, 1981), although it is clear from published descriptions (Stehlin/Graziosi 1935) that its protocone index is low.

TABLE 11

Comparison of selected measurements:
Tall-i-Mushki equid, Can Hasan III equid, and *Equus hemionus*
(data for *E.hemionus* taken from EISENMANN 1979, 1980, 1981)

		E.hemionus mean	Tall-i-Mushki equid mean (wear stages 2–4)	Can Hasan III equid mean (wear stages 2–4)
P^2	OL	34.4	35.8 (2)	31.3 (1)
$P^{3/4}$	OL	25.2	26.6 (6)	23.9 (5)
$M^{1/2}$	OL	22.4	22.8 (21)	21.4 (4)
M^3	OL	24.4	25.2 (4)	-
P_2	OL	29.1	28.1 (5)	24.7 (2)
$P_{3/4}$	OL	25.5	25.8 (12)	22.8 (7)
$M_{1/2}$	OL	23.1	23.4 (13)	21.7 (8)
M_3	OL	28.7	28.5 (5)	27.5 (1)
Metapodia	Bda	38.7 (Mc) 37.5 (Mt)	40.2 (7)	38.1 (7)
	Dd	29.3 (Mc) 30.0 (Mt)	29.5 (5)	29.2 (4)
		(range)		
P^2	protocone index	23.5 18.1–30.3	28.1 (1)	16.0 (1)
$P^{2/3}$	protocone index	45.2 35.6–66.7	45.6 (6)	32.9 (4)
$M^{1/2}$	protocone index	49.1 38.6–70.0	49.8 (20)	43.0 (4)
M^3	protocone index	49.4 40.7–59.2	49.1 (4)	-
P_2	postflexid index	47.9 34.5–56.7	51.1 (5)	52.1 (2)
$P_{3/4}$	postflexid index	47.8 34.6–58.8	47.3 (12)	43.4 (7)
$M_{1/2}$	postflexid index	40.4 28.5–47.9	42.2 (13)	32.7 (8)
M_3	postflexid index	- -	34.9 (5)	27.3 (1)

The Tall-i-Mushki equid closely resembles *E.hemionus,* the onager, as described by EISENMANN on the basis of a sample of about 55 animals from a variety of locations and subspecies (1979, 1980, 1981), and there seems little reason to doubt that it is the early Holocene fore-runner of the modern *E.hemionus onager,* the Iranian onager. Measurements are compared in Table 11. The Tall-i-Mushki equid averages slightly larger in most measurements than EISENMANN's *E.hemionus* sample, but dental indices are, in general, closely similar. The postflexid index is somewhat larger in the Tall-i-Mushki molars (43.5/42.2) than in EISENMANN's sample (40.4), but this index probably decreases with age. (EISENMANN comments (1980:21) that many of the individuals in her sample were rather old.) Caballine folds and ptychostylids are similarly scarce (compare EISENMANN 1980: 90 with Table 2, and EISENMANN 1981: 45 with Table 5), and penetration of the buccal sulcus is similarly weak in the premolars and moderate in the molars (Table 12). Protocone and postflexid indices are both higher than would be expected in wild asses and domestic donkeys (Figure 10).

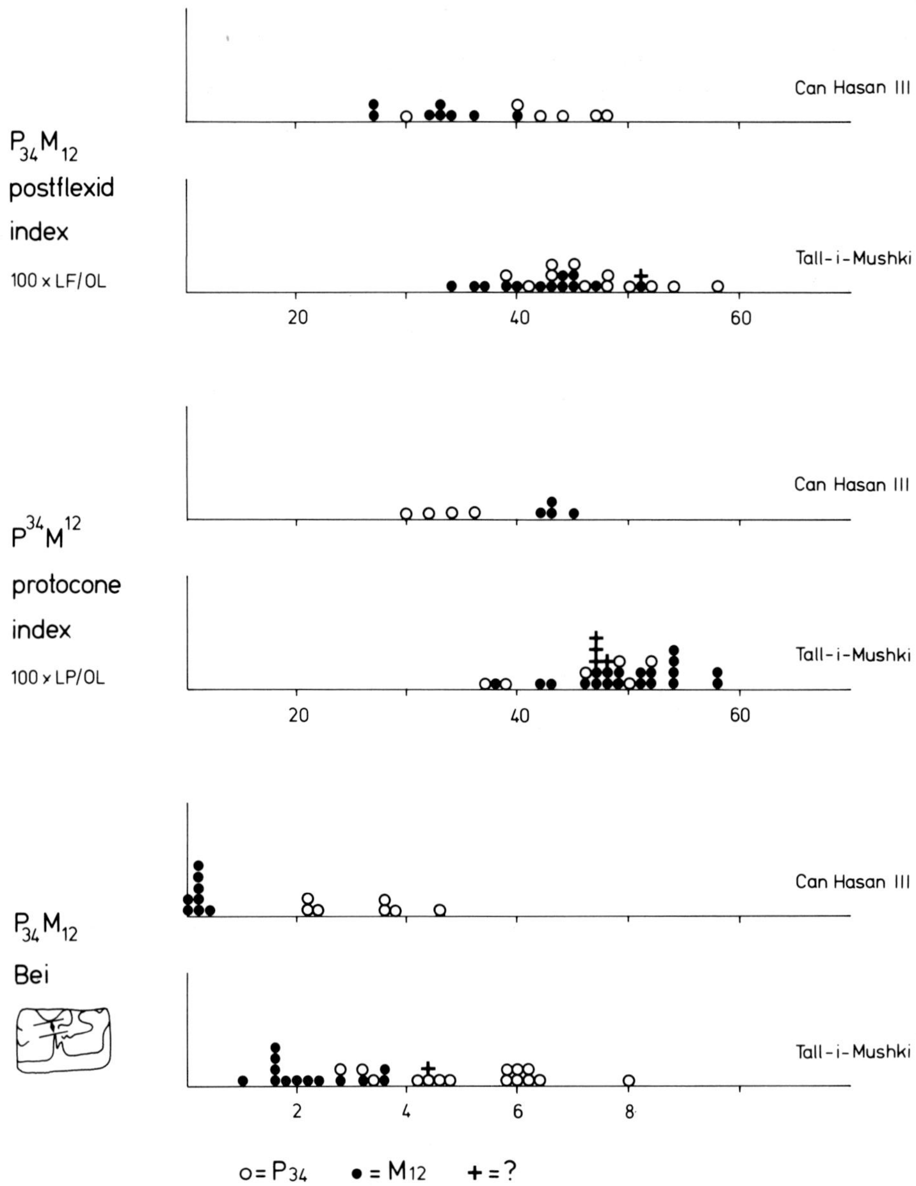

Fig. 11: Comparison between the Tall-i-Mushki and Can Hasan III equids: postflexid index, protocone index, and Bei. In each case the two populations clearly differ, but there is considerable individual overlap.

The Can Hasan III equid, on the other hand, is clearly different. Measurements are compared in Table 11. The Can Hasan III equid is clearly smaller than the Tall-i-Mushki equid. More significantly, protocone and postflexid indices are both low, especially the protocone index in the upper premolars and the postflexid index in the lower molars (Figure 11). This, together with the very deep penetration of the buccal sulcus in the lower molars (Fig. 11 Bei, and Table 12), is a pattern found in modern species of zebra (EISENMANN, 1980, 1981), and to an even

TABLE 12

Comparison of penetration of buccal sulcus in lower cheekteeth:
Tall-i-Mushki equid, Can Hasan III equid, and *Equus hemionus*
(data for *E.hemionus* taken from EISENMANN 1981)

1. Can Hasan III and Tall-i-Mushki

Grade	P_2		$P_{3/4}$		$M_{1/2}$		M_3	
	T-i-M	CH III	T-i-M	CH III	T-i-M	CH III	T-i-M	CH III
1	10	3	14	.	1	.	1	.
2	1	.	7	7	15	.	8	.
3	.	.	.	.	8	.	1	1
4	.	.	.	.	.	8	1	2

2. *Equus hemionus*

Class	P_2	$P_{3/4}$	$M_{1/2}$	M_3
0	100%	100%	72%	88%
1/2	0%	0%	17%	6%
I	0%	0%	11%	3%
c	0%	0%	0%	3%

Notes: For Tall-i-Mushki and Can Hasan III, data are for teeth in wear stages 2–4 only. Data from broken crown-bases have been included when available. The classes used by EISENMANN are similar to, but not identical with, those used in this paper. In particular, the boundary drawn between EISENMANN's classes '0' and '1/2' appears to be drawn at a point at which penetration is greater than the boundary between grades 1 and 2 as defined in this paper.

more marked extent (Fig. 10), in a group of extinct Pleistocene circum-Mediterranean equids that includes the south European *E.stenonis,* the north African *E.tabeti* and *E.mauritanicus,* and *E.hydruntinus.* The last was first described by REGALIA (1907, see also STEHLIN/GRAZIOSI 1935) from the Late Pleistocene of southern Italy and has since been identified from the Upper Pleistocene of most of the Mediterranean coast of Europe and also from Israel (BATE, in GARROD/BATE 1937, DAVIS 1980). Small size, late date, and geographical location might all be taken to suggest that the Can Hasan III equid is *E.hydruntinus,* but this group of equids is relatively poorly known, and further work is needed before this identification can be confirmed.

These suggested identifications are, however, secondary to the basic purpose of this paper, which has been to describe two assemblages of equid teeth and bones in a way that allows them to be compared with each other and with other assemblages. A second purpose of the paper has been to provide some framework for studying variability in equid teeth, which are neither so uniform that single characteristics can be used to identify single teeth nor so variable that there is no purpose in trying to work with and identify collections of isolated teeth.

As Figure 11 shows, there is considerable overlap in individual cases, even between these two very different equids. Nonetheless, when assemblages are studied rather than individual teeth, and when variation is examined in relation to tooth wear, patterns emerge which can be objectively described and which can be used to characterize, compare, and distinguish different populations.

5. Literature

Bohlken, H., 1961, Haustier und Zoologische Systematik. Zeitschrift für Tierzüchtung und Züchtungsbiologie **75**, 107–113.

Corbet, G.B., 1978, The mammals of the Palaearctic region: a taxonomic review; London/Ithaca (NY): British Museum/Cornell University Press.

Davis, S.J., 1980, Late Pleistocene and Holocene equid remains from Israel. Zoological Journal of the Linnean Society **70**, 289–312.

Driesch, A.v.d., 1976, A Guide to the Measurement of Animal Bones from Archaeological Sites; (Peabody Museum Bulletin **1**); Cambridge (MA): Peabody Museum, Harvard University.

Ducos, P., 1970, The Oriental Institute excavations at Mureybit, Syria: preliminary report on the 1965 campaign, Part 4: Les restes d'équidés. Journal of Near Eastern Studies **29**, 273–289.

Eisenmann, V., 1979, Les métapodes d'*Equus sensu lato* (Mammalia, Périssodactyla). Géobios **12**(6), 863–886.

Eisenmann, V., 1980, Les Chevaux (*Equus sensu lato*) fossiles et actuels: crânes et dents jugales supérieures; (Cahiers de Paléontologie); Paris: Éditions du C.N.R.S.

Eisenmann, V., 1981, Étude des dents jugales inférieures des *Equus* (Mammalia, Périssodactyla) actuels et fossiles. Palaeovertebrata **10**(3–4), 127–226.

Eisenmann, V., 1986, Comparative osteology of modern and fossil horses, half-asses, and asses. In: Meadow, R.H./Uerpmann, H.-P., eds., Equids in the Ancient World; (Beihefte zum Tübinger Atlas des Vorderen Orients, Reihe A **19/1**); Wiesbaden: Dr. Ludwig Reichert Verlag, 67–116.

French, D.H., 1972, Excavations at Can Hasan III, 1969–1970. In:Higgs, E.S., ed., Papers in Economic Prehistory; Cambridge: Cambridge University Press, 181–190.

Fukai, S./Horiuchi, K./Matsutani, T., 1973, Marv-Dasht III: the excavation at Tall-i-Mushki 1965; (Tokyo University Iraq-Iran Archaeological Expedition Report **14**); Tokyo: Tokyo University.

Garrod, D.A.E./Bate, D.M.A., 1937, The Stone Age of Mount Carmel: Excavations at Wady el-Mughara: 1; Oxford: Oxford University Press.

Gray, A.P., 1954, Mammalian Hybrids: a Check-list with Bibliography; Farnham Royal, Bucks: Commonwealth Agricultural Bureau.

Gray, A.P., 1966, Mammalian Hybrids: Supplementary Bibliography; Edinburgh: Commonwealth Bureau of Animal Breeding and Genetics.

Groves, C.P., 1971, Request for a declaration modifying Article 1 so as to exclude names proposed for domestic animals from zoological nomenclature. Bulletin of Zoological Nomenclature **27**, 269–272.

Harrison, D.L., 1972, Mammals of Arabia, Volume 3: Lagomorpha, Rodentia; London: Ernest Benn.

Regalia, E., 1907, Sull'*Equus* (*Asinus*) *hydruntinus* Regalia della grotta di Romanelli (Castro, Lecce). Archivio per l'Anthropologia, Etnologia e Psicologia Comparata **37**, 375–390.

Stehlin, H.G./Graziosi, P., 1935, Ricerche sugli Asinidi fossili d'Europa. Mémoires de la Société Paléontologique Suisse **56**(3), 1–73.

Zeder, M.A., 1986, The Equid Remains from Tal-e Malyan, southern Iran. In: Meadow, R.H./Uerpmann, H.-P., eds., Equids in the Ancient World; (Beihefte zum Tübinger Atlas des Vorderen Orients, Reihe A **19/1**); Wiesbaden: Dr. Ludwig Reichert Verlag, 366–412.

Sebastian Payne, 9 Wilberforce Road, Cambridge, CB3 OEQ, England

For definitions of measurements, wear stages, and morphological classifications see Section 2 (Methods). Equivalents are also given to the measurements described by EISENMANN (1985). Epiphyses are fused unless stated. A measurement is given in parentheses when approximate but within ± 2%; "+" is added when the specimen is slightly chipped or abraded but the loss is less than 2%; and "–" when the measurement is slightly too large but again within 2% (*e.g.*, when the bone has split and warped). Morphological data taken from the broken crown-bases of teeth are given in square parentheses, as is the wear stage equivalent to the position of the break.

APPENDIX 1

Tall-i-Mushki *Equus*: measurements and morphological data.

P^2

Specimen number	2MS 25–5	2MS 201–7	2MS 208–21	2MS 233–28
OL	–	36.0	–	(35.6)
LP	(8.8)	10.1	*	–
Wear stage	?4	?3	4	?3
Caballine fold	0	0	0	tr

* Too asymmetrical to be measured.

$P^{3/4}/M^{1/2}$

Specimen number	2MS 1–5	2MS 2–7	2MS 2–5	2MS 10–1	2MS 12–7	2MS 21–3	2MS 25–2
OL	21.7	–	–	–	(22.5)	–	21.9
Be	22.8	(23.5)	–	–	(24.5)	–	–
Bapf	0.8	0.4	1.6	–	0.8	0.2	0.6
LP	11.2	11.4	–	–	(11.1)	–	10.4
Wear stage	2	2	3	1[4]	2[4]	4	4
Caballine fold	+ +	tr	+	-[0]	0[0]	0	0
P/M	M	M	?P	–	M	–	M
Specimen number	2MS 30–11	2MS 36–3	2MS 38–2	2MS 38–3	2MS 47–5	2MS 50–5	2MS 52–5
OL	(26.8)	–	–	–	–	21.9	(23.6)
Be	–	–	–	–	–	(23.0)	–
Bapf	2.8	–	–	–	–	0.8	2.0
LP	13.2	–	–	–	–	10.6	11.2
Wear stage	2	1[4]	1[4]	1[4]	1[4]	4	4
Caballine fold	+	-[0]	-[0]	-[0]	-[0]	0	tr
P/M	P	–	–	–	–	M	–
Specimen number	2MS 56–1	2MS 61–3	2MS 61–2	2MS 66–2	2MS 89–2	2MS 92–4	2MS 99–1
OL	–	24.3	(22.4)	22.4	(27.7)	–	–
Be	–	23.5	–	(25.1)	–	–	–
Bapf	–	0.4	0.8	0.0	1.2	–	–
LP	–	14.1	(12.0)	11.6	12.9	–	–

Wear stage	1[4]	2	3	4	2[4]	1[2–3]	1[4]
Caballine fold	-[0]	+	0	0	?0[0]	-[0]	-[0]
P/M	-	?M	M	?M	?	-	-

Specimen number	2MS	2MS	2MS	2MS	2MS	2MS	2MS
	104–3	110–5	127–2	127–8	128–2	129–8	142–11
OL	-	21.5	-	-	25.1	22.8	(23.2)
Be	-	-	-	-	27.7	(24.3)	23.7
Bapf	-	0.6	-	-	1.4	0.8	0.8
LP	-	(10.4)	-	-	9.9	(11.7)	10.6
Wear stage	1[3]	4	1[4]	1[2–3]	4	2	3
Caballine fold	-[0]	0	-[0]	-[0]	0	?tr	0
P/M	-	M	-	-	P	M	M

Specimen number	2MS	2MS	2MS	2MS	2MS	2MS	2MS
	142–8	155–4	155–5	163–2	167–2	177–3	181–2
OL	(23.5)	21.4	22.3	25.8	-	24.3	22.1
Be	22.6	(23.9)	23.3	-	-	(23.0)	(23.3)
Bapf	0.8	2.0	0.6	1.8	-	1.0	1.0
LP	9.0	10.0	11.3	12.5	-	12.0	(10.4)
Wear stage	2	4	3	2	1[4]	2	3
Caballine fold	0	0	0	tr	-[0]	0	0
P/M	M	?	M	-	-	?M	M

Specimen number	2MS	2MS	2MS	2MS	2MS	2MS	2MS
	194–17	196–1	196–4	196–7	197–3	209–5	218–2
OL	25.4	-	22.4	-	23.0	23.5	27.4
Be	23.2	-	(24.8)	-	24.4	24.6	-
Bapf	0.6	-	0.8	1.0	0.0	0.6	2.2
LP	10.6	-	(13.0)	(14.0)	12.4	12.7	(10.1)
Wear stage	2	1[4]	2	2	4[4]	2	3
Caballine fold	tr	-[0]	0	+ +	0[0]	tr	0
P/M	?M	-	M	-	M	M	P

Specimen number	2MS	2MS	2MS	2MS	2MS	2MS	2MS
	220–1	221–33	221–25	234–11	245–19	262–4	242–16
OL	-	26.1	27.4	23.1	26.7	(22.5)	23.1
Be	-	25.5	-	-	-	(22.5)	(23.7)
Bapf	0.6	1.4	2.0	1.0	1.6	0.8	0.4
LP	-	13.0	12.5	(12.5)	14.0	-	10.0
Wear stage	2[4]	4	2	3	2[4]	2	3
Caballine fold	0[0]	0	0	+	?0[0]	0	0
P/M	?M	P	P	?M	P	M	M

M^3

Specimen number	2MS	2MS	2MS	2MS	2MS	2MS	2MS
	38–4	52–2	66–1	106–2	142–4	155–8	163–3
OL	-	(25.4)	*	-	24.4	25.8	-
Be	-	(20.5)	*	-	-	-	-
LP	-	11.7	*	12.8	(11.7)	12.9	-
Wear stage	1[4]	2	3	2	2	4	1[4]
Caballine fold	-[?0]	0	0	tr	0	0	-[0]

Specimen number	2MS 233-27	2MS 234-5	2MS 261-21
OL	-	*	25.4
Be	-	*	(21.9)
LP	-	*	13.3
Wear stage	1[3]	4	4
Caballine fold	-[0]	0	0

* Not measured because occlusal surface very oblique to axis of tooth-crown.

P_2

Specimen number	2MS 13-11	2MS 30-6	2MS 35-3	2MS 61-16	2MS 77-2	2MS 106-8	2MS 160-5
OL	26.9+	(27.2)	-	-	29.0	-	28.6
B3	9.5	12.3	-	-	11.3	-	10.8
B4	(13.0)	13.5	-	-	-	-	-
Lnd	11.5	12.9	-	-	14.0	-	12.4
LF	11.4	15.5+	-	-	15.5	-	14.9
Bei	5.8	7.7	-	-	7.5	-	5.9
Wear stage	4	2[4]	1[2]	1[3]	3[4]	1[2]	2
Pli hypoconid	tr	+[0]	-[tr]	-[0]	+[?0]	-[+]	++
Ptychostylid	+	++[0]	-[tr]	-[?tr]	+[0]	-[+]	++
Buccal sulcus	1	1[2]	-[1]	-[1]	1[1]	-[1]	1

Specimen number	2MS 254-4
OL	28.8
B3	(12.0)
B4	14.5
Lnd	13.0
LF	14.6
Bei	(8.6)
Wear stage	2-3[4]
Pli hypoconid	+[0]
Ptychostylid	++[0]
Buccal sulcus	1[1]

$P_{3/4}/M_{1/2}$

Specimen number	2MS 1-3	2MS 1-2	2MS 4-18	2MS 4-24	2MS 4-22	2MS 5-	2MS 21-8
OL	(23.6)	24.5	(28.5)	(26.6)	-	-	-
B3	13.3	14.4	11.5	16.5	-	-	-
B4	12.0	13.5	10.5	14.8	16.0	10.5	11.4
Lnd	13.4	14.8	13.9	15.6	16.5+	13.7	15.4
LF	11.1	11.1	11.6	13.8	12.4	10.9	13.8
Bei	2.8	6.3	2.9	5.7	6.1	2.8	3.0
Wear stage	2	2	2*[4]	2[4]	4	2*[5]	2*[4]
Pli hypoconid	0	++	++[0]	++[0]	+	+[0]	++[0]
Ptychostylid	0	0	++[0]	tr[0]	0	++[0]	+[0]
Buccal sulcus	2	1	1[2]	1[2]	1	1[-]	2[?3]
P/M	M	?P	M	P	P	M	M

Specimen number	2MS 25–3	2MS 29–8	2MS 30–7	2MS 30–5	2MS 39–10	2MS 39–1	2MS 39–9a
OL	(24.8)	(23.5)	23.9	–	23.8	27.5	–
B3	–	13.3	11.4	–	15.3	15.5	–
B4	12.6	12.5	9.7	–	(16.0)	16.6	–
Lnd	14.2+	14.2+	13.0	–	15.9	16.5	–
LF	12.7	11.9	8.9	–	9.7	13.1	–
Bei	4.3	2.2	1.6	–	4.6	6.0	–
Wear stage	2[5]	2–3[?4]	2[4]	–[2–3]	4	3	0[4]
Pli hypoconid	++[–]	0[0]	tr[0]	–[0]	tr	+	–[0]
Ptychostylid	++[0]	tr[0]	0[0]	–[0]	0	0	–[0]
Buccal sulcus	1[–]	2[2]	2[3]	–[3]	2	1	–[2]
P/M	?	M	M	–	P	P	–

Specimen number	2MS 39–9b	2MS 40–2	2MS 47–2	2MS 50–7	2MS 50–4	2MS 50–3	2MS 67–1
OL	–	(26.8)	–	–	–	25.0	24.8
B3	–	14.4	–	–	–	15.9	–
B4	–	(15.6)	–	16.7	–	(15.3)	–
Lnd	–	14.2	16.5	–	–	16.9	12.8
LF	–	12.8	10.5	12.8	–	11.5	11.0
Bei	–	5.7	4.1	3.4	–	3.1	3.5
Wear stage	0[4]	2[4]	3	4	–[?4]	3	2
Pli hypoconid	–[?0]	++[0]	tr	0	–[0]	0	tr
Ptychostylid	–[?tr]	tr[0]	0	0	–[0]	0	0
Buccal sulcus	–[2]	1[1]	1	2	–[–]	2	2
P/M	–	P	?P	P	–	P	?M

Specimen number	2MS 68–2	2MS 88–1	2MS 94–1	2MS 121–1	2MS 127–3	2MS 128–5	2MS 148–4
OL	(27.0)	27.2	–	(24.3)	26.2	(22.5)	23.1)
B3	15.0	16.2	–	(12.0)	–	–	15.1
B4	15.8	15.3	–	11.6	(15.6)	12.3	13.9
Lnd	16.3	15.2	–	(14.8)	15.6	13.2	14.0
LF	14.7	11.8	–	(10.3)	11.3	9.7	(9.1)
Bei	8.0	6.1	–	3.6	4.8	1.9	2.7
Wear stage	2[4]	2–3[4]	1[2–3]	2	3	2–3	4
Pli hypoconid	++[0]	+[0]	–[?0]	+	tr	tr	0
Ptychostylid	0[0]	0[0]	–[?tr]	tr	0	0	?tr
Buccal sulcus	1[2]	1[1]	–[?2]	1	1	2	2
P/M	P	P	–	M	P	M	?P

Specimen number	2MS 150–4	2MS 155–6	2MS 166–17	2MS 173–2	2MS 175–5	2MS 189–39	2MS 196–12
OL	–	22.6	–	(23.0)	–	25.9	–
B3	–	13.0	–	11.7	–	15.7	–
B4	–	(12.5)	–	10.6	–	16.4–	(11.2)
Lnd	–	(14.1)	–	14.0	–	17.7–	–
LF	–	(9.0)	–	8.2	10.7	15.0	9.3
Bei	–	1.6	–	0.9	2.9	6.0	4.2
Wear stage	–[4]	3	1[2–3]	?4[?4]	2*	3	2*[?4]

Pli hypoconid	-[-]	0	-[0]	0[0]	0	+	tr[0]
Ptychostylid	-[-]	?tr	-[0]	0[0]	+	0	0[0]
Buccal sulcus	-[3]	3	-[2]	3[3]	2	1	1[?2]
P/M	-	M	-	M	M	P	M

Specimen number	2MS 203-11	2MS 203-2	2MS 208-16	2MS 208-20a	2MS 208-20b	2MS 212-36	2MS 218-22
OL	(26.0)	24.2	(30.6)	-	(26.9)	21.0	-
B3	14.3	(12.8)	11.2-	-	11.6	13.2	-
B4	14.7	12.3	10.8-	-	11.6	12.2	-
Lnd	17.6	13.5	13.2	-	14.7	13.5	-
LF	12.9	11.0	(11.9)	-	(11.1)	7.1	-
Bei	4.3	3.1	4.3	-	2.9	1.7	-
Wear stage	4	2	2*	1[3]	2*[3]	4	-[2-3]
Pli hypoconid	?tr	+	(+ +)	-[?0]	(+ +)[0]	0	-[0]
Ptychostylid	?tr	?tr	0	-[?0]	tr[?tr]	0	-[0]
Buccal sulcus	2	2	1	-[?1]	2[2]	3	-[2]
P/M	P	M	M	P_4	M_1	M	-

Specimen number	2MS 218-16	2MS 227-2	2MS 229-5	2MS 233-20	2MS 236-4	2MS 228-5	2MS 260-9
OL	23.3	-	(24.1)	29.1	-	-	-
B3	13.6	-	-	11.1	11.7	-	-
B4	12.2	-	11.9	10.9	10.4	(15.6)	-
Lnd	14.1	-	-	14.2	13.2	(17.8)	
LF	9.2	-	10.7	12.4	11.5	14.4	-
Bei	1.6	-	1.6	3.6	3.7	-	-
Wear stage	2[?4]	1	2-3[4]	2*[3]	2*[3]	2	0[2]
Pli hypoconid	0[0]	-	0[0]	+[0]	+[0]	+	-[?tr]
Ptychostylid	0[0]	-	0[0]	tr[0]	0[0]	0	-[?0]
Buccal sulcus	2[3]	-	2[?3]	2[2]	2[2]	1	-[?1]
P/M	M	-	M	M	M	P	-

Specimen number	2MS 262-5
OL	23.5
B3	12.4
B4	12.2
Lnd	13.0
LF	10.5
Bei	2.4
Wear stage	2
Pli hypoconid	tr
Ptychostylid	tr
Buccal sulcus	2
P/M	M

M_3

Specimen number	2MS 1-4	2MS 38-9	2MS 46-1	2MS 54-3	2MS 61-1	2MS 67-2	2MS 106-4
OL	-	-	28.3	(26.3)	(29.2)	-	-
B3	9.0	-	12.3	11.2	12.9	13.9	13.0

B4	7.9	-	11.2	9.6	12.8	12.7	11.3
Lnd	10.9	-	12.8	11.2	13.4	13.1	11.6
LF	7.9	-	9.5	9.7	11.5	9.5	11.3
Bei	2.9	-	2.6	3.4	2.8	4.3	4.1
Wear stage	2	-[2-3]	2-3	2	2-3	2-3	2
Pli hypoconid	?tr	-[0]	0	0	0	0	0
Ptychostylid	0	-[0]	tr	0	tr	0	0
Buccal sulcus	2	-[4]	2	2	2	2	2

Specimen number	2MS 125-5	2MS 133-2	2MS 148-9	2MS 218-11
OL	(30.4)	-	28.5	-
B3	13.4	12.2	12.9	-
B4	11.0	10.8	11.0	-
Lnd	12.9	11.7	12.6	-
LF	9.4	9.7	9.6	11.0
Bei	3.0	2.8	1.6	-
Wear stage	2	2-3	2	2
Pli hypoconid	0	0	0	tr
Ptychostylid	0	0	0	tr
Buccal sulcus	2	2	3	1

Scapula

Specimen number	2MS 48-3	2MS 152-1	2MS 235-49	2MS 264-5
SLC (Eisenmann 2)	-	48.5	-	-
GLP (Eisenmann 3)	75.8	-	77.8	-
LG (Eisenmann 4)	(47.7)	-	(49.7)	-
BGmin (Eisenmann 5)	-	-	-	38.5

Humerus

Specimen number	2MS 5-9	2MS 250-4
BT	(65.7)	70.6-

Radius

Specimen number	2MS 241-12	2MS 170-4	2MS 193-1	2MS 217-4
Bp (Eisenmann 4)	71.9+	-	-	-
BFp (Eisenmann 5)	(65.9)	-	-	-
Bd (Eisenmann 7)	-	64.5	66.2+	-
BFd (Eisenmann 8)	-	(50.9)	(53.9)	(52.4) unfused

Metacarpus

Specimen number	2MS 36-18	2MS 214-1
Bp	-	47.5
Dp	31.3+	33.0+

Femur

Specimen number	2MS 48-2
Bd (EISENMANN 7)	(79.0)

Tibia

Specimen number	2MS 224-13
Bd (EISENMANN 8)	61.3
Dd (EISENMANN 9)	40.9

Astragalus

Specimen number	2MS 39-5	2MS 89-1	2MS 212-49	2MS 221-73	2MS 235-37	2MS 248-3
LmT (EISENMANN 2)	-	51.9	-	50.4	59.5	-
LlT	49.7+	47.7	45.8+	47.4	54.6	-
BFd (EISENMANN 5)	-	42.2	-	-	(44.5)	-
GH	(53.0)	(50.0)	-	(50.0)	(55.5)	-
GB	(52.0)	(50.5)	-	(51.0)	(61.0)	(52.5)

Metapodial

Specimen number	2MS 1-1	2MS 38-8	2MS 38-1	2MS 41-2	2MS 128-36	2MS 198-4
Bdfp (EISENMANN 10)	-	40.2+	39.8	40.8	39.7	41.1+
Bda (EISENMANN 11)	-	38.0	41.1	40.2+	38.2	41.3
Dd (EISENMANN 12)	30.6	(29.6)	30.4	-	28.6	28.4+
SD	-	-	27.3	-	-	-
					recently fused	

Specimen number	2MS 221-166	2MS 222-6
Bdfp (EISENMANN 10)	-	-
Bda (EISENMANN 11)	41.9	41.0+
Dd (EISENMANN 12)	-	-
SD	-	-

Phalanx 1

Specimen number	2MS 2-9	2MS 4-2	2MS 4-13	2MS 38-6	2MS 127-1	2MS 157-4
Bp (EISENMANN 4)	45.5+	43.9	-	47.0	39.6	-
SD (EISENMANN 3)	-	-	-	26.4+	25.1	27.6+
Bd (EISENMANN 6)	-	-	-	-	36.7	39.0
BFd (EISENMANN 14)	-	-	-	-	33.8	34.6
GLcal	-	80.2+	76.5+	-	77.2+	79.9+
BFp	(40.9)	(40.3)	-	(41.2)	(37.1)	-
GL	-	(80.0)	-	-	(77.0)	(79.5)

Specimen number	2MS 166-6	2MS 177-1	2MS 182-2	2MS 228-11	2MS 232-250	2MS 256-1
Bp (EISENMANN 4)	-	45.3+	38.9+	44.5+	41.7+	44.3
SD (EISENMANN 3)	27.6	28.7	24.5	27.2	26.4	28.6+
Bd (EISENMANN 6)	-	40.1+	-	41.1	37.7	40.6
BFd (EISENMANN 14)	-	37.5+	33.6+	37.6	35.2	37.4
GLcal	-	83.3+	77.8+	79.5+	79.9+	79.1+
BFp	(41.0)	-	(36.5)	(41.7)	(39.8)	(41.5)
GL	-	(83.0)	(77.5)	(79.5)	(80.0)	(79.0)

Specimen number	2MS 258-5
Bp (EISENMANN 4)	44.7
SD (EISENMANN 3)	25.3
Bd (EISENMANN 6)	-
BFd (EISENMANN 14)	-
GLcal	(74.8)
BFp	(40.1)
GL	(75.0)

Phalanx 2

Specimen number	2MS 5-	2MS 10-3	2MS 67-3	2MS 176-4	2MS 203-9	2MS 228-7
Bp (EISENMANN 4)	41.8	41.0	42.9	-	42.1	43.8
SD (EISENMANN 3)	37.0	35.3	35.6	34.7+	(37.9)	(37.6)
Bd (EISENMANN 6)	38.9+	36.8+	-	36.9+	41.7	39.8+
GLcal	40.5+	42.4	44.2+	39.5	-	39.8
BFp	-	(35.6)	(37.2)	-	(38.0)	(39.0)
GL	(39.5)	(41.0)	(43.5)	(38.5)	(43.5)	(39.0)

Specimen number	2MS 232-112	2MS 258-6
Bp (EISENMANN 4)	40.2	41.1
SD (EISENMANN 3)	35.0	34.5
Bd (EISENMANN 6)	38.3	35.2
GLcal	38.7	39.3+
BFp	(36.7)	(35.4)
GL	(37.5)	(38.5)

Phalanx 3

Specimen number	2MS 152-5	2MS 228-2	2MS 258-4
Ld (EISENMANN 1)	-	46.7	-
BF (EISENMANN 6)	(41.4)	-	(35.1)

APPENDIX 2

Can Hasan III *Equus*: measurements and morphological data.

P^2

Specimen number	49L 109.9 255T1.1	49L 109.9 256T1.2	49L 109.15 272T1.2
OL	31.3	-	27.0
LP	(5.0)	-	(6.9)
Wear stage	3	5	5
Caballine fold	0	0	0

$P^{3/4}/M^{1/2}$

Specimen number	49L 105.11 110T1.1	49L 105.12 121T1.1	49L 109.13 268T1.2	49F 928.3 20T1.1	49T 957.2 30T1.1
OL	23.7	21.6	23.8	20.5	23.2
Be	23.0	22.6	(23.3)	-	(22.9)
LP	7.1	9.3	8.6	(9.2)	7.4
Wear stage	4	2	2	3	2
Caballine fold	0	0	++	0	0
P/M	?P	?M	?P	?M	?P

Specimen number	49L 105.17 132T1.1	49L 109.1 243T1.1	49T 950.4 4T1.1	49T 951.2 8W1.2	49T 950.2 2T1.2
OL	23.7	20.7	21.4	25.2	25.0+
Be	(22.4)	-	(21.3)	-	-
LP	(8.0)	8.6	9.1	9.3	-
Wear stage	2	3	2	2*	2
Caballine fold	+	0	0	tr	0
P/M	?P	?M	?M	-	?P

Specimen number	49L 110.1 314T1.3	49L 105.17 132T1.5	49F 932.7 65W1.4	49F 922+3.1 4W1.2
OL	21.4	(19.1)	(18.8)	-
Be	-	-	20.3	-
LP	-	8.7	8.2	-
Wear stage	5	5	5	2*
Caballine fold	0	0	0	tr
P/M	-	-	-	-

M^3

Specimen number	49L 105.6 51T1.2	49L 109.10 260T1.2	49F 929.1 32T1.1	49L 109.9 255T1.2
LP	9.3	10.4	11.3	9.4
Wear stage	4	4	3	3
Caballine fold	0	0	tr	0

P_2

Specimen number	49L 105.30 171T1.2	49F 929.3 37T1.1	49F 920.15 83W1.1	49F 927.2 14T1.2	49F 929.1 32T1.3
OL	–	23.7	25.6	–	22.5
B3	(10.5)	–	11.1	–	(9.0)
B4	–	12.4	12.4	–	(11.2)
Lnd	12.5	(12.6)	12.1	–	–
LF	(11.6)	10.9	14.9	–	–
Bei	(4.3)	(5.0)	6.6	–	3.9
Wear stage	2	4	?2	5	5
Buccal sulcus	1	1	1	(1)	(1)

$P_{3/4}/M_{1/2}$

Specimen number	49L 102.15 7S1.1	49F 932.12 72T1.1	49L 109.10 260T1.1	49L 109.4 247T1.2	49L 105.45 205T1.4
OL	22.1	(22.7)	(21.7)	22.3	24.0
B3	(11.3)	11.1	10.4	13.8	14.7
B4	(11.2)	–	(9.7)	(13.8)	14.1
Lnd	12.7	–	11.0	13.4	(14.6)
LF	6.0	7.8	7.9	10.4	9.6
Bei	0.0	0.1	0.1	2.2	2.4
Wear stage	2	2	2	2	2
Buccal sulcus	4	4	4	2	2
P/M	M	M	M	P	

Specimen number	49F 932.10 69T1.1	49F 928.2 15T1.2	49L 107.37 236T1.1	49L 109.15 272T1.1	49L 109.14 271T1.1
OL	22.2	22.6	20.1	23.4	21.8
B3	13.7	13.7	12.0	12.5	13.1
B4	13.8	(13.2)	(11.2)	–	12.1
Lnd	13.6	14.0	11.9	13.6	12.8
LF	6.7	12.0	5.4	10.4	7.1
Bei	2.1	3.5	0.3	4.6	0.0
Wear stage	4	3	2–3	2	2
Buccal sulcus	2	2	4	2	4
P/M	P	P	M	P	M

Specimen number	49L 105.12 121T1.2	49B 900.11 11T1.1	49F 932.2 56T1.2	49F 927.3 17T1.1	49L 109.1 241T1.3
OL	22.4	22.1	22.4	21.0	21.7
B3	14.2	(11.7)	–	12.4	11.9
B4	13.5	(11.3)	14.0	11.4	10.8
Lnd	13.3	(12.6)	–	11.5	11.4
LF	9.3	7.0	10.8	6.9	8.6
Bei	3.6	0.1	3.8	0.2	0.2
Wear stage	4	2	3	3	2
Buccal sulcus	2	4	2	4	4
P/M	P	M	P	M	M

Specimen number	49F 928.3 20T1.2	49B 900.11 11W1.4	49L 107.4 209T1.2	49L 109.17 281T1.1	49L 109.14 271T1.2
OL	(26.0)	(28.5)	18.1	17.1	19.5
B3	(9.1)	9.5	–	–	10.6
B4	(8.9)	(9.2)	11.2	–	9.8
Lnd	(11.1)	(13.3)	–	–	(9.9)
LF	(11.3)	9.7	–	–	–
Bei	(0.0)	0.0	5.5	–	2.4
Wear stage	2*	2*	5	5	5
Buccal sulcus	4	4	1	–	1
P/M	M	M	–	–	–

Specimen number	49F 929.2 35T1.1
OL	19.6
B3	11.0
B4	10.0
Lnd	(9.5)
LF	–
Bei	(4.1)
Wear stage	5
Buccal sulcus	1
P/M	–

M_3

Specimen number	49F 932.13 74T1.1	49F 929.1 32T1.2	49F 932.8 66T1.1	49F 932.8 66W1.1
OL	–	26.7	(27.5)	–
B3	–	9.7	10.7	11.2
B4	–	9.5	(9.5)	(10.5)
Lnd	–	10.0	11.4	(11.0)
LF	7.3	–	7.5	6.6
Bei	0.2	1.0	1.0	0.3
Wear stage	2	5	2	4
Buccal sulcus	4	3	3	4

Scapula

Specimen number	49L 105.6 46T1.7	49L 103.9 31T1.1	49L 105.35 185T1.1	49L 107.38 237T1.2
SLC (Eisenmann 2)	–	45.3	36.3	33.5
GLP (Eisenmann 3)	69.4	72.7–	66.2–	65.9
LG (Eisenmann 4)	43.1	46.9–	42.4	(40.5)
BGmin (Eisenmann 5)	37.8	39.1	35.5+	34.6
			fusing	unfused

Pelvis

Specimen number	49L 105.46 206T1.9
LAR	46.8+

Tibia

Specimen number	49L 105.46	49L 110.9
	206T1.6	332T1.2
Bd (Eisenmann 8)	57.5	61.1
Dd (Eisenmann 9)	39.8	42.2

Astragalus

Specimen number	49L 110.3	49L 105.24
	315T1.2	144T1.1
LmT (Eisenmann 2)	48.5	52.4
LlT	44.2	47.9
BFd (Eisenmann 5)	39.4	41.1

Metapodial

Specimen number	49L 109.4	49L 110.19	49L 109.14	49L 100.20
	247T1.4	361T1.7	270T1.5	367W1.1
Bdfp (Eisenmann 10)	-	37.0	-	-
Bda (Eisenmann 11)	39.3+	37.5	38.8-	39.2
Dd (Eisenmann 12)	-	27.8	-	29.1

Specimen number	49L 109.10	49L 105.41	49L 105.17
	262T1.2	198T1.1	132T1.2
Bdfp (Eisenmann 10)	35.3+	38.9	39.5
Bda (Eisenmann 11)	35.3	37.7	38.6
Dd (Eisenmann 12)	-	31.1	28.8
	recently fused		

Phalanx 1

Specimen number	49F 932.13	49L 110.15	49L 109.7	49L 109.14
	74T1.2	355T1.1	253T1.2	270T1.6
Bp (Eisenmann 4)	38.1	40.4+	-	-
SD (Eisenmann 3)	21.3	24.6-	-	Bd (Eisenmann 30.5 6)
35.8	34.9	34.9		
BFd (Eisenmann 14)	30.7	35.0	35.0	33.0
Glcal	71.2	76.8+	-	75.6+

Specimen number	49L 105.6	49F 932.10	49L 103.2
	46T1.6	69T1.3	10W1.1
Bp (Eisenmann 4)	39.4	41.0	42.9
SD (Eisenmann 3)	24.0	25.7-	-
Bd (Eisenmann 6)	31.7+	34.6	-
BFd (Eisenmann 14)	32.0	34.8	-
GLcal	70.1	78.1	-

Phalanx 2

Specimen number	49L 109.1 241T1.4	49L 105.17 132T1.3	49F 932.12 72T1.3	49B 900.10 10T1.1
Bp (Eisenmann 4)	40.0	–	37.9	38.9
SD (Eisenmann 3)	33.9	32.8	32.2	31.8
Bd (Eisenmann 6)	37.2+	34.8	32.5	34.4+
GLcal	39.5	41.2	39.7	40.2

MARIE-FRANÇOISE BONIFAY

Equus hydruntinus Regalia *minor* n. ssp. from the Caves of Lunel-Viel (Herault, France)

Contents

1. Introduction

During the Tübingen symposium, two fundamental needs became apparent. First was the need for better knowledge of the living forms of equid like the hemione to which Holocene species can be connected. Second was the need for a better understanding of the phylogenetic position of fossil forms like *Equus hydruntinus* that were present during the Holocene but extinct today.

The interest of the small European equid *Equus hydruntinus* Regalia lies principally in what it suggests about the evolution of the Quaternary fossil community. Even if this equid is no longer part of modern communities, it nevertheless occupied an important ecological niche in the ancient landscape - that of a small equid adapted to temperate zone environments.

The Pleistocene equid *Equus hydruntinus* was described for the first time by REGALIA in 1907 and cited later by VAUFREY (1928), but it was only following the monograph of STEHLIN/GRAZIOSI, published in 1935 and based on European (principally Italian) material, that the species became relatively well known. Because fossils of this equid are relatively rare and are most often isolated teeth and incomplete long bones, it has usually been necessary to group available data and frequently to compare it with material from widely separated geographical areas and chronological horizons. In spite of these shortcomings, *E.hydruntinus* has been a fossil form easy to recognize because of its principal characteristic, namely, its small size as reflected both in the teeth (microdont) and in the skeleton. Even at sites where it is often not possible to

identify most fossil remains to species, *E.hydruntinus* is easily recognizable. One need merely note the absence of a *pli caballin* from the upper cheekteeth and, in the lower cheekteeth, penetration of the neck of a double knot of stenonine type by the ectoflexid.
Finds of *Equus hydruntinus* exhibit two frequency gradients, the first chronological, the second geographic. Following a sporadic presence in association with Paleolithic assemblages, fossils of this equid become increasingly abundant with time and exceptionally even reach forty percent of the analyzed assemblages, for example from sites dating to the end of the last glaciation in the Mediterranean regions of France (as in Level 10 of the Cornille rockshelter at Istres: M.-F. Bonifay/Lecourtois-Ducgoninaz 1976). Furthermore, just as the sites richest in *E.hydruntinus* remains are the most recent, assemblages from Eastern Europe (several of which are presently being studied) seem to yield the largest collections of this fossil equid.
In sum, populations of *Equus hydruntinus* are rare and when they exist, they are already evolved and morphologically distinct from ancestral forms. Furthermore, as is true of all "Mesolithic" faunas, the fossil remains are fragmentary, the cheektooth rows broken up and the long bones incomplete. This is the reason why the population of this small equid from Lunel-Viel presents two features of interest: it can be considered as the most ancient population of *E.hydruntinus* known from France and, in spite of its age, one of the most complete and best preserved in Europe.

2. The Population of *Equus hydruntinus minor* n. ssp. from Lunel-Viel

2.1. The Sites

The caves of Lunel-Viel, under excavation since 1962, have yielded one of the most important paleontological and prehistoric cultural assemblages of the Mindel-Riss interglacial of Western Europe (E. Bonifay/M.-F. Bonifay 1965). The stratigraphy of Cave I (lower network) includes 14 levels covering a depth of 4.5 meters and yielding a reasonably homogeneous temperate region fauna (M.-F. Bonifay 1971, Mourer 1975, Brugal 1983). The dominant species are the large bovids, cervids (at least three species), and the equids. It is rapidly becoming evident that certain levels include the remains of a small equid significantly different from the most abundant form, the large equid *Equus* (*Equus*) *mosbachensis palustris,* described previously (M.-F. Bonifay 1980).
The caves of Lunel-Viel I served as habitations for prehistoric humans. The faunal remains show signs of having been butchered and the bones are often broken (Guilhaumon 1982). The industry which accompanies this fauna is of the Archaic Middle Acheulean type. Features of human origin including pavings, hearths of various kinds, and small dry-stone walls are among the most ancient structures known from any cave in the world (E. Bonifay 1981).
The stratigraphy of Cave IV (upper network) has been excavated for only a limited period of time and is stratigraphically much less complex. Yellow sands rich in the remains of *Equus hydruntinus* overlie red clayey deposits. This part of the karstic network seems not to contain remains made by humans.

TABLE 1

Inventory of Cranial Remains of *Equus hydruntinus minor* from Lunel-Viel, Caves I (LVI) and IV (LVIV)

CRANIA:

LVI 12059	-	cheektooth row complete but skull flattened and significantly deformed
LVIV 10698	-	incomplete skull, broken in front of second premolar, skull case incomplete, cheektooth row complete
LVIV 13780	-	cheektooth row complete

PARTIAL UPPER CHEEKTOOTH ROWS:

LVI 4-3090	-	M^1, M^2, M^3
LVI 6-1237	-	P^4, M^1, M^2, M^3
LVI 6-2809	-	P^3, P^4, M^1, M^2, M^3
LVI 6-2688	-	P^4, M^1, M^2, M^3

ISOLATED UPPER CHEEKTEETH:

P^2	-	LVI 2-1497
P^3, P^4	-	LVI 4-4362, 6-1287 rt. & lft., 6-2810, 7-13257
	-	LVIV 13809, 13828, 14466, 14467, 14469, 14470
M^1, M^2	-	LVI 2-46 rt. & lft., 4-1924, 4-3079, 5-662, 9-1905
	-	LVIV 13911, 13737, 13995, 15629
M^3	-	LVI 6-1287, 9-1781
	-	LVIV 13743, 14387, 15566
DECIDUOUS	-	LVI 3-8099, 5-4710, 6-4223, 9-12798, 9-12784, 9-12833
	-	LVIV 13967, 14286, 14409, 14916, 15267

MANDIBLES: PARTIAL AND COMPLETE LOWER CHEEKTOOTH ROWS:

LVI 4-3074	-	complete cheektooth row
LVI 5-3601	-	P_3, P_4, M_1, M_2, M_3
LVI 6-3838	-	I_1, I_2, I_3, P_2, P_3, P_4, M_1
LVI 9-1011	-	left cheektooth row complete, right P_2-M_1
LVIV 13891	-	I_2, I_3, P_4, M_1, M_2, M_3
LVIV 14637	-	P_2, P_3, P_4, M_1, M_2
LVI 6-2691	-	D_3, D_4, M_1 (erupting)
LVI 5-12292	-	D_2, D_3, D_4
LVI 6-2298	-	right D_2, D_3, D_4, M_1; left D_2, D_3

ISOLATED LOWER CHEEKTEETH:

P_2	-	LVI 5-3236
	-	LVIV 13740, 13811, 15611
P_3, P_4	-	LVI 5-4735, 9-13236
	-	LVIV 13708, 13808, 13973, 13974, 15609, 15610
M_1, M_2	-	LVI 4-7063, 6-8636
	-	LVIV 13741, 13767, 13807, 13917, 13918, 15598, 15599, 15613, 15614

	M_3	-	LVI 5-4445, 7-2906, 9-1011
		-	LVIV 13810, 15612
	DECIDUOUS	-	LVI 4-3048, 5-7602, 9-12938
		-	LVIV 13744, 13889, 14324, 14538, 15169

ISOLATED INCISORS:

I^2		-	LVI 7-4164
I^3		-	LVI 3-5107, 10-6246
I_1		-	LVI 6-4402, 6-1584
I_2		-	LVIV 16730, LVIV 14988, 14281
	DECIDUOUS	-	LVI 1-156, 7-4164, 10-7075

CANINE:

LVIV 14308

TOOTH BUDS:

LVI 3051, 5-3941, 5-4319, 6-346
LVIV 13742, 13746, 13747, 15600

NOTE: LVI: Lunel-Viel Cave I; the first digit indicates the level, except for LVI V which is the lower part of Lunel-Viel I outside of the area of excavations.
LVIV: Lunel-Viel Cave IV; the number reflects the location of the fossil in the cave; most of the material comes from the level of yellow sands.

2.2. Cranial Remains and Teeth (Table 1)

2.2.1. Cranial Remains (Table 2, Plate 1)

A truly comprehensive study of the cranial remains is not possible because there exist no complete skulls (Table 1). Nevertheless, it is essential to note important aspects of this collection which is, in spite of all limitations, the best preserved available to date.
The skull No. 18698 from Lunel-Viel IV (Plate I) was the subject of a 1977 publication because of its interest from the point of view of ancient cultural practices:

> The rear part of the skull shows a break quite different from the way that the cranial vault is usually and naturally broken. The skull cavity (had been) broken by a blow ... (which) had produced a hole perfectly round in section. The skull of Lunel-Viel IV had been broken intentionally by prehistoric people with the probable purpose of extracting the brain. The dorsal surface of the skull also reveals traces which do not seem natural ... (In particular) a sharp blow above and in front of the orbit led to the formation of a trefoil-shaped cavity with rounded edges ... Likewise, two round and symmetrical holes lead

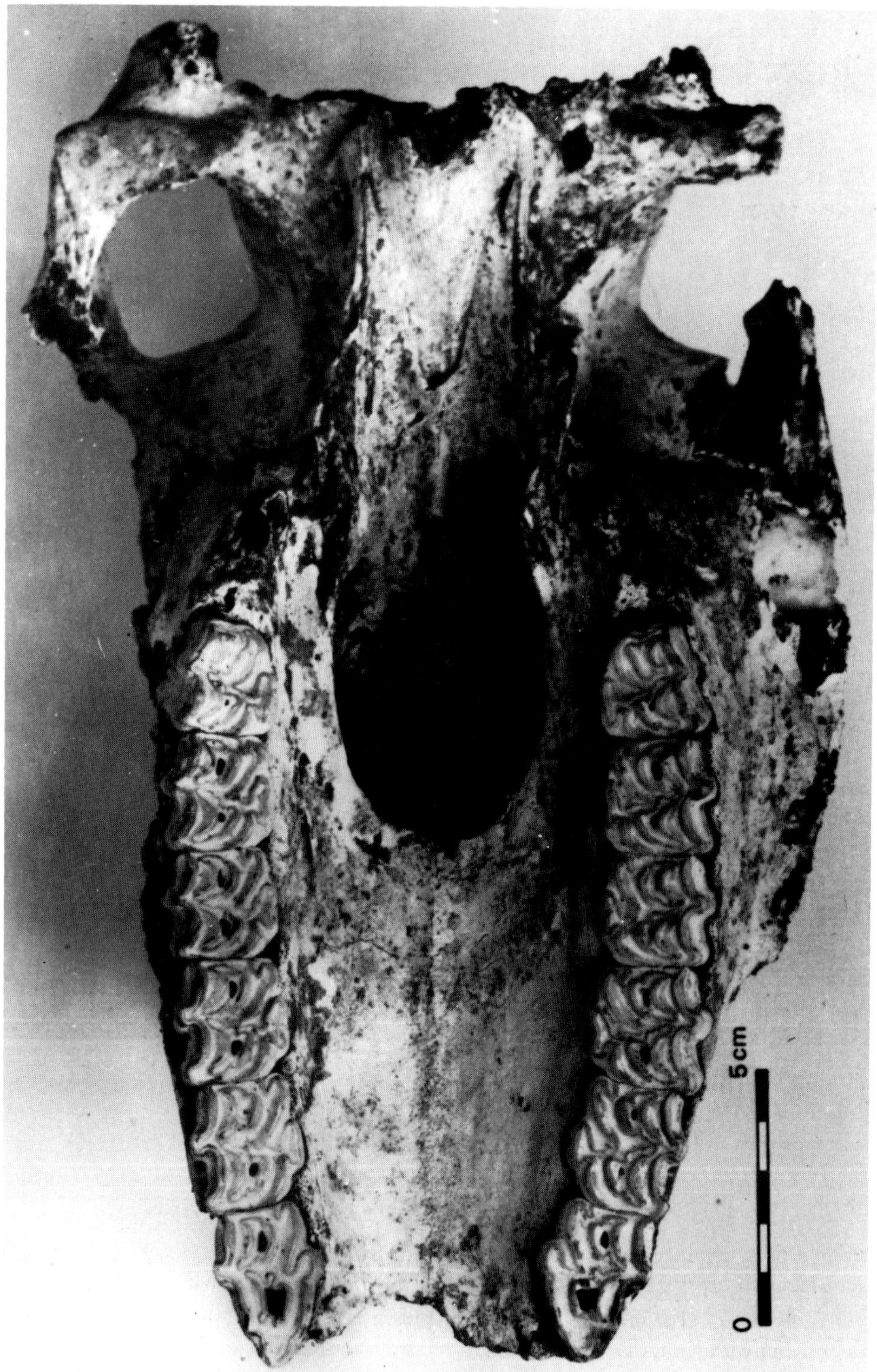

Plate 1: Cranium LVIV 18698, occlusal (basal) view.

TABLE 2

Measurements of Cranial Specimens from Lunel-Viel, Caves I & IV (in mm)

	LVI-12059	LVIV-13780	LVIV-10698
Breadth between M^3	(65)*	–	68
Length P^2–M^3	144.4	147.5	142
Length P^2–P^4	82	81	80
Length M^1–M^3	65	64	62
Breadth between aboral part of the P^2	(49)*	(55)	88
Breadth between oral part of the M^1	(55+)	68	71
Length C–P^2	–	46	–
Length C–I^3	–	12	–

* significant deformation: value is only indicative of approximate size.

one to believe that prehistoric people armed with spears, probably of wood, had speared the animal and pinned it to the spot in order to immobilize it (M.-F. Bonifay 1977).

To the values listed in Table 2, it is possible to add (for LVIV-18698 alone) several other measurements. These include: greatest choanal breadth, 41.1 mm; diameter of the orbit measured perpendicular to the palate, 51 mm; the height of the face measured from the midline at a point between the alveoli of the P^4 and M^1 to the point situated perpendicularly on the internasal suture, 113 mm. It is also possible to estimate the I^2–P^2 distance of skull LVIV-13780 as being about 99 mm.

The total basal length of skull is not measurable on any specimen. Since all indices are calculated as a function of this dimension, we are limited to working only with absolute lengths. This, however, is not a serious drawback because it seems from the work of Forsten (1982) that the indices have little phylogenetic value in equids. According to her, equids evolved either along several axes that vary independently of one another or along a single axis the observed variations of which are due to allometric growth. Furthermore, study of the upper cheekteeth provides a large amount of data and particularly interesting results.

2.2.2. Upper Cheekteeth (Tables 3 and 4, Figures 1–5)

Tooth rows are complete on the two more or less complete skulls from Cave IV and on a deformed skull found in Cave I. Enamel patterns are distinct and have few folds (features which facilitate study) and the folding of the fossettes is poorly developed. I have not observed any isolated hypoglyphs (Figure 1).

The P^2 is a triangular-shaped tooth one of the points of which is directed orally. Only the P^3 reveals a suggestion of a *pli caballin,* while both the P^3 and P^4 have short, slipper-shaped protocones. Among the loose teeth, only premolars showed any signs of a *pli caballin.* The M^1 and

TABLE 3

Measurements of Upper Cheekteeth found in connection at Lunel-Viel Caves I (LVI) and IV (LVIV)
(Occlusal Length, Occlusal Breadth, Protocone Length in millimeters)
(Protocone Index = Pr. L./Oc. L. × 100)

	LVIV 13780	LVIV 10698	LVI 12059	LVI 6–2809	LVI 6–1237	LVI 6–2688	LVI 4–3090	TOT n	mean	s.d.
P^2										
Oc. L.	31.0	31.8	31.7	–	–	–	–	3	31.5	0.44
Oc. B.	22.1	22.5	23.5	–	–	–	–	3	22.7	0.72
Pr. L.	6.2	6.0	7.0	–	–	–	–	3	6.4	0.53
Pr. In.	20.0	18.9	22.1	–	–	–	–	3	20.3	1.63
P^3										
Oc. L.	26.4	25.4	27.0	24.4	–	–	–	4	25.8	1.14
Oc. B.	23.0	23.7	24.5	22.5	–	–	–	4	23.4	0.87
Pr. L.	7.9	8.8	7.0	7.0	–	–	–	4	7.7	0.86
Pr. In.	29.9	34.7	25.9	28.7	–	–	–	4	29.8	3.64
P^4										
Oc. L.	25.0	24.0	23.4	22.0	24.1	24.0	–	6	23.7	0.99
Oc. B.	23.2	24.4	23.0	24.1	24.2	24.9	–	6	24.0	0.73
Pr. L.	9.4	9.1	8.0	8.0	7.4	8.0	–	6	8.3	0.77
Pr. In.	37.3	37.9	34.2	36.4	30.7	33.3	–	6	35.0	2.74
M^1										
Oc. L.	20.5	20.9	21.6	21.8	20.0	19.8	22.6	7	21.0	1.02
Oc. B.	22.2	22.7	23.7	23.4	23.8	23.8	23.7	7	23.3	0.63
Pr. L.	8.3	8.8	7.9	8.0	7.1	7.7	8.9	7	8.1	0.63
Pr. In.	40.5	42.1	36.6	36.7	35.5	38.9	39.4	7	38.5	2.37
M^2										
Oc. L.	22.6	21.7	21.6	21.6	21.6	21.4	24.0	7	22.1	0.94
Oc. B.	21.4	22.6	22.0	22.3	22.7	23.7	22.5	7	22.5	0.70
Pr. L.	9.0	9.0	8.4	7.5	7.6	8.8	9.3	7	8.5	0.71
Pr. In.	39.8	41.5	38.9	34.7	35.2	41.1	38.8	7	38.6	2.68
M^3										
Oc. L.	20.5	20.9	19.2	21.0	22.7	22.6	erupting	6	21.2	1.33
Oc. B.	17.8	19.7	16.7	20.2	20.6	19.8	"	6	19.1	1.53
Pr. L.	10.0	9.8	8.0	8.5	8.8	8.8	"	6	9.0	0.77
Pr. In.	48.8	46.9	41.7	40.5	38.8	38.9	"	6	42.6	4.24
P^2–M^3 Length	147.5	142.0	144.4	–	–	–	–	3	144.6	2.76
P^2–P^4 Length	81.0	80.0	82.0	–	–	–	–	3	81.0	1.00
M^1–M^3 Length	64.0	62.0	65.0	64.0	64.0	63.5	–	6	63.75	0.99

TABLE 4

Measurements of Isolated Cheekteeth from Lunel-Viel, Caves I & IV
(Occlusal Length, Occlusal Breadth, Protocone Length in millimeters)
(Protocone Index = Pr. L. / Oc. L. × 100)

Upper Cheekteeth	n	Range	Mean	s.d.	Lower Cheekteeth	n	Range	Mean	s.d.
P^2					P_2				
Oc. L.	1	–	31.3	–	Oc. L.	4	25.5–27.5	26.7	0.93
Oc. B.	1	–	21.0	–	Oc. B.	4	14.1–15.2	14.5	0.49
Pr. L.	1	–	5.5	–					
Pr. I.	1	–	17.5	–					
P^3, P^4					P_3, P_4				
Oc. L.	11	20.7–26.4	23.4	1.79	Oc. L.	8	23.3–25.9	24.7	0.81
Oc. B.	11	21.8–24.8	23.4	0.84	Oc. B.	8	13.0–17.1	15.9	1.24
Pr. L.	11	6.2– 9.4	8.0	0.89					
Pr. I.	11*	25 –40	34.1	5.16					
M^1, M^2					M_1, M_2				
Oc. L.	10	22.0–27.6	25.1	1.91	Oc. L.	10	21.1–25.8	22.8	1.44
Oc. B.	10	19.8–24.9	22.5	1.78	Oc. B.	10	11.0–15.6	13.8	1.40
Pr. L.	10	8.2–10.5	9.3	0.83					
Pr. I.	10**	32 –42	36.4	2.67					
M^3					M_3				
Oc. L.	4	21.0–24.3	22.4	1.40	Oc. L.	5	25.8–27.5	26.4	0.74
Oc. B.	3	19.1–19.8	19.3	0.41	Oc. B.	5	11.3–13.6	12.2	0.83
Pr. L.	4	7.2–12.0	9.1	2.06					
Pr. I.	4	33 –49	39.7	6.80					

* partitioned: LVIV n = 6, m = 36.8, s.d. = 4.57
LVI n = 5, m = 30.8, s.d. = 3.9
** partitioned: LVI n = 6, m = 36.6, s.d. = 3.50
LVIV n = 4, m = 36.0, s.d. = 0.81

M^2 are particularly small, squarish teeth with well-developed protocones. In contrast, the M^3 is more triangular in form, has an elongated protocone, and sometimes an opening in the post-fossette. A *pli caballin* is extremely rare on molars.

Metrical data for the upper cheekteeth are listed in Table 3 (teeth in maxillaries) and Table 4 (isolated teeth).

The mean occlusal length of the upper cheekteeth of the Lunel-Viel animal decreases in a constant fashion from the P^2 to the M^1 with the greatest value being that of the P^2. The two last molars provide an interesting feature in that the length of the M^3 is not greater than that of the M^2, as in *Equus asinus* and *E.stenonis,* but instead is of equal or sometimes lesser length as in *E. burchelli* (Figure 2).

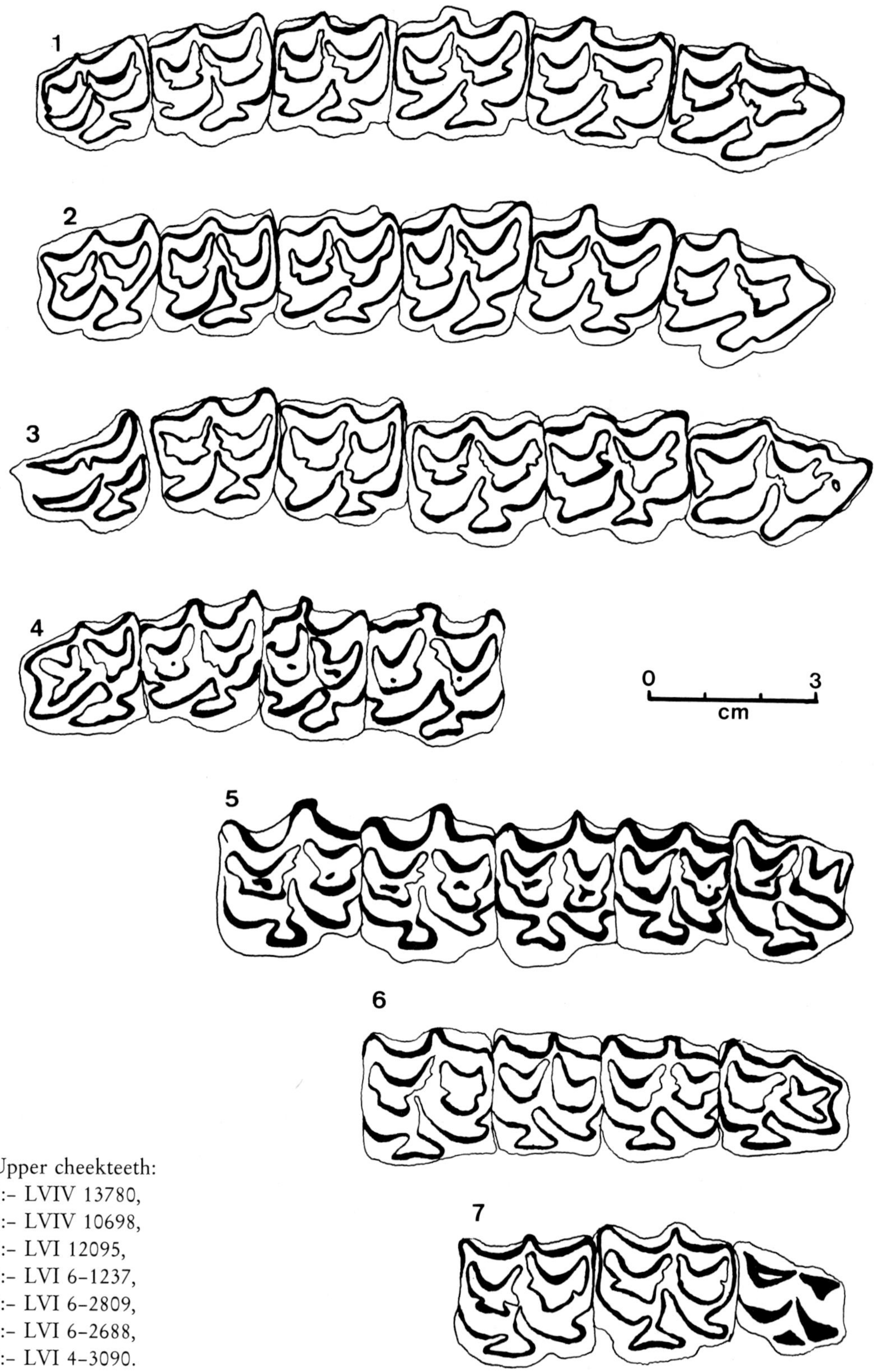

Fig. 1: Upper cheekteeth:
1:- LVIV 13780,
2:- LVIV 10698,
3:- LVI 12095,
4:- LVI 6-1237,
5:- LVI 6-2809,
6:- LVI 6-2688,
7:- LVI 4-3090.

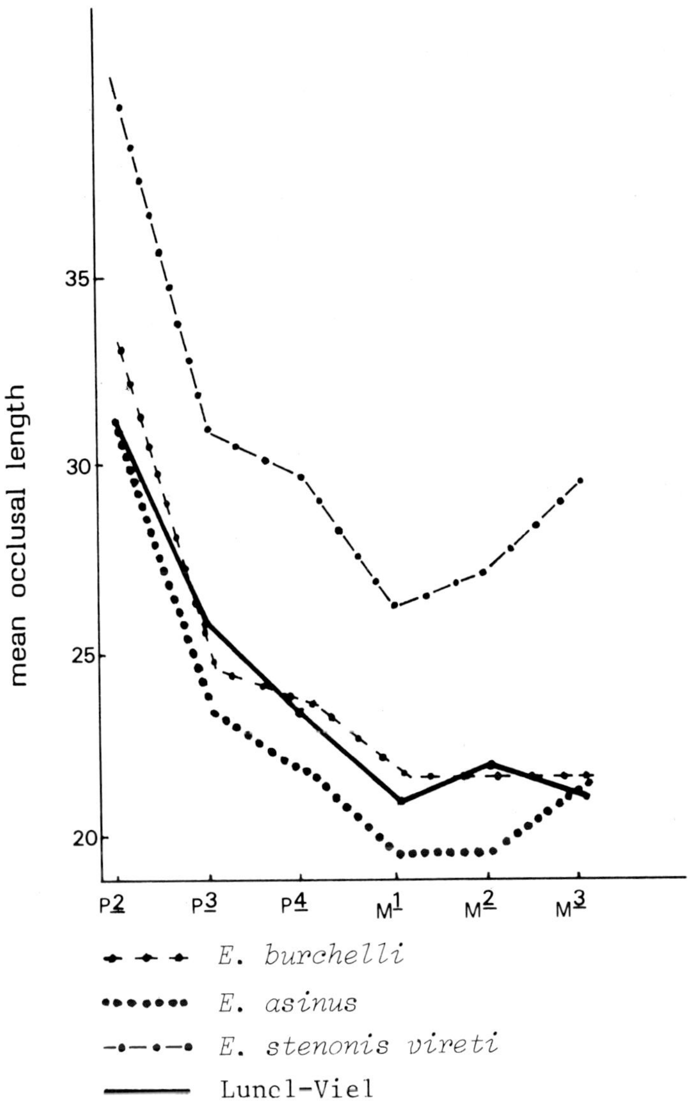

Fig. 2: Mean occlusal lengths of equid upper cheekteeth (cf. EISENMANN 1980: figs. 44, 45, 53).

The length of the protocone in the Lunel-Viel specimens increases progressively from the P^2 through the P^3 to the P^4 but then remains about the same in the M^1 in contrast to the situation in *Equus asinus* where the protocone length of the M^1 is smaller than that of the P^4. Also in the Lunel-Viel material there is not the marked difference between the protocone lengths of the M^2 and M^3 that exists in quagga and that is characteristic of *E. burchelli* (Figure 3). Instead of being much longer, the protocone length of the M^3 is only slightly greater than that of the M^2.

The protocone index (I. P.) of the average Lunel-Viel M^1 is greater than for the P^4 although, overall, it is low having a value less than 43 (Figure 4). In both modern equids and the Lunel-

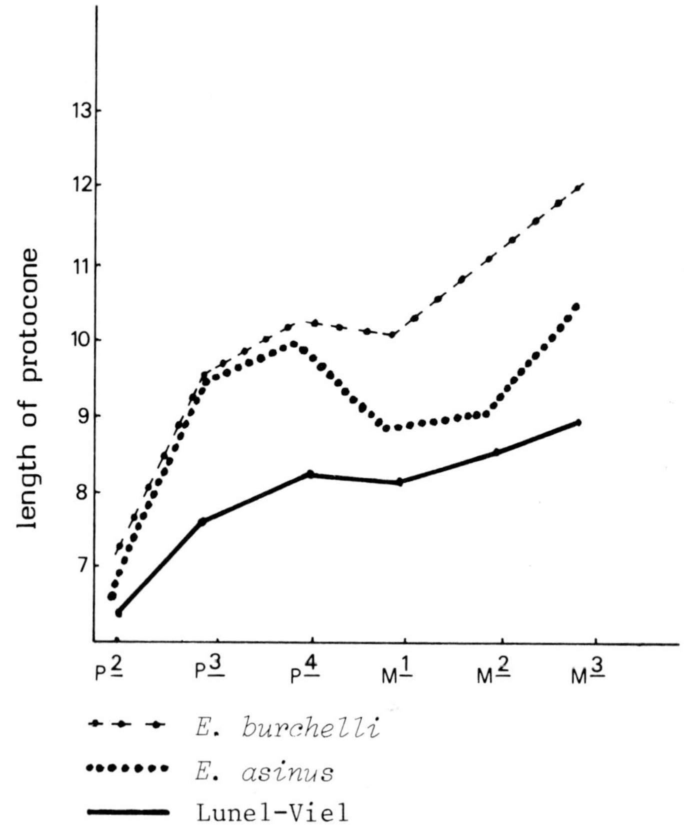

Fig. 3: Mean protocone lengths of equid upper cheekteeth (cf. EISENMANN 1980: figs. 48, 49).

Viel material, the protocone index of the P^3 is greater than that of the P^2 in all individuals, while in more than 85 percent of the cases the protocone index of the P^4 is greater than the protocone index of the P^3. Differences arise with the molars where the protocone index of the M^1 increases in *Equus grevyi, E.burchelli, E.quagga,* the caballines, and the Lunel-Viel fossil while it decreases in hemiones, ass, and *E.zebra* in comparison with the P^4.

The curve closest in shape to that obtained for Lunel-Viel is the one for *Equus mauritanicus* (Figure 5) with the exception that the M^2 to M^3 segment of the latter curve slopes down whereas the analogous segment for the Lunel-Viel material rises. In comparison with other fossil equid material, it seems that the curve for Lunel-Viel is most similar to that for *E.stenonis.*

Among the sub-species of *Equus stenonis,* however, some have a protocone index which decreases from the M^2 to M^3 (e.g., *E.s.senezensis,* 1.5 million years BP) while others have, like *E.mauritanicus,* a protocone index that is the same for the M^2 and M^3 (e.g., *E.s.vireti,* 1.8 m.y. BP). Only in the case of *E.stenonis* cf. *vireti* from Puebla de Valverde (2 m.y. BP), the least evolved of the stenonines, does one find an increase in the protocone index from M^2 to M^3 like what we see in the Lunel-Viel material.

Following EISENMANN (1980), it is possible to establish a "protocone formula" to characterize a form. Class 1 is defined to correspond to protocone indices between 15 and 20, Class 2

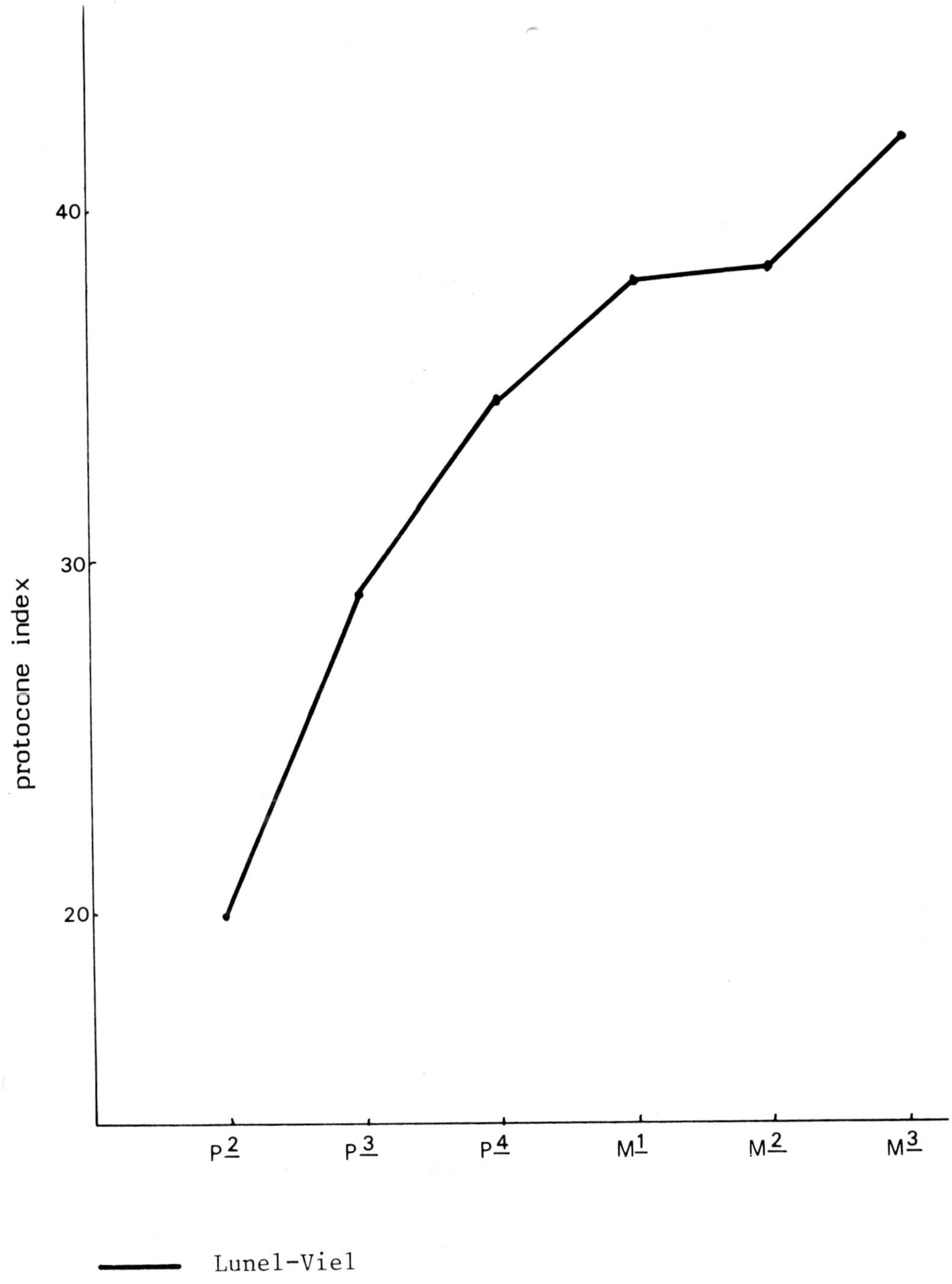

Fig. 4: Mean protocone indices for upper cheekteeth of *Equus hydruntinus minor* from Lunel-Viel.

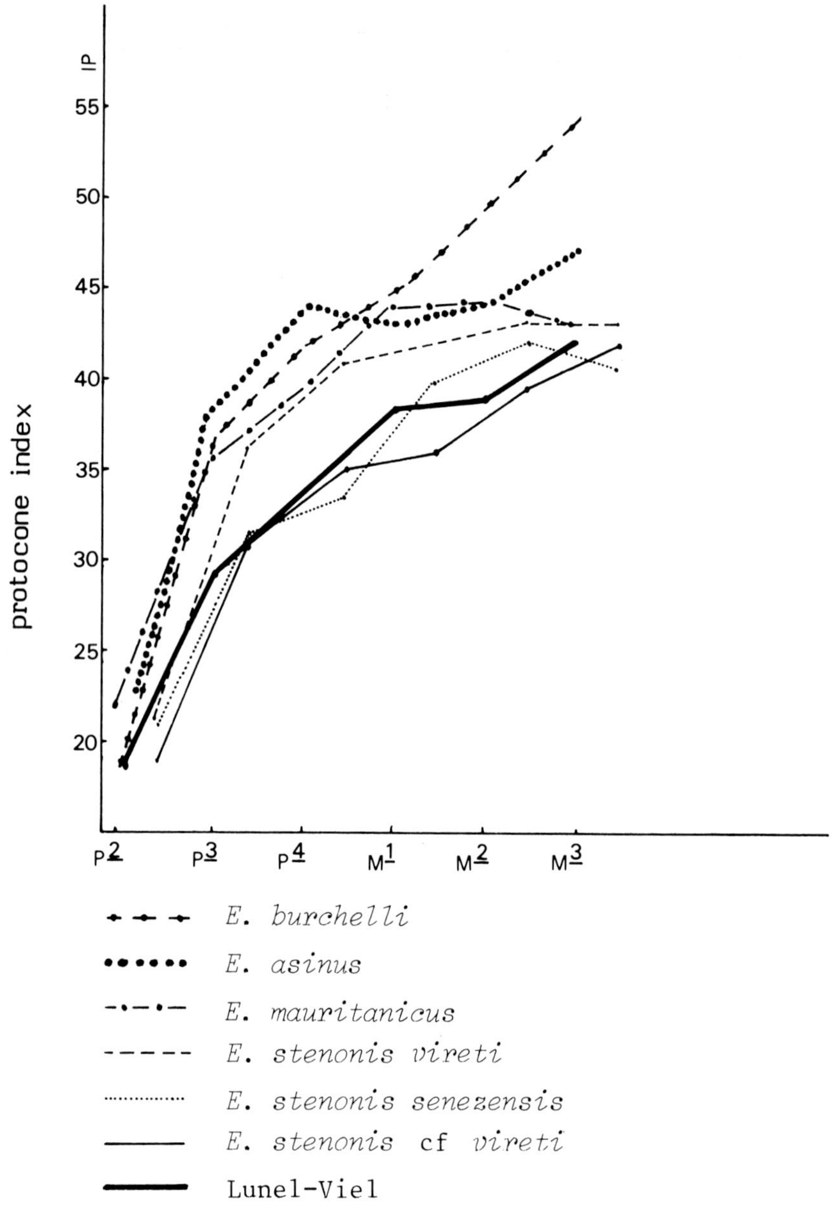

Fig. 5: Mean protocone indices of equid upper cheekteeth (cf. EISENMANN 1980: figs. 50, 51, 55, 64).

between 20 and 25, Class 3 between 25 and 30, Class 4 between 30 and 35, and so on. For *Equus hydruntinus* of Lunel-Viel, the following formula can be defined, arrayed from P^2 to M^3:

2 2 3 3 3 4

which is close to that of *E.mauritanicus* of Ternifine:

2 2 3 3 3 3

and not too distant from modern *E.burchelli*:

2 3 4 4 5 6.

The ratio of length of P^2 to length of P^3 is 1.22:1 (index = 122). This low index tends to confirm the archaic nature of the Lunel-Viel equid. Among the lowest indices in modern species is that for *Equus africanus* (127) and in fossil forms those for the Würm equid *E.germanicus* (124.2) and *E.mauritanicus* (126).

The ratio of P^3 protocone index to P^2 protocone index is 1.46:1 (index = 146). This is also a low index as can be seen by noting that the lowest index in modern material is 168.3 for *Equus grevyi* and in fossil material 148.3 for the equids from Kotheny.

Finally, the ratio of M^1–M^2 protocone index to P^3–P^4 protocone index is 1.06:1 (index = 106). This value is comparable to that found in various modern equids including *Equus africanus* (105.3), *E.asinus* (105.6), and *E.kiang* (106.2). This index for the Lunel-Viel material is close to that for *E.stenonis vireti* (110.7) but less than that for Würm horses, clustering instead with various pre-Würm forms (from Pech de l'Aze and Arago, for example).

2.2.3. The Mandible (Figure 6)

No mandible is preserved complete with ascending ramus. On the most complete mandible from Lunel-Viel (LVI 6–3838), the narrow part of the interalveolar region is situated orally to and at the level of the mental foramen. Furthermore, there are no canines nor sockets for them on this mandible. Breadth across the incisors is 48.9 mm, and length of the symphysis 68 mm. More interesting is the ratio of minimum breadth of symphysis (19.5 mm) to length of the diastema (I_3–P_2: 66 mm) which is 0.29:1 (index = 29). According to PRAT (1980), horses and modern equids have indices greater than 40. With a value less than 30, we again approach those for *Equus stenonis senezensis* and *E.stenonis vireti.*

2.2.4. Lower Cheekteeth (Tables 4 and 5, Figure 7)

As with the upper cheekteeth, the "microdont" character of the small Lunel-Viel equid is evident on the two complete tooth rows which come from Cave I.

On the triangular-shaped P_2, the double knot is deformed. P_3 and P_4 are rectangular in shape and the double knot is rounded; the ectoflexid does not penetrate the neck of the double knot. M_1 and M_2 are almost square. On the M_1, the ectoflexid often slants orally and, in all cases, deeply penetrates the neck of the double knot. The same is true on the M_3, a tooth triangular in form with an ampule-shaped double knot elongated aborally. As a rule, the double knot is of stenonine type, the metastylid and metaconid having rounded borders (Figures 6 and 7). The linguaflexid is pointed. There is no caballine fold. The neck of the double knot is either

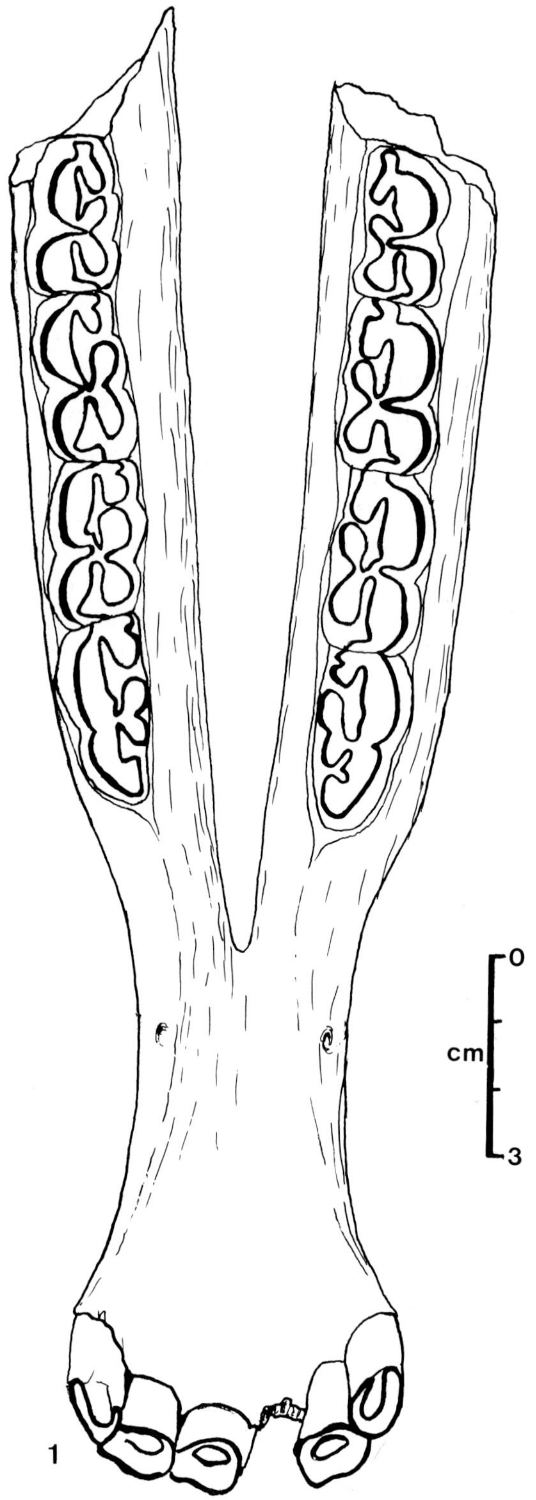

Fig. 6: Mandible LVI 6-3838, occlusal (dorsal) view.

TABLE 5

Measurements of Lower Cheekteeth found in connection at Lunel-Viel Caves I (LVI) and IV (LVIV)
(Occlusal Length, Occlusal Breadth in millimeters)

	LVIV 13891	LVIV 14637	LVI 9-1011	LVI 5-3601	LVI 4-3074	LVI 6-3838	TOT n	mean	s.d.
P_2									
Occlusal L.	–	27.0	28.4	–	26.1	27.0	4	27.1	0.95
Occlusal B.	–	15.0	14.4	–	14.2	13.2	4	14.2	0.75
P_3									
Occlusal L.	–	25.5	26.0	25.6	24.8	24.8	5	25.3	0.53
Occlusal B.	–	16.3	15.5	17.3	15.5	15.1	5	15.9	0.88
P_4									
Occlusal L.	24.4	24.4	25.5	24.5	24.7	24.6	6	24.7	0.42
Occlusal B.	17.6	17.0	16.2	18.3	16.0	17.2	6	17.1	0.86
M_1									
Occlusal L.	21.0	21.8	23.8	21.8	20.8	21.9	6	21.9	1.06
Occlusal B.	16.5	15.5	18.2	17.7	14.8	14.5	6	16.2	1.53
M_2									
Occlusal L.	22.0	21.9	22.8	21.5	20.5	–	5	21.7	0.84
Occlusal B.	14.0	13.8	12.8	15.9	13.2	–	5	13.9	1.19
M_3									
Occlusal L.	24.3	–	27.0	27.5	26.2	–	4	26.3	1.41
Occlusal B.	12.5	–	10.2	13.4	12.2	–	4	12.1	1.35
Length P_2-M_3	–	–	151.0	–	138.7	–	2	144.85	–
Length P_2-P_4	–	–	79.0	–	74.2	–	2	76.6	–
Length M_1-M_3	68.0	–	71.2	69.0	66.2	–	4	68.6	2.08

open (as on the molars) or narrow (as on the premolars); this relates to the nature of the ectoflexid which penetrates the neck deeply in the M_1-M_2 and M_3 (the more when the tooth is more heavily worn) but which remains at a distance from the linguaflexid in the premolars. There is sometimes contact between the lingua- and ectoflexids in the M_1 and M_2.

From the point of view of measurements, the mean occlusal length of the lower teeth decreases from the P_2 toward the M_1, is nearly the same on the M_1 and the M_2 (as with *Equus burchelli*) and increases markedly from the M_2 to M_3 (as with *E.quagga* and the caballines). The ratio of P_2 length to P_3 length is 1.07 (index = 107) and approaches that of the ass, Burchell's zebra, and quagga (114–118). But even more clearly than for the upper cheekteeth, the small size of the Lunel-Viel teeth contrasts markedly with the situation in *E.burchelli.* Having had the opportunity to study a relatively abundant collection of upper cheekteeth, it seems unneces-

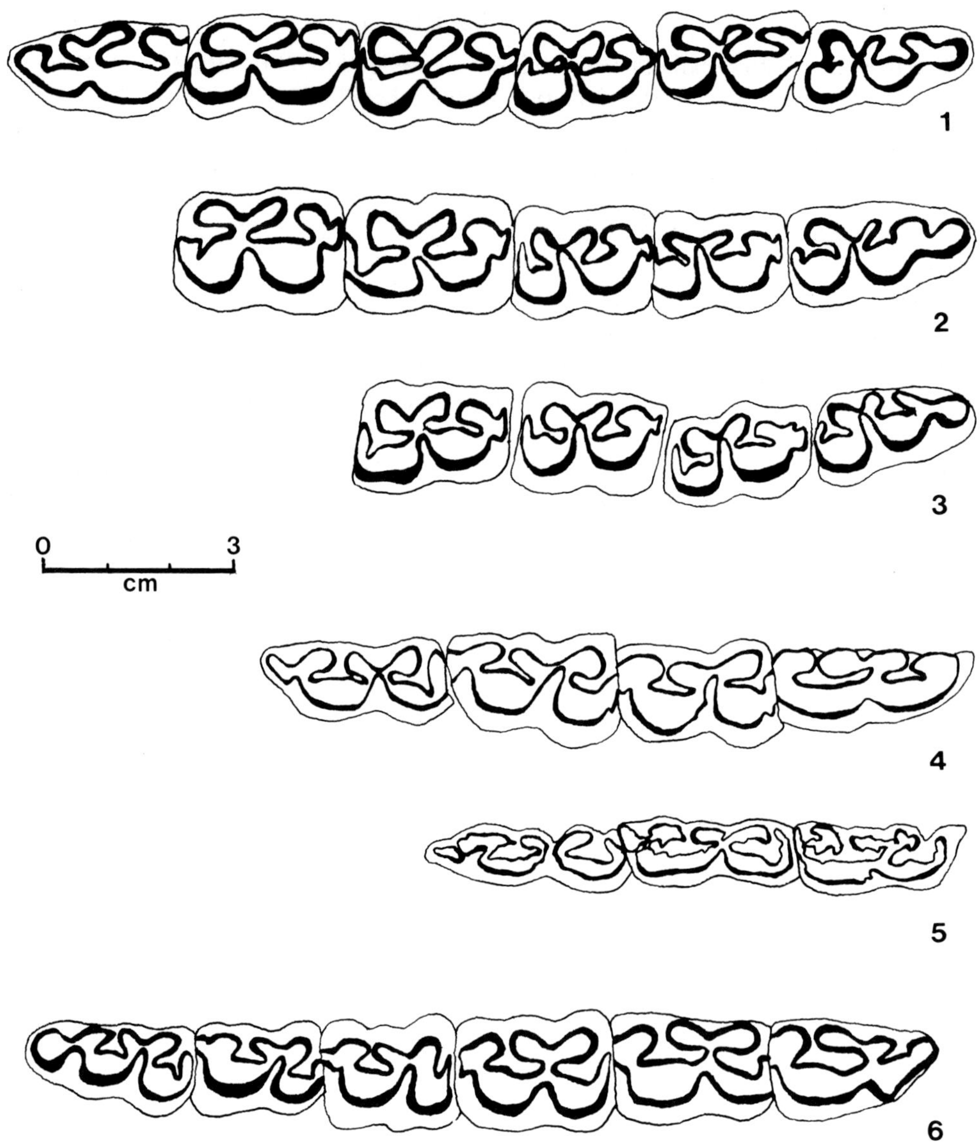

Fig. 7: Lower cheekteeth:
1:- LVI 8-1011,
2:- LVI 5-3601,
3:- LVIV 13891,
4:- LVI 6-2298 (milk teeth + M_1),
5:- LVI 5-12292 (milk teeth),
6:- LVI 4-3074,
7:- LVI 8-1011.
For tooth order, see Table 1.

sary to repeat analogous calculations for the lower teeth and to analyze the post-flexid index since these data will not provide any new information and will only serve to confirm the results obtained from analysis of the protocone index of the upper cheekteeth.

2.2.5. Incisors, Canines, Deciduous Teeth

Isolated incisors could be separated into upper and lower teeth. Medial incisors are worn to a shape round in section, middle incisors to a shape oval in section. Both conditions correspond to a horse of about six years of age and indicate that all the teeth came from young adults. On no isolated tooth is there a cornet present (EISENMANN 1979). In contrast, mandibles Lunel-Viel LVI 6–3638 and LVIV–13891 have lateral incisors with perfectly formed cornets; they are closed in the former and open in the latter indicating that the first mandible came from a younger animal with more rounded incisors than the second.
Deciduous teeth from foals are not numerous in the collection. The cheekteeth have simple enamel patterns only exceptionally folded in a complex fashion (Figure 7, no. 5). The examples illustrated in Figure 7 as nos. 4 and 5 seem quite different. In fact, one illustration (no. 5) shows the lightly worn teeth of a very young animal while the other (no. 6) shows the teeth of an older animal with the first (permanent) molar erupted; the older animal was also in the process of getting its permanent incisors. If one adds tooth buds to deciduous teeth, the total makes up nearly 50 percent of the total for permanent teeth. As with the true horses from Lunel-Viel, this ratio reveals the presence of a relatively high proportion of foals in comparison to adult animals. When this phenomenon is added to the fact that most of the permanent teeth come from young animals, it is clear that the fossil population of *Equus hydruntinus* from Lunel-Viel is heavily dominated by young adults, that is to say individuals about to lose their deciduous teeth and in the course of acquiring their permanent dentition or individuals with permanent teeth in an early stage of wear.
Canines are rare in the Lunel-Viel collection. The only specimen from Cave IV is relatively poorly developed.

2.3. **Post-cranial Remains** (Table 6)

In order to avoid too lengthy descriptions, I will only deal here with anatomical and metrical data which appear to me to be particularly characteristic of the small *Equus hydruntinus.*

2.3.1. Humerus (Table 7, Figure 8)

There is only one almost complete humerus from a small equid; its length (not noted in Table 7) was approximately 200 mm.
From an anatomical point of view, I will compare the bones of the small *Equus hydruntinus* of Lunel-Viel on the one hand with the population of *E.caballus* from the same assemblage and on the other hand with the population of *E.stenonis* from Ceyssaguet (Haute-Loire, excavations of BONIFAY in progress). There are many classic characters to be considered both on the diaphysis and on the distal extremity.
As in *Equus stenonis,* the breadth of the humerus at the deltoid tuberosity (46.5 mm) is considerably less than twice the mid-shaft breadth (25.9 mm). Even though the bone was broken at

TABLE 6

Inventory of Post-Cranial Remains of *Equus hydruntinus minor* from Lunel-Viel, Caves I (LVI) and IV (LVIV)

HUMERUS:	Incomplete:	LVI 6–4567
RADIUS and ULNA:		
	Complete:	LVI 9–1646
	Radius only:	LVI 9–2418
METACARPAL III:		
	Complete:	LVI 2/4–3798 (with one stylet) LVI 7–2883 (with both stylets) LVI 9–2715 (with both stylets)
		LVIV 13750 (with one stylet) LVIV 13930 (with one stylet) LVIV 13790, 13816, 13817, 13969
	Incomplete:	LVI 9–866 LVIV 15339, 13673
METACARPAL II:		LVIV 14191
METACARPAL IV:		LVIV 14193
TIBIA, distal ends:		LVI 3–3973, 5–4465, 7–1598 LVIV 13704, 13708, 13759, 13938, 13991, 14137, 14174, 15037
METATARSAL III:		
	Complete:	LVI 6–2535, 9–2409 LVIV 13620, 13763, 13765, 13916, 13983 (with one stylet), 13984, 13985, 14419, 14420, 14421, 14445, 14547, 15340
	Incomplete:	LVIV 14118, 14209, 13707, 13758
METATARSAL II:		LVI 9–2442, 9–2931 LVIV 13622, 14187, 14190
METATARSAL IV:		LVIV 13621, 14189
CALCANEUM:		LVI 7–2852 LVIV 13628, 13705, 13791, 14528, 14925
ASTRAGALUS:		LVI 4–3949, 7–2933 LVIV 13627, 13755, 13768, 13769, 13781, 13959, 14091, 14443, 14446 (eroded)

CENTRAL TARSAL:		LVI 0–3015, 6–5057 LVIV 13626, 13784, 13815, 14527, 14832, 15494
OS TARSALE III:		LVI 9–1342, 9–2414 LVIV 13623, 14054, 14186, 14422, 14423
OS TARSALE II:		LVI 9–3519 LVIV 13625
OS TARSALE IV:		LVI 9–2445 LVIV 13674
SESAMOIDS:		LVI 6–5055 LVIV 15536
PHALANX 1 –	Anterior:	LVI 7–1645 LVIV 13630
	Posterior:	LVI 6–5058 LVIV 13619, 13795, 13978, 13798, 14027, 14100
PHALANX 2 –	Anterior:	LVIV 13806, 13895
	Posterior:	LVI 6–5049, R–2903 LVIV 13618, 13706, 13787, 13796
	Position n. d.:	LVIV 13898, 14082
PHALANX 3 –	Complete:	LVI 7–1568 LVIV 13617, 13757, 15007
	Incomplete:	LVI 9–2735 LVIV 13685, 14092, 14158

TABLE 7

Measurements of Long Bones from Lunel-Viel Caves I (LVI) and IV (in mm)

	Humerus LVI 6–4567	Radius & Ulna LVI 9–1646	Radius & Ulna LVI 9–2418	Tibia n	Tibia Range	Tibia Mean	Tibia s.d.
1. Total Length	–	(300)	–	–	–	–	–
1 bis. Length of Radius	–	250.8	257.0	–	–	–	–
2. Proximal Breadth	64.0	52.2	–	–	–	–	–
3. Proximal Depth	58.2	24.0	31.0	–	–	–	–
4. Mid-Shaft Breadth	25.9	27.2	29.2	–	–	–	–
5. Mid-Shaft Depth	30.2	20.0	22.6	–	–	–	–
6. Distal Breadth	55.0	52.1	55.3	9	49.2–53.1	51.5	1.49
7. Distal Depth	57.4	29.3	33.2	11	32.0–35.8	34.4	1.23

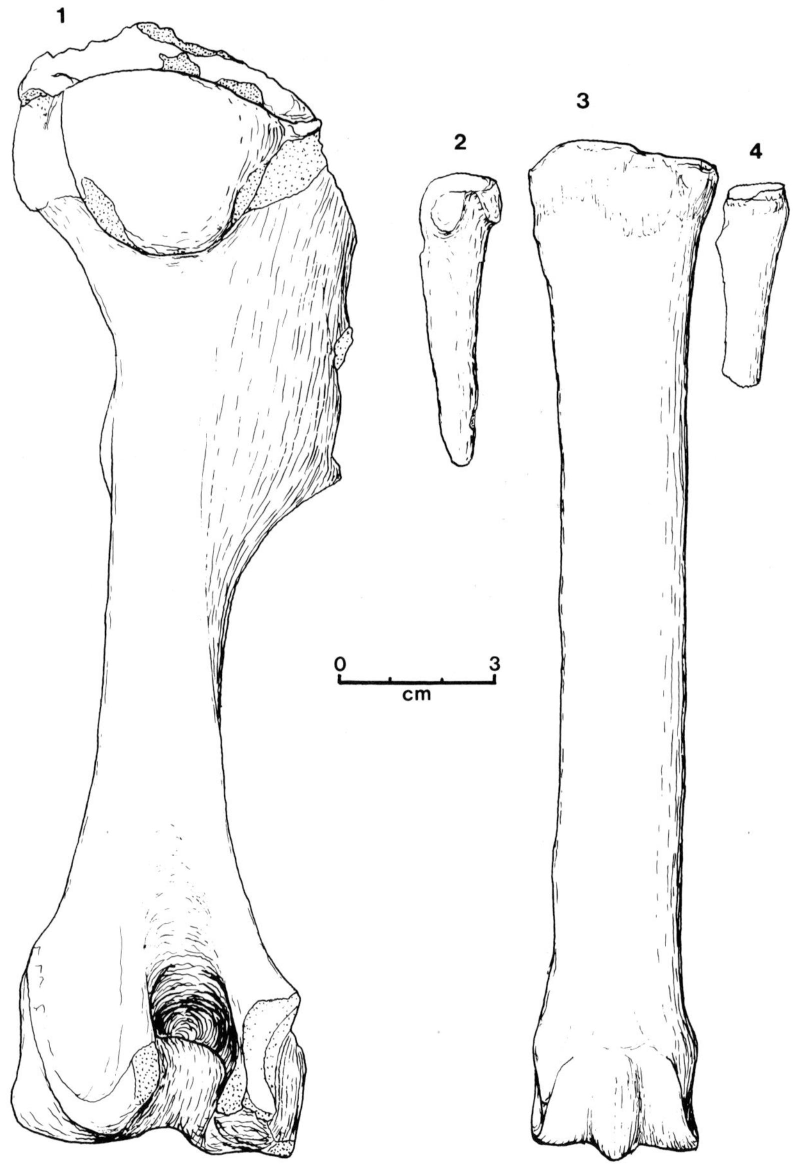

Fig. 8: Bones of the front leg:
1:- Humerus: LVI 6-4567, caudal view;
2:- Metacarpal II: LVIV 14193;
3:- Metacarpal III: LVIV 13790, dorsal view;
4:- Metacarpal IV: LVIV 13191.

this point, it seems evident that the deltoid tuberosity could not have been as much as 60 mm, the value necessary to be twice the shaft breadth of the smallest *E.caballus.*

In cranial view, the best character is the shape of the coronoid (radial) fossa which extends transversely and, as in *Equus stenonis,* runs along and clearly delimits the upper edge of the distal articular surface. It is easy to distinguish the humerus of *E.caballus* using this character since in that animal the fossa is more scooped out and confined to a central position over the trochlear groove. Also, the small lateral swelling visible on stenonine humeri limiting the upper part of the coronoid (radial) fossa is absent on *E.caballus.*

In caudal view, the olecranon fossa is triangular in shape, narrow, and not deeply scooped out in *Equus caballus*; triangular, wide, and deep in *E.stenonis*; and oval, wide, and deep in the small equid of Lunel-Viel.

As far as measurements are concerned, since the small size of *Equus hydruntinus* serves to completely isolate it from other forms, it seems worthwhile to consider only indices and not absolute values. The ratio of proximal breadth to length is c. 0.30:1 (index = 30), a value which is close to those determined for the stenonine of Seneze (31–33: PRAT 1980) and clearly less than those for the horse of Lunel-Viel (35–39). The ratio of mid-shaft breadth to length (robustness index) is 0.12:1 (index = 12) in *E.hydruntinus* which is similar to ratios for *E.stenonis* from Seneze but not so high as the index for the robust *E.caballus* (index = 15).

In sum, the humerus from the small equid of Lunel-Viel has clear morphological similarities with *Equus stenonis* although it has a somewhat differently shaped distal articulation, a smaller proximal end, and a generally less robust overall character.

2.3.2. Radius and Ulna (Table 7, Figure 9)

From an anatomical point of view, there is one useful distinguishing feature easily visible on the radius. On the volar (palmar) side of the distal end of the Lunel-Viel specimens in volar view, there is a relatively deep longitudinally oriented median trough between the lateral (ulnar) and medial (radial) sides of the bone. This depression is completely absent in horses and only vaguely outlined in the stenonines. The boney swelling which marks the point of insertion for the *flexor digitorum superficialis* and *f. d. profundus* is relatively better developed on the Lunel-Viel specimens than in the horse, whereas it is absent or almost so in the *Equus stenonis* of Ceyssaguet. This feature, which reflects the presence of powerful musculature, is confirmed by the prominence of areas for muscle insertion arranged as distinct protrusions on the two small radii from Lunel-Viel. This, however, does not imply an enlargment of the long bones or a skeleton which one might term "stocky"; as for the humerus, the mean index of robustness which is about 12.35 for *E.stenonis* from Seneze and 13.9 for *E.caballus* from Lunel-Viel is significantly lower in our *E.hydruntinus* (10 and 11).

2.3.3. Tibia (Table 7, Figure 10)

The tibia is represented only by two distal ends. The sole index which it is possible to calculate is the ratio between the dorso-plantar depth and the medio-lateral breadth of the distal end. For the Lunel-Viel *Equus hydruntinus* the mean value of the index is 66 (ratio = 0.66:1); for *E.stenonis* the value centers on 64; for *E.caballus* from Lunel-Viel, the index is between 57 and 60. This index indicates a tendency for the distal tibia of *E.hydruntinus* to be more rectangular, this also being a characteristic of *E.stenonis.*

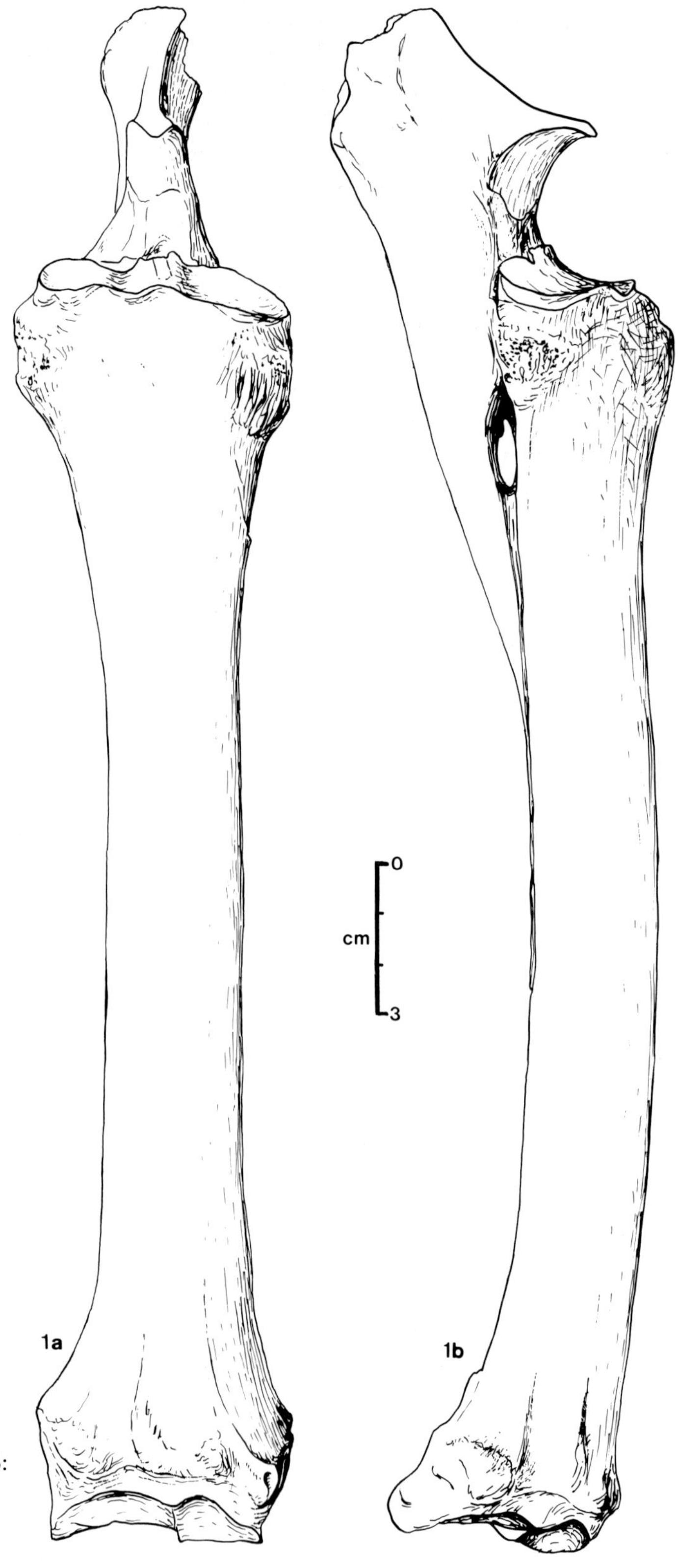

Fig. 9: Radius and Ulna: LVI 9-1646:
1a:- dorsal view,
1b:- lateral view.

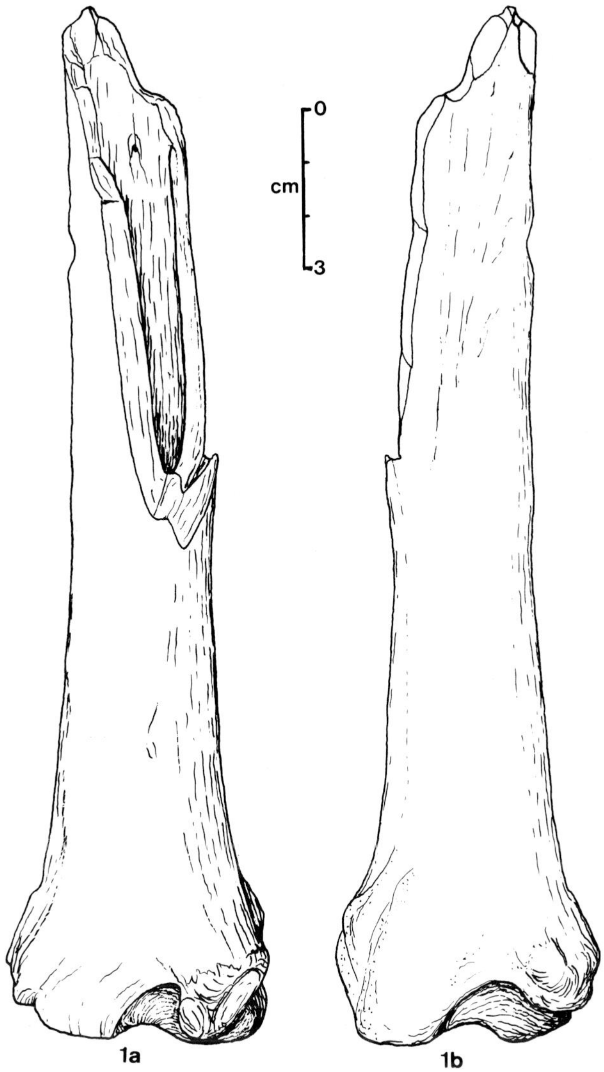

Fig. 10: Distal tibia: LVI 7–1598:
1a:– dorsal view,
1b:– plantar view.

2.3.4. Metapodials (Tables 8–10, Figures 8, 11, 12)

From the complete metapodials it is possible to calculate height at the withers:

190.26 mm (mean length of McIII) × 6.41 = 1219 mm (1.219 m);
229.32 mm (mean length of MtIII) × 5.33 = 1222 mm (1.222 m);
mean height at withers: c. 1.2 m.

From an anatomical point of view, when one considers the wealth of comparative material, few distinguishing characters are evident and the third metacarpals in particular seem to have a

TABLE 8

Measurements of Third Metapodials from Lunel-Viel, Caves I and IV (in mm)
(L. = Length, B. = Breadth (Medial-Lateral), D. = Depth)

	METACARPAL III				METATARSAL III			
	n	Range	Mean	s.d.	n	Range	Mean	s.d.
1. Total L.	9	186.3–193.2	190.2	2.43	14	224.9–234.5	229.3	3.10
2. Proximal B.	8	34.7– 38.5	36.2	1.11	15	33.8– 38.7	36.5	1.52
3. Proximal D.	8	23.5– 26.5	25.1	1.07	15	28.2– 33.7	30.8	1.53
4. Mid-Shaft B.	9	22.4– 25.2	24.6	1.16	14	22.3– 26.7	24.9	1.17
5. Mid-Shaft D.	9	19.5– 22.0	21.0	1.00	13	22.8– 25.4	24.2	0.84
6. Supra-Artic. Distal B.	9	30.2– 34.6	32.4	1.32	15	31.9– 34.3	33.0	0.94
7. B. Distal Articulation	12	30.8– 35.0	33.1	1.07	16	31.7– 34.6	32.6	2.29
8. D. Distal Articulation	11	23.2– 26.6	25.4	1.18	17	24.5– 26.9	25.8	0.58
Robustness Index (4./1. × 100)	8	12.3– 13.5	13.0	3.53	12	10.1– 11.6	10.9	4.14
Caballine Index (6./7. × 100)	9	94 - 98	97.0	1.50	14	96 -103	98.5	1.78

TABLE 9

Measurements of Metapodials II and IV from Lunel-Viel, Caves I (LVI) and IV (LVIV) (in millimeters)

		Total Length	Proximal Depth
Metacarpal IV:	LVIV 14191	–	12.7
Metacarpal II:	LVIV 14193	–	13.5
Metatarsal II:	LVI 9–2442	96	13.5
	LVI 9–2931	75	14.2
	LVIV 14187	73	14.4
	LVIV 14190	67	13.8
	LVIV 13622	93	14.0
Metatarsal IV:	LVIV 13621	99	14.7
	LVIV 14189	63	14.5

TABLE 10

Tabulation of Third Metapodials according to numbers of articulating accessory metapodials: Lunel-Viel, Caves I (LVI) and IV (LVIV)

Cave	McIII n	with no fused acces. Mp: n	with 1 fused acces. Mp: n	with 2 fused acces. Mp: n
LVI	3	0	1	2
LVIV	6	4	2	0
TOTAL	9 (100%)	4 (44%)	3 (33%)	2 (22%)

Cave	MtIII n	n	n	n
LVI	2	2	0	0
LVIV	14	13	1	0
TOTAL	16 (100%)	15 (94%)	1 (6%)	0 (0%)

rather homogeneous morphology. Only for the proximal end does there seem to be a valid feature characteristic of *Equus hydruntinus* (following Prat 1980), namely that the upper articular surface has volar-lateral and volar-medial notches which are less marked on *E.caballus* in relation to the size of the stylets.

As for the stylets, McII and McIV are numerous, whether fused to the McIII or loose (Table 10). A ratio of one stylet fused for every one unfused (50 percent) is the same for both the large and small fossil equids of Lunel-Viel and represents a value between that observed for modern horses in which most of the stylets are loose and *E.stenonis* where they are fused. In contrast, the MtII and MtIV were found loose in 93 percent of the cases (15 out of 16) at Lunel-Viel. The length of the accessory metatarsals varies between 27 and 43 percent of the length of the average MtIII in the *E.hydruntinus* of Lunel-Viel (Table 9). For comparison, the variation is between 36 and 50 percent in contemporary caballine equids.

As far as indices for the third metapodials are concerned, the caballine index (distal supra-articular breadth/distal articular breadth × 100) is generally less than 100 (range 94–103, see Table 8). An index below 100 is a relatively evolved trait; the index for *Equus stenonis* is above 100 as, in fact, is the index for part of the population of *E.caballus* at Lunel-Viel.

The index of robustness (mid-shaft breadth/total length × 100) is relatively low, averaging 12.97 for McIII and 10.85 for MtIII. These figures are some 4 points below the average calculated for the metapodials of *Equus caballus* from Lunel-Viel and more than 2 points below the average indices for *E.stenonis* from Seneze.

As far as absolute size is concerned, the *Equus hydruntinus* of Lunel-Viel is closest to *E.burchelli.* The average length for our third metacarpals is 190.26 mm and for our third metatarsals

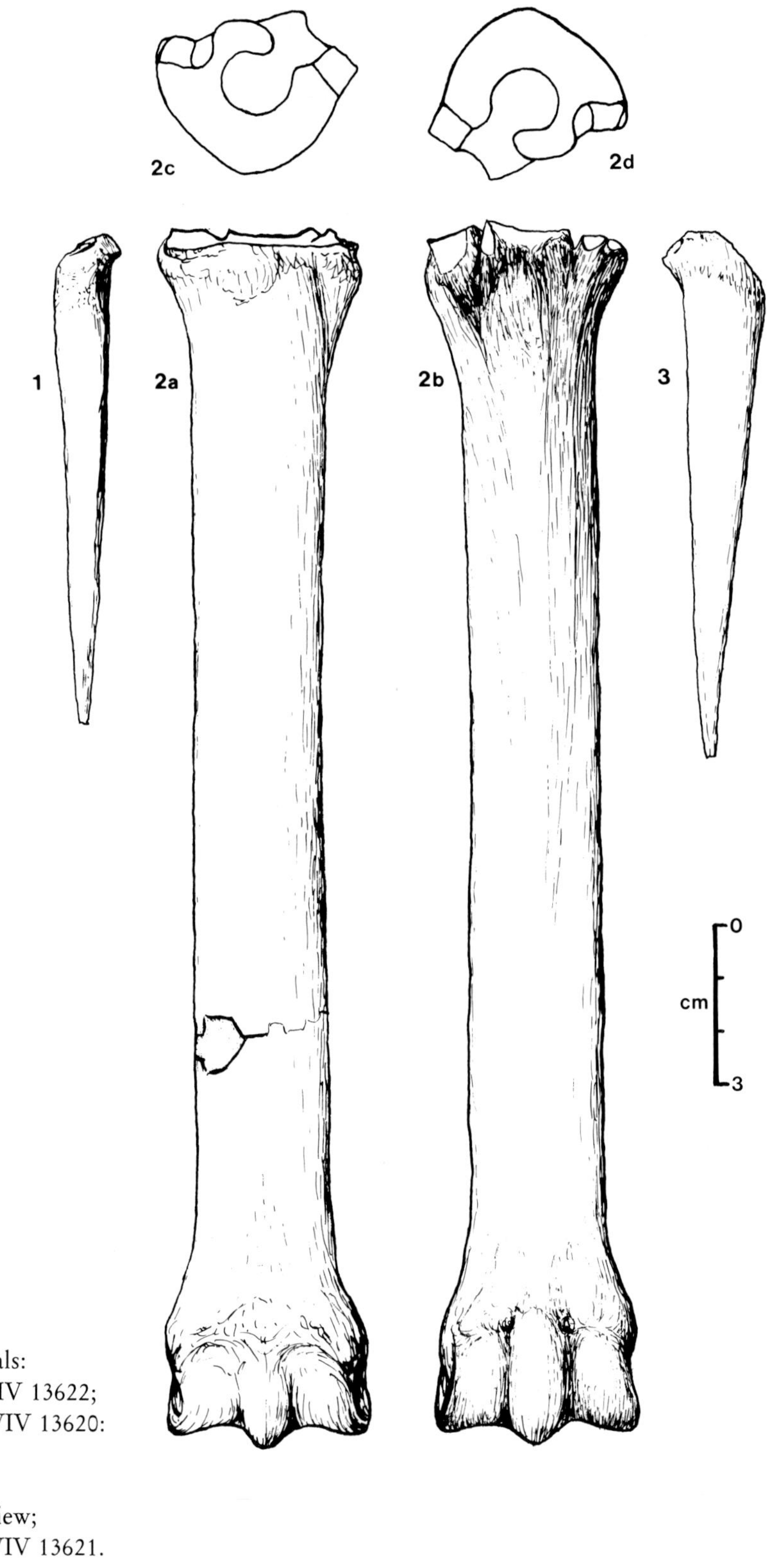

Fig. 11: Articulating metatarsals:
1:- Metatarsal II: LVIV 13622;
2:- Metatarsal III: LVIV 13620:
2a:- dorsal view,
2b:- plantar view,
2c & 2d:- proximal view;
3:- Metatarsal IV: LVIV 13621.

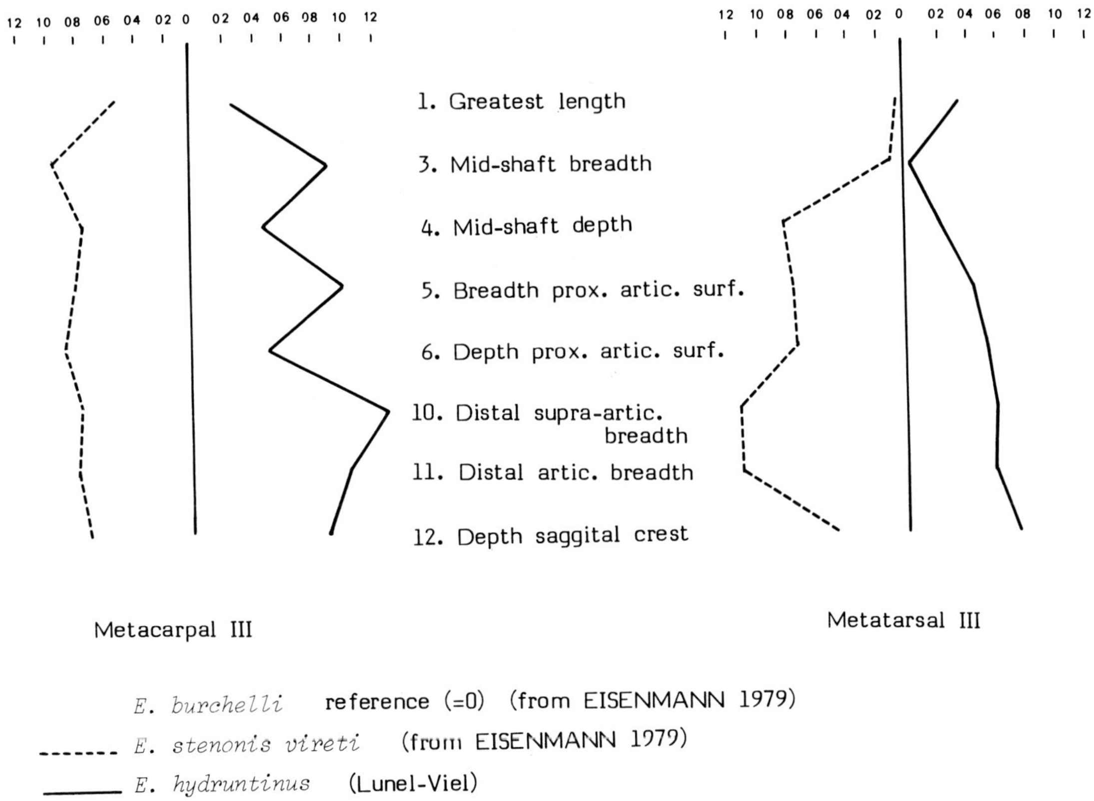

Fig. 12: Ratio diagrams for Metacarpal III and Metatarsal III using the difference of logarithms technique.

229.32 mm. Corresponding lengths for *E.burchelli* are 201.7 mm and 226.4 mm. Also as in Burchell's zebra, the proximal breadth of the Lunel-Viel MtIII is slightly greater than that of the McIII.

It is possible to push metapodial comparisons a little further, but not like EISENMANN/KARCHOUD (1982). If one takes *Equus burchelli* as the reference taxon, uses the measurements published by EISENMANN (1979), and constructs curves employing the difference of logarithms technique (Figure 12), it is interesting to note that while the metacarpals of *E.stenonis* are similarly proportioned to those from *E.burchelli,* for the metatarsals it is those of *E.hydruntinus* (and not *E.stenonis*) which are more similar to the reference. As far as the metacarpals are concerned, it is the three breadth dimensions (mid-shaft, proximal, and distal) which are the most aberrant; for the metatarsals it is only the mid-shaft diameter which seems out of line. In any event, it seems that the third metacarpals of both *E.stenonis* and *E.hydruntinus* are relatively less gracile than those of *E.burchelli,* but this character is less evident for third metatarsals.

2.3.5. Phalanges (Tables 12–14, Figure 13)

Posterior phalanges are more numerous than anterior ones which are rare. Anterior first phalanges are a little longer and less robust than are posterior ones; anterior second phalanges are a little shorter but also less robust than posterior ones (Tables 13–14). The posterior first pha-

TABLE 11

Measurements of Astragali and Calcanea from Lunel-Viel, Caves I and IV (in millimeters)

	n	Range	Mean	s.d.
ASTRAGALUS				
1. Greatest Height	10	40.5–45.2	43.5	1.57
2. Medial Height	9	41.5–46.0	44.4	1.40
3. Lateral Height	9	41.5–47.0	43.5	1.57
4. Greatest Breadth	10	38.5–44.1	42.3	1.95
5. Breadth of Distal Articulation	10	36.1–40.0	37.4	1.39
6. Depth of Distal Articulation	10	23.5–27.2	25.5	1.04
Ratio: 4./1.	10	0.85–1.05	0.97	0.06
Ratio: 5./1.	10	0.79–0.90	0.85	0.03
CALCANEUM				
1. Greatest Length	2	84.6–84.8	84.7	–
2. Greatest Breadth	6	32.0–33.8	33.1	0.70
3. Least Breadth of body of bone	6	14.7–15.4	15.1	0.29
4. Depth at 3.	5	35.0–36.7	35.6	0.64
Ratio: 2./1.	2	0.39	0.39	–

TABLE 12

Measurements of Central and Third Tarsals and Third Phalanges from Lunel-Viel, Caves I and IV (in millimeters)

	n	Range	Mean	s. d.
CENTRAL TARSAL				
1. Greatest Breadth	8	33.7–38.3	36.3	1.59
2. Greatest Depth	8	29.1–32.2	30.8	1.13
THIRD TARSAL				
1. Greatest Breadth	7	30.8–35.8	33.6	1.85
2. Greatest Depth	7	29.8–34.4	30.9	1.63
PHALANX 3				
1. Greatest Median Length	4	37.7–39.6	38.8	0.81
2. Greatest Articular Breadth	7	27.5–30.2	29.2	1.05
3. Median Articular Depth	8	17.0–19.5	18.5	0.91

TABLE 13

Measurements of Anterior First and Second Phalanges from Lunel-Viel, Caves I (LVI) and IV (LVIV) (in millimeters)

	LVI 7–1645 PH. 1	LVIV 13630 PH. 1	LVIV 13896 PH. 2	LVIV 13806 PH. 2
1. Greatest Length	68.8	70.3	35.0	33.8
2. Proximal Breadth	35.4	36.2	32.0	29.7
3. Proximal Depth	22.0	27.6	24.0	21.7
4. Mid-Shaft Breadth	21.4	23.0	27.3	26.2
5. Mid-Shaft Depth	19.5	19.8	–	–
6. Distal Breadth	(34.0+)	31.9	28.6	26.2
7. Distal Depth	15.0	15.2	16.7	14.5
Robustness Index (4./1. × 100)	31.1	32.7	78.0	77.5
Index: (2./1. × 100)	51.5	51.5	91.4	87.9
Index: (6./1. × 100)	(49.4+)	45.4	81.7	77.5

TABLE 14

Measurements of Posterior First and Second Phalanges from Lunel-Viel, Caves I and IV (in millimeters)

FIRST PHALANX	n	Range	Mean	s.d.
1. Greatest Length	7	60.8–66.2	63.9	2.03
2. Proximal Breadth	7	35.8–38.4	37.1	0.93
3. Proximal Depth	7	21.7–28.6	24.8	2.71
4. Mid-Shaft Breadth	7	20.8–22.7	21.8	0.71
5. Mid-Shaft Depth	7	18.0–20.2	19.5	0.73
6. Distal Breadth	7	29.7–31.8	30.8	0.87
7. Distal Depth	7	14.2–15.6	15.1	0.43
Robustness Index (4./1. × 100)	7	32.8–35.1	34.0	0.86
Index: (2./1. × 100)	7	56.5–59.2	58.1	1.12
Index: (6./1. × 100)	7	46.8–49.1	48.2	0.91
SECOND PHALANX				
1. Greatest Length	6	34.9–37.2	36.0	0.92
2. Proximal Breadth	6	33.8–35.8	34.5	0.80
3. Proximal Depth	6	23.4–24.9	21.3	0.53
4. Mid-Shaft Breadth	6	28.7–30.3	29.4	0.69
6. Distal Breadth	6	29.3–31.6	30.1	1.02
7. Distal Depth	6	15.0–17.6	16.3	0.86
Robustness Index (4./1. × 100)	6	79.5–85.3	81.5	2.09
Index: (2./1. × 100)	6	92.7–98.5	95.5	2.18
Index: (6./1. × 100)	6	80.6–87.7	83.2	2.64

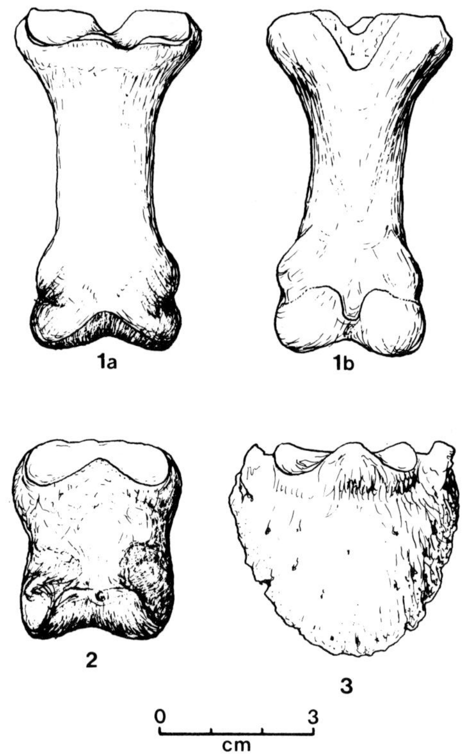

Fig. 13: Phalanges:
1:- posterior Phalanx 1: LVIV 13619:
1a:- dorsal view,
1b:- plantar view;
2:- posterior Phalanx 2: LVIV 13618, dorsal view;
3:- Phalanx 3: LVIV 13617, dorsal view.

langes are gracile with an average smallest breadth to greatest length robustness index of 34 as compared with 42.2 for *Equus stenonis* from Seneze and 44 or 45 for *E.caballus* from Lunel-Viel. The supra-articular tubercles are also well developed on the Lunel-Viel *E.hydruntinus* posterior first phalanges. As for the posterior second phalanges, the Lunel-Viel small equids have an index of robustness averaging 81.45. This can be compared with indices between 92 and 95 for the horse from Lunel-Viel and about 85 for the stenonine from Seneze. The index of distal breadth to greatest length for the second phalanges from the Lunel-Viel *E.hydruntinus* averages just over 83 in comparison with more than 100 for horse and about 92 for the stenonine equid.

2.3.6. Other Post-Cranial Elements (Tables 11–12, Figure 14)

Among the compact bones of the post-cranial skeleton, only the tarsals are represented in the collection from Lunel-Viel. Their study did not reveal any particularly characteristic anatom-

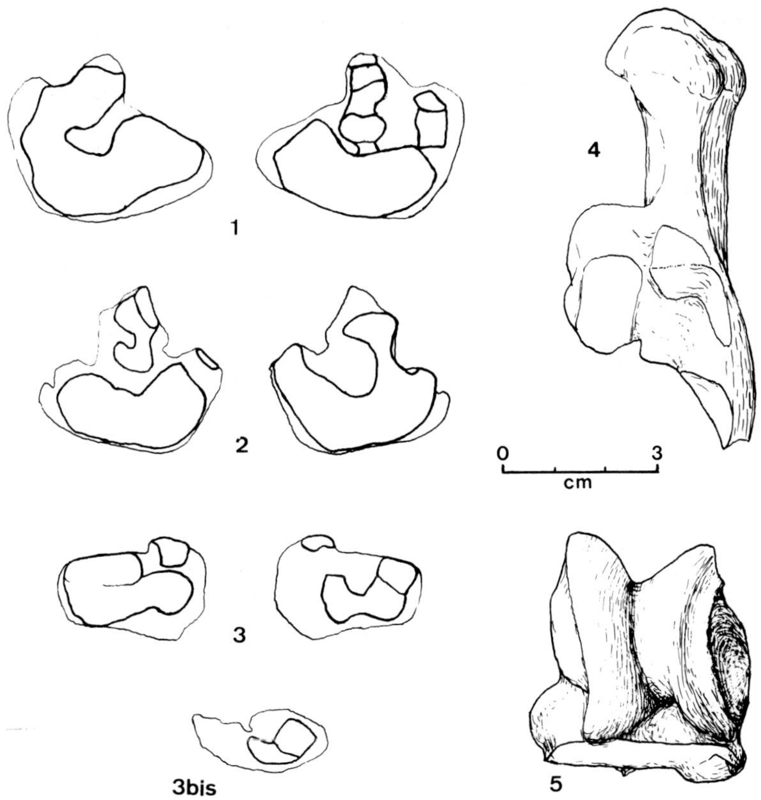

Fig. 14: Carpals and Tarsals (all except Calcaneum found articulated):
1:- Central Tarsal: LVIV 13626, proximal and distal views;
2:- Os Tarsale III: LVIV 13623, proximal and distal views;
3:- Os Tarsale IV: LVIV 13624, proximal and distal views;
3bis:- Os Tarsale II: LVIV 13625, proximal view;
4:- Calcaneum: LVIV 13791, dorsal view;
5:- Astragalus: LVIV 13627, dorsal view.

ical traits. As for dimensions, the astragalus of the small equid is higher than it is wide. This is the opposite of the situation with *Equus caballus* from Lunel-Viel while with the stenonine equid, the two dimensions are more or less equal.
In conclusion, the majority of the indices indicate that the bones of the post-cranial skeleton of the *Equus hydruntinus* of Lunel-Viel came from a relatively gracile animal.

3. Chronological and Geographic Distribution and Climatic Adaptation

In order to compare the population of small equids from Lunel-Viel with fossil remains of *Equus hydruntinus* already studied from other sites, it is necessary to examine the distribution of the species a little more closely.
In Europe, as in the Near East (TCHERNOV 1981), this small equid is known only from the beginning of the Riss glaciation, c. 400,000 years BP. Thus, the Mindel-Riss interglacial finds

of Lunel-Viel are older than most from western Europe. More to the east, the species appeared at nearly the same time (if the correlations are valid) with the Middle Acheulean of the Tejanski Complex of Moldavia (DAVID 1977). In the Near East it survives just into the Holocene (TCHERNOV 1981), in Roumania to the Neolithic (NECRASOV/HAIMOVICI 1959, SAMSON/RADULESCO 1966, SAMSON 1975) and in the Ukraine also to the Neolithic at the site of Selichte in the Orgiejewsk region (DAVID 1977) where it is called the "European ass."

The geographic distribution of *Equus hydruntinus* corresponds, by and large, to that described by STEHLIN/GRAZIOSI (1935) although it is necessary to entend the distribution at least into the Mediterranean littoral regions of the Iberian Peninsula (UERPMANN 1976) and the Peloponnesus (PAYNE 1975). These additions, however, only serve to confirm a generally southern European distribution for this species, there being few fossils of this species identified from northern European countries (HEMMER/JAGER 1969, MUSIL 1978). It is evident that *E.hydruntinus* prefered zones with a temperate climate. STEHLIN/GRAZIOSI (1935) have already remarked on the complementary distributions of *E.hydruntinus* and saiga (*Saiga* sp.), a fact which was confirmed by RADULESCO/SAMSON (1965) for Dobroudja in Roumania. When saiga is abundant, signifying a cold and dry phase often accompanied by the deposition of loess, *E.hydruntinus* is absent. Again in Roumania, *E.hydruntinus* occurs as an abundant member of the fossil community during the Atlantic climatic phase of the Holocene (NECRASOV/HAIMOVICI 1959) when it occurs with red deer, roe deer, and fallow deer. Further confirmation can be found in the observations of PRAT (1968) who notes that at Achenheim in northern France "in the recent loess horizons where the remains of reindeer are found in greater or lesser numbers, it (*E.hydruntinus*) is rare or absent ..." In contrast, the small equid occurs with *Cervus elaphus, Capreolus capreolus, Equus mosbachensis,* and *Elephas trogontherii* at that site.

Study of material from Balauziere Cave (M.-F. BONIFAY 1967) in the central Mediterranean region of France has proved to be particularly interesting in this regard. *Equus hydruntinus* appears in Level VII where it occurs with cave bear but not cold-weather species. In contrast, in Level V where it is most abundant, there are neither any remains of Arctic species like the reindeer nor humid zone species like cave bear. Thus, if *E.hydruntinus* shows a marked preference for temperate climates, it seems to have been able to tolerate cold if not too intense. When abundant, however, it signals a dry climate with particularly low humidity.

A detailed review of the French material by PRAT (1968) showed it to appear homogeneous and able to be grouped as a single species *Equus hydruntinus.* PRAT, however, noted a larger protocone in the more recent *E.hydruntinus davidi* than in the older form *E.hydruntinus hydruntinus* with a smaller protocone. It seems to have been a sporadically occurring species coming principally from prehistoric sites of southwest France like Pair-non-Pair, Combe-Grenal, La Ferrassie, Le Pech de l'Aze or of southeast France like La Balauziere, Rigabe (M.-F. BONIFAY 1963), and Valorgues, but only exceptionally in open-air sites (CREGUT-BONNOURE/GRANIER 1982). A few sites mark its extension to the north (Achenheim, Grenelle) or south (Isturitz). It is a rare fossil during the Riss glacial period, becoming more abundant at the beginning of the Würm where it is found in Mousterian sites. *E.hydruntinus* lasted to the end of the Würm in the southwest (PRAT 1968) and to the Mesolithic (7000 BC) in the southeast of France (LECOURTOIS-DUCGONINAZ 1984, ESCALON DE FONTON/ONORATINI/BONIFAY 1977, POULAIN 1971, M.-F. BONIFAY 1968). It is rarely abundantly represented; thus at Achenheim in Alsace there are only a few isolated teeth, two mandible fragments, some metapodials, a few other long bones, some tarsals, and a few phalanges. Where remains seem more numerous, as in the caves of La Chaise in Charente, the 77 teeth found actually come from three sites and seven different levels.

4. Comparisons with Pleistocene Non-Caballine Equids

4.1. Comparisons with *Equus hydruntinus* in France (Figure 15)

Figure 15 shows that lines representing the mean values for protocone index and upper cheek-tooth length calculated by EISENMANN (from the data of PRAT 1968), lines for the peri-Mediterranean populations of Istres and Valorgues, and lines for the small equid of Lunel-Viel all run generally parallel to each other. The lines for protocone index basically ascend with greatest values being for the third molar. The flattening-out evident for the P^3 and P^4 in the French material (EISENMANN curve) could be the result of grouping measurements from adult and young animals. The lines for mean length of upper cheekteeth do just the opposite of those for the protocone index and basically descend with the lowest values being for the molars. Taken together, this seems to be good evidence that we are dealing with the same animal.

I propose giving *Equus hydruntinus* from Lunel-Viel the subspecific name *minor.* I justify this distinction on the basis of the fact that the absolute lengths of the Lunel-Viel bones are less than lengths for the other French material (Table 15).

4.2. Comparisons with Other Pleistocene Equids

In our European regions, there is no hemione. Even when the presence of this equid has been suggested, it has never been confirmed with certainty.

Various authors have noted the presence of small equids which they believed to be more or less similar to the ass. *Equus wernerti* is represented by a single third metatarsal from Achenheim (France). This unique specimen, clearly not from an hemione, is anatomically not much different from the Lunel-Viel material although it does come from an even smaller animal. It is essential that we have additional material before we can reasonably deal with this problem.

Equus graziosi has been described by AZZAROLI (1979). It comes from Upper Pleistocene levels in Tuscany (Italy). It is characterized by its small size, small teeth, and gracile limbs. It has clear similarities with *E.hydruntinus* from which it differs in having a narrow mandibular symphysis. According to AZZAROLI, it is clearly similar to the ass.

TABLE 15

Comparisons of Greatest Lengths of Long Bones from Lunel-Viel and other sites in France (in mm; averages unless ranges shown)

	Lunel-Viel	*Equus hydruntinus* (from PRAT 1968)
Radius	250.8–257.0	285 –305
Metacarpal III	190.3	210 –222
Metatarsal III	229.3	249 –265
Phalanx 1 posterior	63.9	68.5– 82
Phalanx 2 posterior	35.9	37 – 45

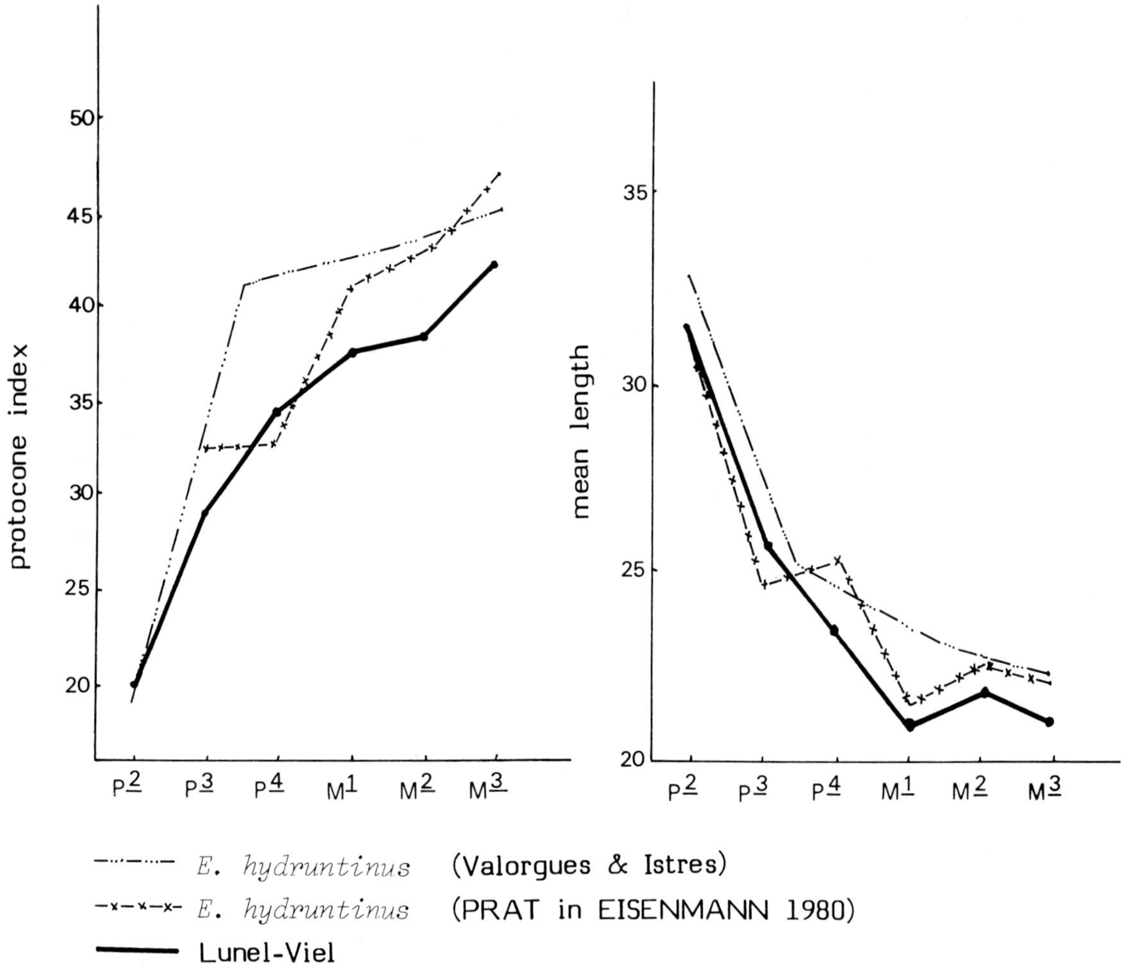

Fig. 15: Comparisons of mean values for protocone index and upper cheektooth length for populations of *Equus hydruntinus* from France (EISENMANN/PATOU 1980), from Mediterranean regions (M.-F. BONIFAY 1968, 1978), and from Lunel-Viel.

Equus altidens also reveals similarities with *E.hydruntinus,* and it is possible that it forms with the latter an "independent subgenus" (MUSIL 1969). The very ancient dating of this form (Süssenborn is contemporary with the Mindel) suggests a phyletic rather than a synonimic relation. In any case, additional study is clearly indicated.

Finally, in Africa, *Equus mauritanicus* seems closely linked to *E.hydruntinus* and, if one follows TCHERNOV (1981), it could be thought to be the same species. It exists in the assemblage from Ternifine (ARAMBOURG 1970) that is very close in time to that of Lunel-Viel.

5. Conclusions

On the basis of study of the upper cheekteeth, it seems logical to relate, phylogenetically, *Equus mauritanicus, E.burchelli, E.quagga,* and *E.stenonis* (EISENMANN 1980). This grouping is confirmed by study of the lower cheekteeth as well (EISENMANN 1981). Incontestably it is to this group that our small equid from Lunel-Viel belongs with affinities particularly to *E.stenonis.* Certain authors have remarked on this, ALIMEN already in 1946 noting "that *Hydruntinus* seems to continue into Europe an archaic stage close to that of *stenonis,* their characters being indistinguishable." In reality, if we assemble the various scattered characters, the fossil of Lunel-Viel seems allied more with quagga group in general than with any single species.
The length of the M^3 is equal to or less than the length of the M^2 as in *Equus burchelli*; the protocone index of the M^1 is greater than that of the P^4 as in *E.burchelli* and *E.quagga.* The curve for protocone indices is comparable to that of the most archaic form of *E.stenonis.* The protocone formula resembles that of *E.mauritanicus* and is similar to that of *E.burchelli*; the ratio of

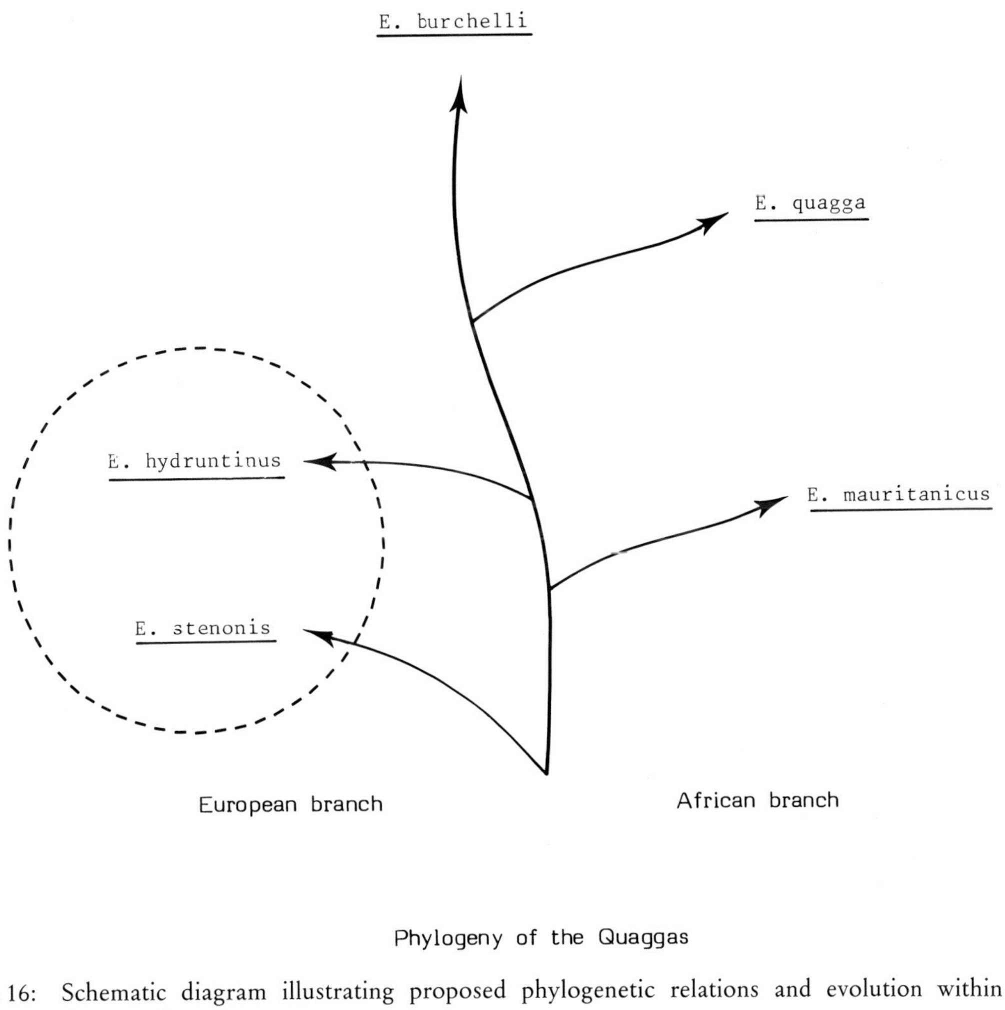

Fig. 16: Schematic diagram illustrating proposed phylogenetic relations and evolution within the Quagga Group.

length of P^2 to length P^3 is comparable to that of *E.mauritanicus*; the ratio of the protocone index for M^1–M^2 to the protocone index of P^3–P^4 is comparable to that of *E.stenonis vireti.* The mean occlusal length of the lower teeth diminishes from the P^2 toward the M^1, is similar in the M^1 and M^2 as in *E.burchelli,* and increases from the M^2 to M^3 as in *E.quagga.* The skeletal characters seem to be closest to those of *E.stenonis* in the anatomy of the distal end of the humerus and in the proportions of the distal end of the tibia. By contrast, those characters which are thought to be more "evolved", e.g., the metapodial index, are less archaic in the Lunel-Viel equid than in *E.stenonis.* For these reasons, I believe that it is possible to suggest a scheme like that presented in Figure 16. *E.stenonis* came to Europe from eastern Asia with the migration of Villafranchian fauna. During the major climatic amelioration of the Mindel-Riss interglacial, coming from the same geographical regions and from the same stock, the small *Equus hydruntinus* (in the *minor* form), spread over western Europe and into North Africa. Thereafter, it evolved locally in those regions into the Holocene.

6. Literature

Alimen, H., 1946, Remarques sur *Equus hydruntinus* Regalia. Bulletin de la Société Géologique de France **16**(5éme série), 585–595.

Arambourg, C., 1970, Les vertébrés du Pléistocène de l'Afrique du Nord. Archives du Muséum National d'Histoire Naturelle, Paris **10**(7ème série), 1–128.

Azzaroli, A., 1979, On a late Pleistocene ass from Tuscany, with notes on the history of asses. Palaeontographia Italica **71**(n.s. 41), 27–47.

Bonifay, E., 1981, Les plus anciens habitats sous grotte découverts à Lunel-Viel (Hérault). Archéologia **150**, 30–43.

Bonifay, E./Bonifay, M.-F., 1965, Age du gisement de Mammifères fossiles de Lunel-Viel (Hérault). Comptes Rendues de l'Academie des Sciences, Paris **260**(9), 3441–3444.

Bonifay, M.-F., 1963, Présence d'*Equus hydruntinus* dans la grotte de Rigabe. Annales de Paléontologie **49**(159–170)

Bonifay, M.-F., 1967, Étude paléontologique de la grotte de la Balauzière. Bulletin du Musée d'Anthropologie préhistorique de Monaco **13**, 91–139.

Bonifay, M.-F., 1968, La faune de l'abri Cornille, Istres, Bouches-du-Rhône. In: La Préhistoire: Problèmes et Tendances; Paris: Éditions du C.N.R.S., 47–57.

Bonifay, M.-F., 1971, Carnivores quaternaires du Sud-Est de la France. Mémoires du Muséum National d'Histoire Naturelle, Paris **21**(Série C), 49–377.

Bonifay, M.-F., 1977, Essai de corrélation entre les os cassés intentionnellement et leur origine anatomique au Paléolithique ancien (grottes de Lunel-Viel, Hérault). In: 1ér Colloque international sur l'Industrie de l'Os dans la Préhistoire; Éditions de l'Université de Provence, 35–46.

Bonifay, M.-F., 1978, Le gisement de la baume de Valorgues à St. Quentin-la-Poterie (Gard): Paléolithique supérieur et final et Épipaléolithique, II: La grande faune. Gallia-Préhistoire **21**, 123–142.

Bonifay, M.-F., 1980, Le cheval du Pléistocène moyen des grottes du Lunel-Viel (Hérault): *Equus mosbachensis palustris* n.ssp. Gallia-Préhistoire **23**, 233–281.

Bonifay, M.-F./Lecourtois-Ducgoninaz, A., 1976, Les faunes des niveaux inférieurs de l'abri Cornille (Istres, Bouches-du-Rhône). In: Congrès Préhistorique de France, Session 20; Provence, 228–247.

Brugal, J.-P., 1983, Applications d'analyses multidimensionnelles à l'étude systématique du squelette des membres des grandes Bovidés Pléistocènes des grottes de Lunel-Viel (Hérault); Thèse, Sciences naturelles (3ème cycle), Marseille-Luminy.

Cregut-Bonnoure, E./Granier, J., 1982, À propos d'une dent d'*Equus hydruntinus* (Mammalia, Perissodactyla) du loess récent de Collias (Gard, France). Bulletin mensuel de la Société Linnéenne de Lyon **51**, 234–240.

David, A. I., 1977, Novie dannie o rasprostranenii iskopaemogo evropejskogo osla v vostochnej Evrope (Nouvelles données au sujet de la répartition de l'âne pléistocene en Europe orientale). Izvestija Akademii Nauk MSSR, Serija biologicheskikh i khimicheskikh **4**, 89–91.

Eisenmann, V., 1979, Les métapodes d'*Equus sensu lato* (Mammalia, Perissodactyla). Géobios **12**(6), 863–886.

Eisenmann, V., 1980, Les Chevaux (*Equus sensu lato*) fossiles et actueles: crânes et dents jugales supérieures; (Cahiers de Paléontologie); Paris: Éditions du CNRS.

Eisenmann, V., 1981, Étude des dents jugales inférieures des *Equus* (Mammalia, Perissodactyla) actuels et fossiles. Palaeovertebrata **10**(3–4), 127–226.

Eisenmann, V./Karchoud, A., 1982, Analyses multidimensionnelles des métapodes d'*Equus* (Mammalia, Perissodactyla). Bulletin du Muséum National d'Histoire Naturelle, Paris, Section C, **4**(4ème série, no. 1–2), 75–103.

Eisenmann, V./Patou, M., 1980, La faune de la grotte de Félinés-Termenès (Aude), resultats préliminaire et étude detaillée des restes d'*Equus hydruntinus* (Mammalia, Perissodactyla). L'Anthropologie **84**, 633–649.

Escalon de Fonton, M./Onoratini, G./Bonifay, M.-F., 1977, Les industries de filiation magdalénienne dans le Sud-Est de la France: leur position géochronologique et les faunes. In: Colloque international du C.N.R.S. **271**; Bordeaux: C.N.R.S., 306–311.

Forsten, A., 1982, Indices in equid systematics and phylogeny. Annalecta Zoologica Fennica **19**, 183–191.

Guilhaumon, D., 1982, Approche méthodologique pour l'étude du matériel osseux fragmenté du gisement de Lunel-Viel (Hérault); Marseille: Université de Provence, Centre Saint-Charles. (Diplome d'Études Supérieures, Polycopie).

Hemmer, H./Jaeger, R., 1969, Ein Radius von *Asinus* cf *hydruntinus* Reg. aus dem Jungpleistozan von Stammheim/Kreis, Freiberg (Hessen). Hessische Landesamt für Bodenforschung, Notizblatt **97**, 38–43.

Lecourtois-Ducgoninaz, A., 1984, La faune du gisement de Mourre-Poussiou à Fos-sur-Mer, Bouches-du-Rhône. Gallia Préhistoire **27**, 67–80.

Mourer, C., 1975, Les Oiseaux du Pléistocène moyen et supérieur de France; Université Claude Bernard, Lyon. (Thèse de Doctorat es-Sciences Naturelles, no. 75–14).

Musil, R., 1969, Die Equiden-Reste aus dem Pleistozän von Süssenborn bei Weimar. Palaeontologische Abhandlungen **A–3**, 367–788.

Musil, R., 1978, Die fossilen Equiden-Reste aus dem Travertinen von Burgtonna in Thüringen. Quartärpaläontologie **3**, 137–138.

Necrasov, O./Haimovici, S., 1959, Sur la présence d'une espèce pléistocène d'*Equus hydruntinus* Reg. dans le Néolithique roumain. Analele Stiintifice ale Universitatii "Al. I Cuza" (n.s., Sect. 2), 137–146.

Payne, S., 1975, Faunal change at Franchthi Cave from 20,000 B.C. to 3,000 B.C. In: Clason, A.T., ed., Archaeozoological Studies; Amsterdam: North Holland, 120–131.

Poulain, T., 1971, Le camp mésolithique de Gramari à Méthamis (Vaucluse), III: Étude de la faune. Gallia-Préhistoire **14**, 121–131.

Prat, F., 1968, Recherches sur les Équidés pléistocènes de France; Université de Bordeaux, Bordeaux. (Thése Doctorat d'État).

Prat, F., 1980, Les Equidés villafranchiens en France (genre *Equus*); (Cahiers du Quaternaire **2**); Paris: Éditions du Centre National de la Recherche Scientifique.

Radulesco, C./Samson, P., 1965, Sur la présence d'*Hydruntinus hydruntinus* (Reg.) en Roumanie. Quaternaria **7**, 219–233.

Regalia, E., 1907, Sull'*Equus (Asinus) hydruntinus* Regalia della grotta di Romanelli (Castro, Lecce). Archivio per l'Anthropologia, Etnologia e Psicologia Comparata **37,** 375–390.

Samson, P., 1975, Les Équidés fossiles de Roumanie (Pliocène moyen – Pléistocène supérieur). Geologica Romana **14,** 165–352.

Samson, P./Radulesco, C., 1966, Asupra prezentei lui *Hydruntinus hydruntinus* (Regalia) in postglaciarul Dobrogii. Lucrarile Institutului de speologie "Emil Racovita" **5,** 251–260.

Stehlin, H.G./Graziosi, P., 1935, Ricerche sugli Asinidi fossili d'Europa. Mémoires de la Société Paléontologique Suisse **56**(3), 1–73.

Tchernov, E., 1981, The biostratigraphy of the Middle East. In: Préhistoire du Levant; Paris: Éditions du CNRS, 67–97.

Uerpmann, H.-P., 1976, *Equus (Equus) caballus* und *Equus (Asinus) hydruntinus* im Postpleistozän der Iberischen Halbinsel (Perissodactyla, Mammalia). Säugetierkundliche Mitteilungen **24,** 206–218.

Vaufrey, R., 1928, Le Paléolithique italien. Archives de l'Institut de Paléontologie Humaine **3,** 1–196.

Marie-Françoise Bonifay, Laboratoire de Géologie du Quaternaire, C.N.R.S. – Luminy, Case 907, 13288 Marseille CEDEX 9, France.

SÁNDOR BÖKÖNYI

The Earliest Occurrence of Domestic Asses in Italy

Contents

1. Introduction

True wild asses belong to a single species *Equus* (*Asinus*) *africanus* FITZINGER, 1857 that can be divided into three subspecies each of which has or had its own well-defined area of distribution, albeit at one time contiguous.

The Nubian wild ass *Equus africanus africanus* FITZINGER formerly lived in large areas of the eastern part of Africa including Egypt although later it withdrew into the region between the Nile, the Red Sea, and the Abyssinian Plateau. In the third decade of this century the free-ranging form became extinct, and it today survives only in zoological gardens (ANTONIUS 1922: 248, ZEUNER 1963: 374).

The second subspecies, the Somalian wild ass *Equus africanus somaliensis* NOACK, 1884, also inhabited a larger area in the past but for the last one hundred years or so has been confined to the territory of Somaliland (ANTONIUS 1922: 248, WERTH 1929: 343).

The vaguely known third subspecies, the northwest African *Equus africanus atlanticus* P. THOMAS, 1884, used to live in the Atlas Mountains but probably became extinct during the first centuries of the Christian era (ZEUNER 1963: 374).

DUCOS has described asses considered by him to be wild from Bronze Age levels of Tell Gat in the Mediterranean zone of Palestine. To these he gave the name of *Equus palestinae* sp. n. DUCOS, 1968. He also identified materials from Muraibit in northern Syria as belonging to the same species (DUCOS 1970, 1975, 1978) but recently has noted that the Tell Gat material was probably domestic and has given the name of *Equus africanus mureybeti* ssp. n. DUCOS, 1986 to the ninth and eighth millennium b. c. material from Muraibit. I agree with DUCOS that the Tell Gat material is domestic ass probably coming from Egypt (EPSTEIN 1971: 394), but I also believe that the Muraibit material is all probably onager (*Equus* (*Hemionus*) *hemionus*; see BÖKÖNYI 1986).

Of the three wild subspecies, the Nubian wild ass was probably the ancestor of the domestic donkey. This view is based on morphological, zoological, and archaeological considerations. Morphologically, the great majority of domestic asses show the characteristics of the Nubian subspecies, particularly the stripe along the back, the cross-stripe on the shoulders, and the "zebra-stripes" on the legs. Further evidence is provided by the fact that the earliest domestic donkeys have been found in or near the area of distribution of the Nubian wild ass. Finally, the oldest Egyptian representations of domestic asses show animals similar to the Nubian wild ass (Antonius 1922: 269–270).

Originally students of the subject believed that the earliest domestication of the ass took place within the area of the known modern distribution of the Nubian wild ass south of Egypt and from there spread north into Egypt itself. We know today, however, that the Nubian wild ass originally included Egypt within its range of distribution, and thus it seems highly probable that its earliest domestication was carried out in Egypt under the stimulus of the domestication of other animals. Although it has not been possible to distinguish the bones of wild and domestic asses, it seems highly probable that the *Asinus* bones unearthed along with other bones of domestic animals from the early fourth millennium B.C. Amratian (Naqada I) culture came from domestic donkey (Boettger 1958: 106) simply because that was the period of the domestication of the first work (as opposed to food) animals. The asses represented on a slate palate from the middle fourth millennium B.C. Gerzean (Naqada II) culture were almost surely domestic because they were depicted by the ancient artist exclusively in the company of domestic animals. The ass remains from Maadi (just south of Cairo) dating to the second half of the fourth millennium B.C. undoubtedly came from domestic asses since the site lies outside of the area of distribution of the Nubian wild ass (Bökönyi, 1985: 496 ff.).

By the second half of the fourth millennium B.C., asses had spread to Southwest Asia, for example, to southern Iran (Zeder 1986) and even to Anatolia (Bökönyi, 1988: 589). The latter identification is based on a metacarpus from Arslantepe near Malatya in eastern Anatolia that is eight millimeters shorter and much broader than any hemione metacarpus. This wide distribution strengthens my argument that the earliest domestication of the donkey goes back at least to the middle of the fourth millennium.

The fourth millennium was the great period for the domestication and spread of horses and donkeys alike. After its first domestication in the southern Ukraine (Bibikova 1967), the horse (like the ass) was quickly taken in all directions and reached Asia Minor before the end of the millennium (Bökönyi 1978, 1980 a, this volume) and even Mesopotamia around 3000 B.C. In fact, domestic horses and donkeys first met in Anatolia and it would not be surprising if the earliest mules were to be found in that region.

Both domestic horses and donkeys had a marked inpact on the economies of the regions into which they were spread. Thus Bibikova (1975) noted marked differences in eastern Europe before and after the introduction of the horse. Both horse and donkey played important parts in agriculture and in transport. With the ass, heavy burdens could be carried; with the horse the additional benefit of speed was gained. In warfare horses also played important roles in pulling chariots and as mounts for warriors (Bökönyi 1980 a: 921). The role of donkeys was confined to that of a pack animal. With the development of lighter wheeled vehicles and systems of roads, the horse could replace oxen with a resulting increase in speed and range of trade (Bökönyi 1980 b).

The domestic ass appeared in Europe rather late. Although donkeys were quite common in Anatolia during the Hittite period (Herre/Röhrs 1958: 60), the earliest European domestic ass bones come from the southern Ukraine in the ninth and eighth centuries B.C. (Pidopličko 1956: 91). These animals undoubtedly reached the region by way of Greece, for the Greeks had close contacts with both Egypt and Anatolia. This supposes, of course, that donkeys were introduced into Greece earlier although for this we do not yet have any direct evidence. We do know, however, that mule races were held at the Olympic Games from 580 B.C. onwards, and

there are representations on vases which show the lame Hephaistos and other elderly or infirm persons riding asses and mules in that period (ANDERSON 1961: 43, BÖKÖNYI 1974: 306).
The Greeks played an important part in spreading donkeys through Europe. In Greek colonies on the Black Sea, asses apparently were kept in large numbers. Ass bones have been found in the sixth to first century B.C. layers of Olbia (PIDOPLIČKO 1956: 91) and in the Graeco-Roman layers of Scythian Neapolis, Pantikapaion, Myrmekion, Tyritake, and Ilurat (ZALKIN 1969: 49).
Asses probably reached the Apennine Peninsula by way of Greece as well, although until now any written or osteological evidence has been lacking. Recent excavations at the site of Tortoreto-Fortellezza in central Italy near the Adriatic coast, however, have provided some new information. The excavations were carried out by the Chieti Museum under the directorship of Prof. G. SCICHILONE. The bulk of the occupation belongs to the Iron Age (tenth to sixth or fifth centuries B.C.) although the Neolithic, Bronze Age, and Roman-Medieval periods are also represented (M. TOSI, personal communication 1985).

2. Material

Along with ass, a wide range of both wild and domestic species were found in the Iron Age levels of Tortoreto-Fortellezza (Table 1). Twenty-one ass bones were excavated but only nine of them came from well-dated levels. These include a right P^3 (Plate 1); fragments of a right

TABLE 1

Taxa identified from Tortoreto-Fortellezza

		domestic animals
cattle	–	*Bos taurus* L.
sheep	–	*Ovis aries.* L.
goat	–	*Capra hircus* L.
pig	–	*Sus scrofa domesticus* L.
horse	–	*Equus caballus* L.
ass	–	*Equus asinus* L.
cat	–	*Felis domestica* Briss.
dog	–	*Canis familiaris* L.
		wild animals
red deer	–	*Cervus elaphus* L.
fallow deer	–	*Dama dama* L.
roe deer	–	*Capreolus capreolus* L.
boar	–	*Sus scrofa ferus* L.
cat	–	*Felis silvestris* Schreber
badger	–	*Meles meles* L.
brown bear	–	*Ursus arctos* L.
fox	–	*Vulpes vulpes* L.
wolf	–	*Canis lupus* L.
brown hare	–	*Lepus europaeus* Pallas
duck	–	*Anas* sp.
birds	–	Aves (species not determined)
turtle	–	*Chelonia* sp.

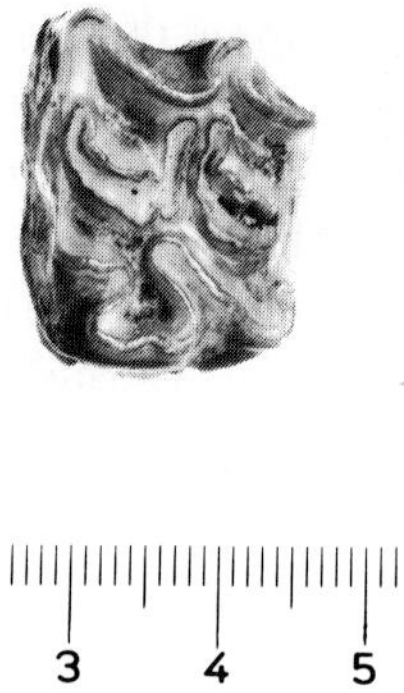

Plate 1 Right P³: Iron Age (7th–6th centuries BC)

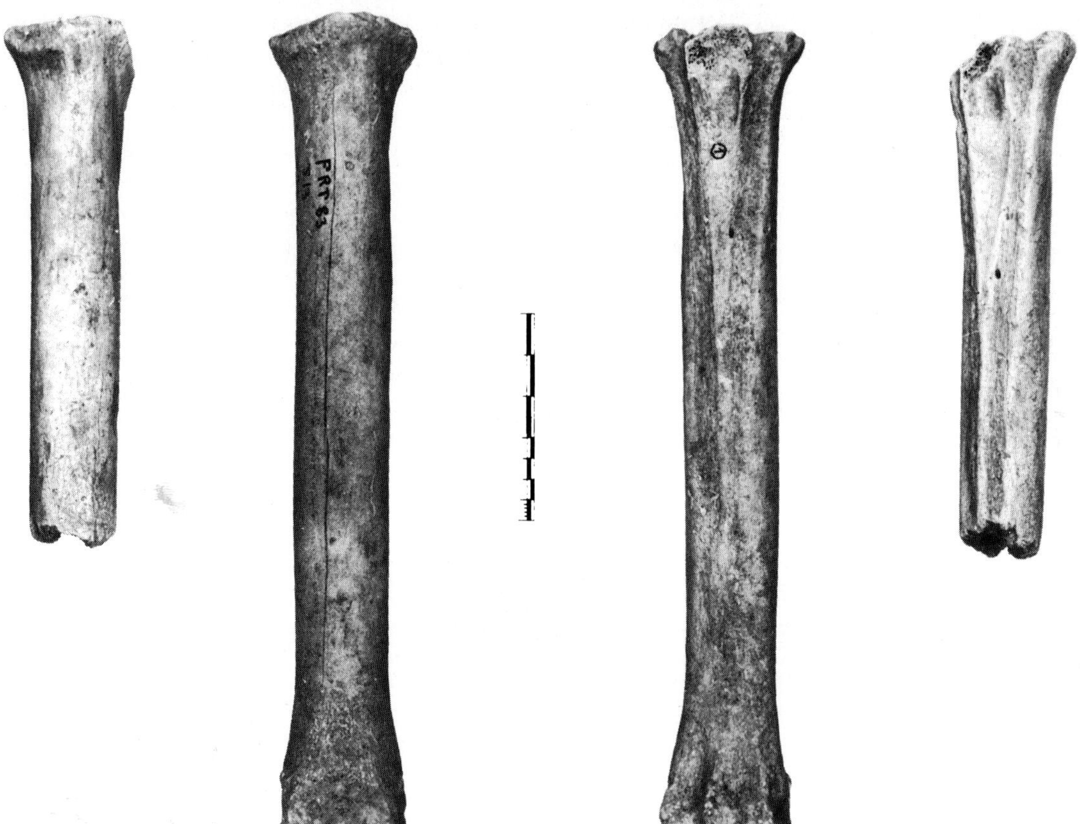

Plate 2 Left complete Metatarsal III, Iron Age (7th–6th centuries BC), left proximal Metatarsal III, mixed deposits, dorsal and plantar views

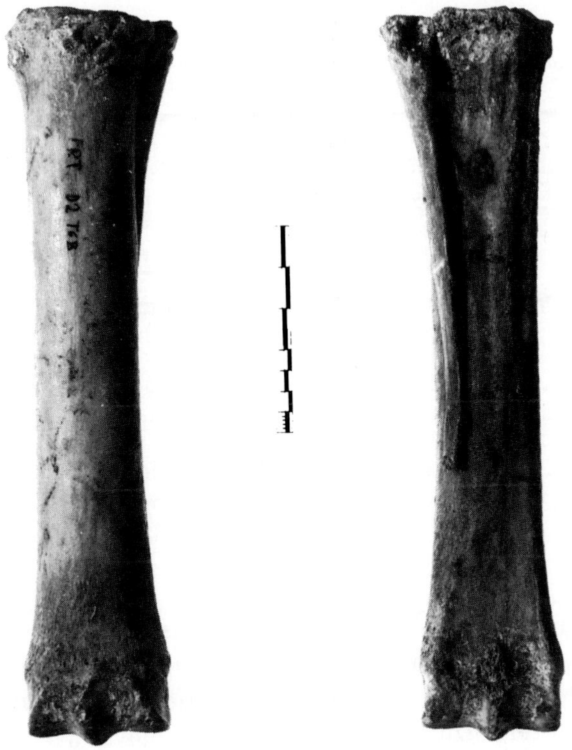

Plate 3 Right Metacarpals II and III, mixed deposits

Plate 4 Right distal Metatarsal III, mixed deposits

TABLE 2

Measurements of ass bones from Tortoreto-Fortellezza (in millimeters)

Element	GL	GB	Bp	Dp	SB	SD	Bd	Dd	Context
Mc III	173.5	–	37.5	27.0	26.0	16.5	34.5	27.0	mixed
Tibia	295.0	–	–	–	32.0	23.0	56.0	41.0	mixed
Tibia	–	–	–	–	32.5	20.5	51.0	35.5	Iron Age
Talus	44.0	42.0	–	–	–	–	–	–	mixed
Mt III	(205)		37.5		23.0	17.5	32.0	26.0	Iron Age
Mt III	–	–	36.0	(30)	22.0	–	–	–	mixed
Mt III	–	–	–	–	–	20.0	37.5	30.0	mixed

GL = Greatest Length, GB = Greatest Breadth, Bp = proximal Breadth, Dp = proximal Depth, SB = Smallest Breadth, SD = Smallest Depth, Bd = distal Breadth, Dd = distal Depth

upper premolar, of an atlas, of a cervical vertebra, of a distal scapula, of a left pelvis, of a right femur, of a left tibia (shaft); and a left Metatarsus III (Plate 2) all from Iron Age (seventh to sixth centuries B.C.) deposits. The remaining twelve bones came from mixed deposits. These include a right temporal fragment, a left upper molar, a right M_3, a right Metacarpus III with Mc II (Plate 3), a right pelvic fragment, a left femur shaft fragment, a left tibia, a left distal tibia fragment, a right astragalus, a left proximal Metatarsus III (Plate 2), and a right distal Metatarsus III (Plate 4). There were no ass bones in well-defined pre- or post-Iron Age levels; this indicates that all the ass bones in fact come from Iron Age deposits.

Of the ass bones, the left upper molar came from an adult/mature individual and the M_3 from a sub-adult/adult. All the other bones of determinable age came from adult individuals. Measurements for eight specimens are listed in Table 2.

3. Discussion

With regard to the ass bones from Tortoreto, how can we tell that they came from domestic donkeys? First, the small size of the bones permits a clear separation from the bones of true horses also found at the site. While the horses of Tortoreto were small (as one would expect in the Iron Age west of the Venice-Vienna line: Bökönyi 1968), the asses were even smaller.

Second, the enamel patterns of the upper cheekteeth clearly distinguish the Tortoreto material from true horses. Thus the absence of a *pli caballin* combined with the presence of abrupt and sharply angled transitions from the interstylar surfaces to the columns on the buccal side of the teeth serve to exclude horse. Among the other possibilities are the hemione which can be excluded by the presence of short protocones. The protocones are not so short, however, as to invite identification as *Equus hydruntinus* (which lived in Italy into the early Neolithic).

Third, the Metacarpus III, unfortunately from a mixed deposit, is more than ten millimeters shorter than and has a slenderness index (14.99) higher than any hemione metacarpus. At the same time, this bone is far smaller than that from any known wild ass metacarpus. The complete Metatarsus III, from a seventh century B.C. level, is also far shorter (at GL = 205 mm) than any known hemione metatarsus while, again, the slenderness index (11.22) is rather high for a hemione.

Fourth, the shapes of the distal ends of the metapodials are also more asinine than hemione. HILZHEIMER was the first to note that the supra-articular area of both metatarsals (HILZHEIMER 1921: 141) and metacarpals (HILZHEIMER 1941: 10) manifest a marked broadening in horses and asses. I also have noted this phenomenon (BÖKÖNYI 1972: 16) as has EISENMANN (1979: 869). Such a broadening can be clearly seen on the Tortoreto metapodials (Plates 2, 3, 4) and indicate that the bones came from asses.

Thus the small equids of Tortoreto-Fortellezza can be identified as donkeys, almost certainly domestic. As such, they are the earliest yet known from Italy. Their presence points to contacts with Greece (via Magna Graecia) or directly with northeastern Africa. Interestingly enough, the remains of domestic cats also occur in the faunal collection from Tortoreto. This animal also came directly or indirectly from northeastern Africa.

Looking at the donkeys of Tortoreto-Fortellezza with a breeder's eye, one can state that they were rather variable in size and proportions. The complete metapodials, astragalus, and some teeth indicate the presence of smaller animals. Other remains such as the complete tibia and the distal metatarsus came from larger animals.

Based on the fact that only one bone came from an individual that was not fully grown, we can suppose that the donkeys were used as work animals at Tortoreto-Fortellezza. At the same time however, the fragmented bones suggest that they were at least occasionally eaten. Asses, like horses, however, were probably domesticated primarily for labor and transport (ZEUNER 1963: 378).

Through direct contacts with Greece or by way of Italy, the domestic ass also reached Gaul where it has been described from at least two La Tène sites (Vielle Toulouse by POULAIN-JOSIEN (1964: 78) and La Croix-des-Sables à Maine also by POULAIN-JOSIEN (1971: 469)). Nevertheless, it was the Romans who were ultimately to spread the ass through their provinces in Europe and into Germania (SCHLOSSER 1888: 19, VOGEL 1940: 110, LÜTTSCHWAGER 1967: 355, MÜLLER 1967: 28, and SWEGAT 1976: 20), Gallia (POULAIN-JOSIEN 1962: 238 and JOURDAN 1976: 223), Britannia (EWART 1911: 362), AND PANNONIA (BÖKÖNYI 1974: 301).

4. Literature

ANDERSON, J.K., 1961, Ancient Greek Horsemanship; Berkeley: University of California Press.

ANTONIUS, O., 1922, Grundzüge einer Stammesgeschichte der Haustiere; Jena: Gustav Fischer.

BIBIKOVA, V.I., 1967, K izucheniu drevneishkh domashnikh loshadei Vostochnoi Evropy (Studies of ancient domestic horses in Eastern Europe). Bulleten Moskovskogo Obshchestva Ispitatelei Prirody, Otdel Biologicheskii **72,** 106–118.

BIBIKOVA, V.I., 1975, Formen der Viehzucht bei den äneolithischen Stämmen Südosteuropas. In: Moderne Probleme der Archäologie; Berlin: Akademie-Verlag, 237–245.

BOETTGER, C.R., 1958, Die Haustiere Afrikas; Jena: Gustav Fischer.

BÖKÖNYI, S., 1968, Data on Iron Age horses of Central and Eastern Europe. (Bulletin of the American School of Prehistoric Research 25); Cambridge (MA): Peabody Museum, Harvard University, 1–71.

Bökönyi, S., 1972, Once more on the osteological differences of the horse, the half-ass, and the ass. In: Firouz, L., comp., The Caspian Miniature Horse of Iran; Miami: Field Research Projects, 12–23.

Bökönyi, S., 1974, History of Domestic Animals in Central and Eastern Europe; Budapest: Akadémiai Kiadó.

Bökönyi, S., 1978, The earliest waves of domestic horses in East Europe. Journal of Indo-European Studies **6**, 17–76.

Bökönyi, S., 1980a, La doméstication du cheval. La Recherche **11**(114), 919–926.

Bökönyi, S., 1980b, The importance of horse domestication in economy and transport. In: Sørbom, P., ed., Transport Technology and Social Change; (Tekniska Museet Symposia **2**); Stockholm, 15–21.

Bökönyi, S., 1986, The Equids of Umm-Dabagiyah, Iraq. In: Meadow, R.H./Uerpmann, H.-P., eds., Equids in the Ancient World; (Beihefte zum Tübinger Atlas des Vorderen Orients, Reihe A **19/1**); Wiesbaden: Dr. Ludwig Reichert Verlag, 302–317.

Bökönyi, S., 1985, The animal remains of Maadi, Egypt: a preliminary report. In: Liverani, M./Palmieri, A./Peroni, R., eds., Studi di palentologia in onore di Salvatore M. Puglisi; Rome: Università di Roma "La Sapienza," 495–499.

Bökönyi, S., 1988, Late chalcolithic and Early Bronze Age animal remains from Arslantepe near Malatya, Anatolia: a preliminary report. Origini **13**(2): 581–598.

Ducos, P., 1968, L'Origine des animaux domestiques en Palestine; (Publications de l'Institut de Préhistoire, Université de Bordeaux **6**); Bordeaux: Institut de Préhistoire.

Ducos, P., 1970, The Oriental Institute excavations at Mureybit, Syria: preliminary report on the 1965 campaign, Part 4: Les Restes d'Équidés. Journal of Near Eastern Studies **29**, 273–289.

Ducos, P., 1975, A new find of an equid metatarsal bone from Tell Mureibet in Syria and its relevance to the identification of equids from the early Holocene of the Levant. Journal of Archaeological Science **2**, 71–73.

Ducos, P., 1978, Tell-Mureybet (Syrie, IXe-VIIe Millenaires): Étude archéozoologique et problèmes d'écologie humaine: 1; Paris: Éditions du CNRS.

Eisenmann, V., 1979, Les métapodes d'*Equus sensu lato* (Mammalia, Perissodactyla). Géobios **12**(6), 863–886.

Epstein, H., 1971, The Origin of the Domestic Animals of Africa, Volume 2; New York: Africana Publishing.

Ewart, J.C., 1911, Animal remains. In: Curle, J., A Roman Frontier Post and its People: the Fort of Newstead in the Parish of Melrose; Glasgow, 362–377.

Herre, W./Röhrs, M., 1958, Die Tierreste aus den Hethitergräbern von Osmankayasi. In: Bittel, K., et al., Die hethitischen Grabfunde von Osmankayasi; Bogazköy-Hattusu 2; (Wissenschaftliche Veröffentlichung der Deutschen Orient-Gesellschaft **71**); 60–80.

Hilzheimer, M., 1921, Ueber den Metatarsus eines *Equus hemionus fossilis* Nhrg. aus Königswusterhausen. Sitzungsberichte der Gesellschaft Naturforschender Freunde zu Berlin, 140–142.

Hilzheimer, M., 1941, Animal Remains from Tell Asmar; (Studies in Ancient Oriental Civilization **20**); Chicago: The University of Chicago Press.

Jourdan, L., 1976, La Faune du Site Gallo-Romain et Paléo-Chrétien de la Bourse (Marseille); Paris: Éditions du C.N.R.S.

Lüttschwager, J., 1967, Zum Vorkommen von Hauseseln zur Römerzeit nördlich der Alpen. Zeitschrift für Säugetierkunde **32**, 355–361.

Müller, R., 1967, Die Tierknochenfunde aus den spätrömischen Siedlungsschichten von Lauriacum 2 – Wild- und Haustierknochen ohne die Rinder; Dissertation, Institut für Palaeoanatomie, Munich.

Pidopličko, I.G., 1956, M. Materiali do vivchenniya minulikh faun URSR, **2**; Kiev: Akademia Nauk URSP.

Poulain-Josien, T., 1962, Les chevaux gallo-romains découverts à Paris rue Henri-Barbusse (IIIe-IVe siècles). Zeitschrift für Tierzüchtung und Züchtungsbiologie **76**, 238–242.

Poulain-Josien, T., 1964, Les animaux domestiques et sauvages en France du Néolithique au Gallo-Romain; Dissertation, Université de Paris, Paris.

Poulain-Josien, T., 1971, Le site gaulois de la Croix-des-Sables à Maine (Charente) – Étude de la faune. Bulletin de la Société Préhistorique Française **68**, 469–471.

Schlosser, M., 1888, Ueber Säugethier- und Vogelreste aus den Ausgrabungen in Kempten stammend. Correspondenzblatt der Deutschen Gesellschaft für Anthropologie, Ethnologie, und Urgeschichte **19**, 17–22.

Swegat, W., 1976, Die Knochenfunde aus dem römischen Kastell Künzing-Quintana; Dissertation, Institut für Palaeoanatomie, Munich.

Vogel, R., 1940, Die alluvialen Säugetiere Württembergs. Jahreshefte des Vereins für Vaterländische Naturkunde in Württemberg **96**, 93–110.

Werth, E., 1930, Zur Abstammung des Hausesels. Sitzungsberichte der Gesellschaft Naturforschender Freunde zu Berlin (1929), 342–355.

Zalkin, V. I., 1960, Domashnie e dikie zhivotnie Severnogo Prichernomorya v epokhi rannego zheleza; Moscow: Materialy i Issledovania Archeologii SSSR **53**.

Zeder, M. A., 1986, The Equid Remains from Tal-e Malyan, southern Iran. In: Meadow, R. H./Uerpmann, H.-P., eds., Equids in the Ancient World; (Beihefte zum Tübinger Atlas des Vorderen Orients, Reihe A **19/1**); Wiesbaden: Dr. Ludwig Reichert Verlag, 366–412.

Zeuner, F. E., 1963, A History of Domesticated Animals; London: Hutchinson.

Sándor Bökönyi, Archaeological Institute of the Hungarian Academy of Sciences, Uri utca 49, 1250 Budapest 1, Hungary

ANNEKE T. CLASON

Horse Remains from Swifterbant, The Netherlands

Contents

1. Introduction

Between 1971 and 1981 the Biologisch-Archaeologisch Instituut of the State University of Groningen excavated a number of Neolithic sites in the new Flevopolder in the former Zuiderzee.

These sites were situated on natural levees of freshwater creeks near the village of Swifterbant (Figures 1 and 2) (DECKER/DE ROEVER/VAN DER WAALS 1980). The sites were ideally situated for fishing and hunting. Among the animal remains, those of beaver, otter, catfish, and other fish species abound (CLASON/BRINKHUIZEN 1978), although bones of domestic cattle, pig, sheep or goat (ZEILER 1986), and dog were also collected in small numbers.

2. The Horse Remains

Four of the mammal remains belonged to a small equid, in all probability a horse – *Equus przewalskii* Poliakov, 1881. Two incisors, a left I^2 (Plate 1) and a right I^3 (Plate 2), of the upper jaw made into a pendant, and a left metacarpus IV (Plate 3), possibly used as a piercer, were found in site S_3. A distal part of a scapula was collected in the creek next to S_3. Five samples of S_3 were collected for radiocarbon dating and gave dates of between 3525 ± 40 B.C. and 3280 ± 40 B.C. (All dates quoted are uncalibrated.) Four radiocarbon measurements are available from the creek; they range between 3390 ± 35 and 3305 ± 35 B.C. (DECKER/DE ROEVER/VAN DER WAALS 1980).

Both incisors may have belonged to the same animal.

The front and back rims of the I^2 are both worn. According to HABERMEHL (1975), this stage of wear is reached at approximately 4.5 years of age by both large horses and small horses such as Shetland ponies, Haflingers, and Norwegian and Icelandic horses. The front rim of the I^3 is worn but the back rim is worn only slightly or not at all. According to HABERMEHL, wear is

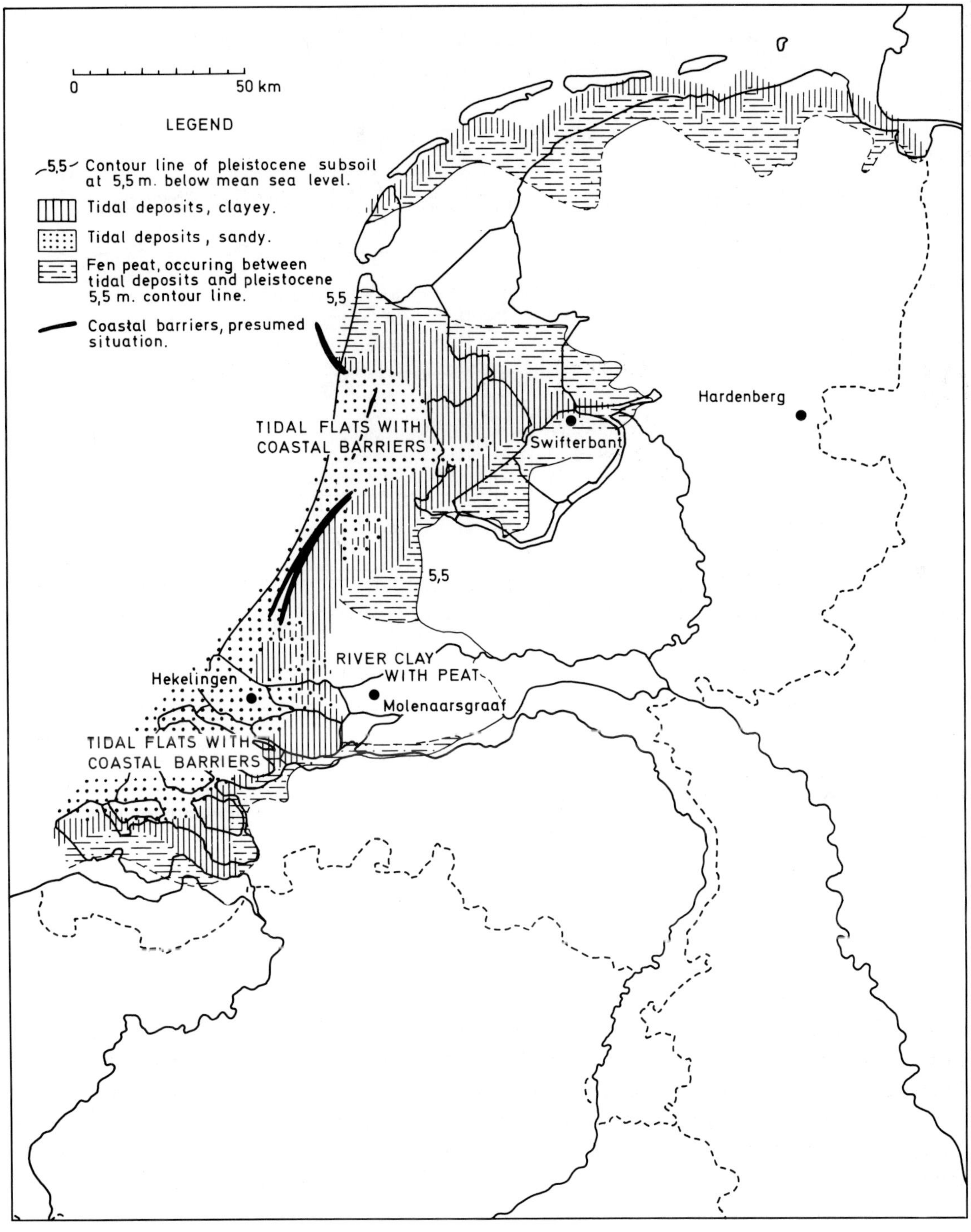

Fig. 1 The situation of Swifterbant and Hardenberg in the fourth millennium B.C. (after DECKER/DE ROEVER/VAN DER WAALS 1980). The location of Hekelingen and Molenaarsgraaf is geographically correct but is not related to the landscape of the fourth millennium B.C. Drawing: J.M. SMIT/H.R. ROELINK, B.-A.I.

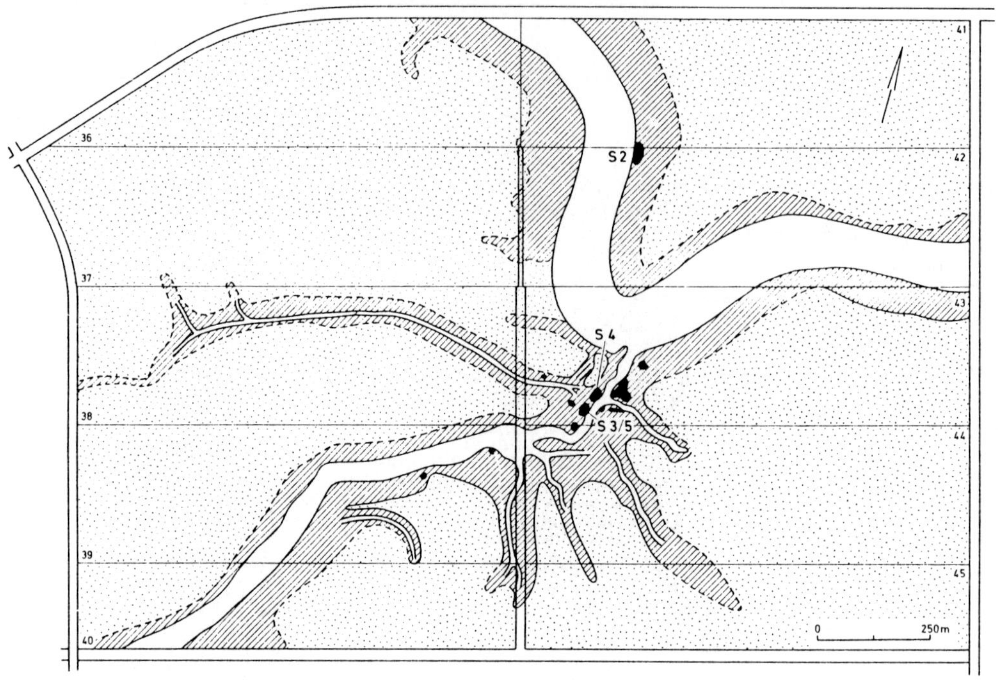

Fig. 2 Swifterbant. Situation of the sites on the levees along the creeks. (after Decker/De Roever/Van der Waals 1980). Drawing: J. H. Zwier/J. Klein, B.-A. I.

noticeable on the front rim of I^3 at an age of approximately five years, and on the back rim at 5.5 years for both large and small horses. This stage of wear of the incisor confirms our impression that they probably belonged to the same animal. The age at death of this animal was between 5 and 6 years of age.

3. Discussion

Theoretically, the two pendant incisors could have been imported from elsewhere or could even have been heirlooms, handed down through the ages since the late Palaeolithic when horses probably were more numerous. The scapula and the metacarpus indicate, however, that horses could have been hunted by the inhabitants of Swifterbant. Whether these animals were living in the tidal area on the levees on which the dwelling sites were also situated is still uncertain. That rather wet biotope would not seem favorable for horses, but it was probably impossible for the inhabitants of Swifterbant by going overland to reach the higher Pleistocene sands in the east. It should be remembered, however, that a P^2 and phalanx 3 of a horse were also found at the site of Hekelingen in the Province of South Holland (Fig. 1), radiocarbon dated between 2225 ± 120 B.C. and 2130 ± 85 B.C. (Clason 1967). This site was situated in the same kind of habitat as those near Swifterbant. The surroundings at the site of Molenaarsgraaf (Fig. 1), dated c. 1800 B.C. (Louwe Kooijmans 1974), whence a horse radius and tibia

Plate 1 Left I^2 from Swifterbant

have been identified, were more favorable. That horses were living on the Pleistocene sands east of Swifterbant seems to be confirmed by the discovery of the proximal part of a metatarsus and a badly preserved phalanx 2 among bones from red deer, wild and domestic pig, wild and domestic ox, and red deer near Hardenberg (Fig. 1).

No radiocarbon dates are available, but a T-shaped axe of red deer antler indicates that at least part of the material is contemporary with Swifterbant where similar T-shaped axes were found.

Horse remains are scarce in Mesolithic and Neolithic settlements in northwestern Europe. In the Mesolithic site of Hohen Viecheln (Mecklenburg, c. 5500 B.C.), one metatarsus, three small metapodials, one scapula fragment, and one molar of the upper jaw were found (GEHL 1961).

One specimen is known from the Mesolithic settlement of Rüde (c. 4000–3500 B.C.) and two from Bondebrück (c. 4000–3500 B.C.) in Schleswig-Holstein (LÜTTSCHWAGER 1967). Horse remains are also known from Mesolithic and Neolithic settlements in the Baltic region (PAAVER 1965). These finds corroborate the observation of NOBIS (1971) that horses were living in northwestern Europe during the Neolithic.

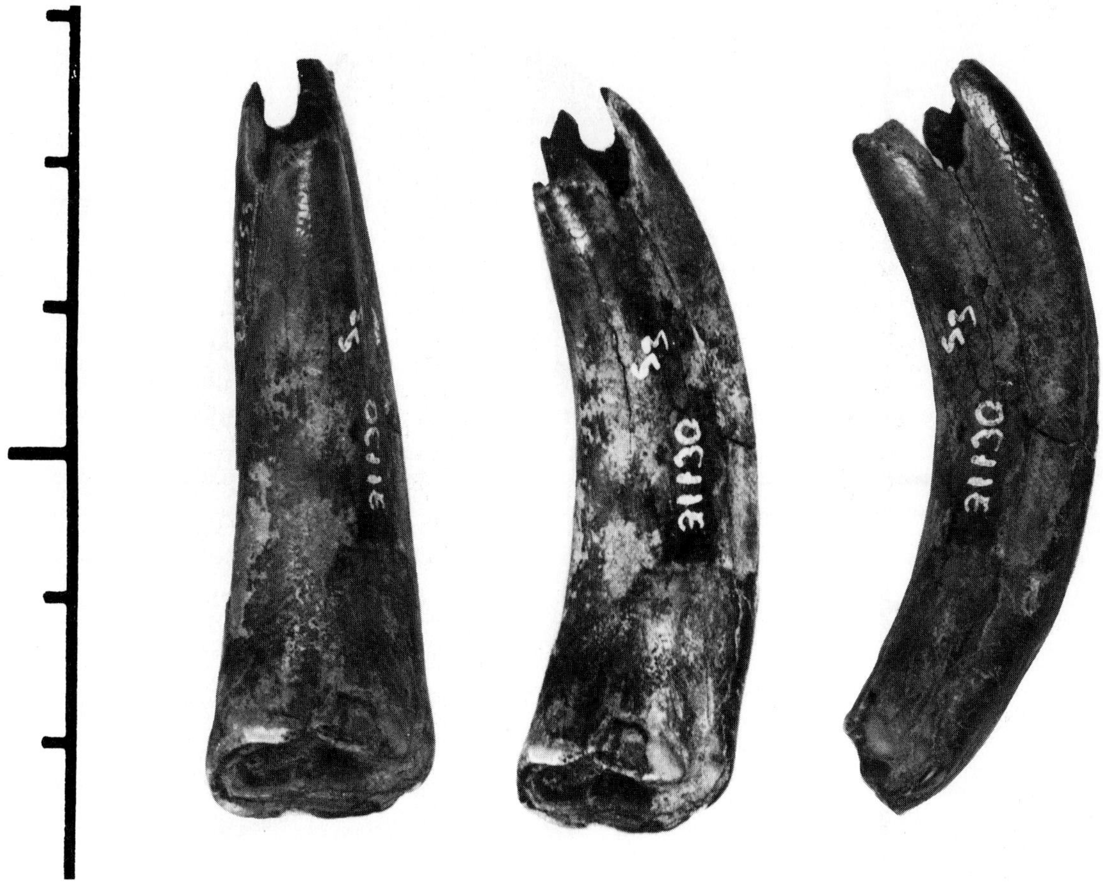

Plate 2 Right I^3 from Swifterbant

The horse is supposed to have been first domesticated in southwestern Russia at c. 3000 B.C., the oldest finds being from the Neolithic site of Dereivka in the Ukraine. If this is really the earliest date for the domestication of the horse, the animal(s) from Swifterbant must have been wild. It has been suggested, however that the horses found in Hekelingen (late third millennium: Clason 1967) and Molenaarsgraaf (early second millennium: Clason 1977) were probably domesticated, although there exists no solid proof of this.
It is still too early to say whether the wild horses still living in The Netherlands in the second half of the fourth millennium B.C. disappeared before the introduction of domestic horses in the second half of the third millennium B.C., or whether they survived long enough to be locally domesticated or crossed with imported domestic animals.
One possibility for getting an answer to these questions could be to carbon-date subfossil horse remains that are found without archaeological context.

Plate 3 Left Metacarpal IV from Swifterbant

4. Literature

Clason, A.T., 1967, Animal and Man in Holland's Past; Groningen: Noordhoff.

Clason, A.T., 1977, Bouqras, Gomolava en Molenaarsgraaf, drie stadia in de ontwikkeling van de veeteelt. Museologia **7**(1), 54–64.

Clason, A.T./Brinkhuizen, D.C., 1978, Swifterbant: mammals, birds, fishes: a preliminary report (Swifterbant Contributions 8). Helinium **18,** 69–82.

Decker, P.H./Roever, J.P. de/Waals, J.D. van der, 1980, Jagers, vissers en boeren in een prehistorisch getijdengebied bij Swifterbant. Z.W.O. Jaarboek, 111–146.

Gehl, O., 1961, Die Säugetiere. In: Schuldt, E., ed., Hohen Viecheln: ein mittelsteinzeitlicher Wohnplatz in Mecklenburg; Berlin: Akademie-Verlag, 40–63.

Habermehl, K.H., 1975, Die Altersbestimmung bei Haus- und Labortieren; Berlin: Paul Parey Verlag. (2nd edition).

Louwe Kooijmans, L. P., 1974, The Rhine-Meuse Delta: Four Studies on its Prehistoric Occupation and Holocene Geology; Leiden: Leiden University Press.

Lüttschwager, H., 1967, Kurzberichte über Tierfunde aus meso- und neolitischen Moorsiedlungen in Schleswig-Holstein. Schriften Naturwissenschaftlicher Verein Schleswig-Holstein, 53–64.

Nobis, G., 1971, Vom Wildpferd zum Hauspferd; (Fundamenta, Reihe B, Band **6**); Köln: Bohlau Verlag.

Paaver, K., 1965, Die Entstehung der Säugetierfauna und die Variabilität der Säugetiere des Ostbaltikums in Holozän; Tartu. (in Russian with German summary).

Zeiler, J. T., 1986, Swifterbant: dwelling place for a season or throughout the whole year? An archaeozoological contribution. In: Fokkens, H./Banqa, P./Birmu, M., eds., Opzock naar mens en matriëk cultuur; Groningen: B.-A. Inst., 85–95.

A. T. Clason, Biologisch-Archaeologisch Instituut, Poststraat 6, 9712 Groningen, The Netherlands.

RICHARD BURLEIGH, ANDREW CURRANT, ELIZABETH JACOBI, AND ROGER JACOBI

A Note on some British Late Pleistocene Remains of Horse (*Equus ferus*)

Contents

1. Introduction

As part of a program for investigating the chronology of the introduction and extinction of larger terrestrial mammals in Britain at the end of the Pleistocene and in the Holocene (CLUTTON-BROCK/BURLEIGH 1983), six bones of wild horse (*Equus ferus*) from Gough's Cave, Cheddar, Somerset, England (51° 15′N, 2° 45′W) have been dated by the British Museum radiocarbon laboratory (AMBERS/MATTHEWS/BURLEIGH 1985).
Gough's Cave from which these equid remains came has yielded a number of important archaeological finds including the contracted human burial known as 'Cheddar Man', artifacts of bone and antler including two bâtons-de-commandement, and more than 7000 stone artifacts of terminal Palaeolithic form (BURLEIGH/JACOBI/JACOBI 1985, GARROD 1926), as well as a variety of animal remains. One of us (APC) has recently re-examined the Pleistocene faunal remains from the cave, excavated by R. F. Parry in 1927–31, held in the Department of Palaeontology at the British Museum (Natural History). The assemblage includes fourteen species of mammal other than *Homo sapiens.* Of these, only three are common, there being over 200 stratified specimens of horse (*Equus ferus*) and some 40 stratified specimens each of red deer (*Cervus elaphus*) and hare (*Lepus timidus*). There is direct evidence of human utilization of all three forms. Other species tend to be poorly represented; a complete list is given below.

Homo sapiens	human
Lepus timidus	hare
Ochotona pusilla	pika
Castor fiber	beaver
Lemmus lemmus	lemming
Arvicola terrestris	water vole
Canis lupus	wolf

Alopex lagopus	arctic fox
Vulpes vulpes	red fox
Ursus arctos	brown bear
Equus ferus	horse
Cervus elaphus	red deer
Rangifer tarandus	reindeer
Bos primigenius	aurochs
Saiga tatarica	Saiga antelope

The abundance of horse and the marked scarcity of reindeer at Gough's Cave is of particular interest given the overwhelming dominance of the latter at many other sites of Late Glacial human activity. The age determinations given below (Table 1) show that the principal phase of bone accumulation at Gough's Cave corresponded to the climatic amelioration of the Late Glacial interstadial, and it is reasonable to infer that this was a time when reindeer were rare, or even periodically absent from Britain, rather than interpreting their low numbers as being due to selective hunting bias.

The six horse bones selected for dating came from the lower part of a fill of cave earth with angular limestone and pebbles (spits 1–16) overlying an older Pleistocene deposit (spits 17–25). (Parry's excavation comprised twenty-five 6-inch [15 cm] spits numbered from the top downwards and conforming with the natural dip of the deposits.) None of the bones except the atlas vertebra from spit 10 (Table 1) was observed to carry cut-marks or other unambiguous evidence of modification by humans. While the bones could possibly have reached the cave other than by human agency, none showed signs of damage inflicted by carnivores. The conjoining of anciently broken stone tools from spits 8–17 shows that the individual spits do not correspond to discrete archaeological events, and the faunal remains from spits 10–18 were probably accumulated over a relatively short period. As will be seen, the radiocarbon dates obtained for the horse bones from these spits provide strong support for this assertion.

2. Measurements and results

The BM-radiocarbon dates for the bones of horse from Gough's Cave are listed in Table 1 and were obtained by the conventional liquid scintillation technique (see Burleigh 1985 for details of sample preparation and measurement). These dates form a remarkably tightly clustered group with an unrounded weighted mean value of 12,318 radiocarbon years bp, and corresponding standard error on the mean of ± 33 years (based on the variation of the dates about the mean rather than the individual errors based on counting statistics given in the table). At the 95% confidence level the limits of age about the mean are 12,250–12,390 bp.

TABLE 1

Radiocarbon dates for bones of wild horse (*Equus ferus*) from Gough's Cave, Cheddar

Spit	Material	Radiocarbon age (5570 yr half-life)	Lab. no.
10	atlas vertebra (collagen)	12,350 ± 160 bp	BM-2183 R
12	calcaneum (collagen)	12,250 ± 160 bp	BM-2184 R
13	metapodial (collagen)	12,200 ± 250 bp	BM-2185 R
14	metapodial (collagen)	12,470 ± 240 bp	BM-2186 R
16	metapodial (collagen)	12,300 ± 200 bp	BM-2187 R
18	metapodial (collagen)	12,380 ± 230 bp	BM-2188 R

Subsequently, four more bones from spits 13–18 (a right calcaneum of Saiga antelope, a distal right metatarsal of red deer, and a distal right metacarpal and 2nd right phalanx of horse) were dated at the Research Laboratory for Archaeology and the History of Art at Oxford, by the much less destructive accelerator (AMS) technique (Burleigh/Hall 1984), the weighted mean of the four determinations (OxA-463-466) being 12,500 ± 100 bp. Three of these bones (the metatarsal of red deer, from spit 13, the metacarpal of horse, from spit 18, and the phalanx of horse, from spit 14) had distinctive cut-marks left by stone tools during disarticulation. The mean of the OxA-dates is somewhat older than that of the BM-dates, but the two series are not discordant and may reflect the true spread of age within the deposit.

3. Discussion

In our previous publication (Burleigh/Jacobi/Jacobi 1985) we were concerned mainly with the indirect evidence that these remains of horse provide for the early return of human hunting groups to southern Britain during the warm episode following the last glacial maximum. Here our interest is more in the remains themselves. Wild horse, together with other open-country, grazing species such as the mammoth (*Mammuthus primigenius*) and woolly rhinoceros (*Coelodonta antiquitatis*), was present in Britain during of the Devensian, although (and here further research is needed) perhaps not during the coldest stage from about 18,000–15,000 bp. After this, the sparse fossil and radiocarbon evidence so far available suggests that the latter two species did not return (Stuart 1982).

That horses were hunted for food by humans in the late glacial period, and perhaps even in preference to other species, is well attested. In Britain, with the eventual eustatic rise in sea-level around 8000 bp, stocks of semi-migratory species such as the wild horse and migratory species such as the reindeer could no longer be replenished. It seems probable, however, that the wild horse became extinct in Britain and northwest Europe before the British Isles were isolated from the European mainland. The latest date we have obtained for remains of horse from Britain during the earlier Holocene is 9840 ± 120 bp (BM-1619) for a metapodial from the Darent river gravels, Sevenoaks, Kent (Burleigh/Ambers/Matthews 1982, Clutton-Brock/Burleigh 1983, Harrison/Clutton-Brock/Burleigh 1981). An engraved horse mandible (diastema) from Kendrick's Cave, Llandudno, North Wales, in the British Museum collection (B.M.Reg. No.1959, 12-3,1: Sieveking 1971), dated by the accelerator technique, has given a very similar result of 10,100 ± 200 bp (OxA-111). After this period, the horse was absent from Britain until reintroduced as a domestic animal in the Beaker period (Clutton-Brock/Burleigh, this volume: Grime's Graves).

Although extinctions can have multiple causes (Martin/Klein 1984) the abundance of lithic scatters dating to the earlier Holocene (Wymer 1977) strengthens the hypothesis that overhunting by humans was the most important factor in the ultimate demise of this as well as other large mammal species, such as the reindeer and the elk (*Alces alces*), in the British Isles.

4. Acknowledgements

We thank Dr J.A.J.Gowlett and Dr R.E.M.Hedges and colleagues of the University of Oxford Radiocarbon Accelerator Unit, supported by Serc, for providing accelerator dates.

5. Literature

Ambers, J./Matthews, K./Burleigh, R., 1985, British Museum natural radiocarbon measurements XVIII. Radiocarbon **27**(3).

Burleigh, R., 1986, Chronology of some early domestic equids in Egypt and Western Asia. In: Meadow, R.H./Uerpmann, H.-P., eds., Equids in the Ancient World; (Beihefte zum Tübinger Atlas des Vorderen Orients, Reihe A **19/1**); Wiesbaden: Dr. Ludwig Reichert Verlag, 230–236.

Burleigh, R./Ambers, J./Matthews, K., 1982, British Museum natural radiocarbon measurements XV. Radiocarbon **24**(3), 262–290.

Burleigh, R./Hall, E.T., 1984, Accelerator dating (AMS). Antiquity **58**(224), 205–206.

Burleigh, R./Jacobi, E.B./Jacobi, R.M., 1985, Early human resettlement of the British Isles following the last glacial maximum: new evidence from Gough's Cave, Cheddar. Quaternary Newsletter **45**, 1–6.

Clutton-Brock, J./Burleigh, R., 1983, Some archaeological applications of the dating of animal bone by radiocarbon with particular reference to post-Pleistocene extinctions. In: Mook, W.G./Waterbolk, H.T., eds., Proceedings of the First International Symposium on C-14 and Archaeology; (Pact Journal **8**); Strasbourg: Council of Europe, 409–419.

Garrod, D.A.E., 1926, The Upper Palaeolithic Age in Britain; London: Oxford University Press.

Harrison, D.L./Clutton-Brock, J./Burleigh, R., 1981, Remains of mammals from the Darent river gravels at Sevenoaks Reserve, Kent. Archaeology of Canterbury **97**, 27–52.

Martin, P.S./Klein, R.G., 1984, Quaternary Extinctions; Tucson: University of Arizona Press.

Sieveking, G. de G., 1971, The Kendrick's Cave mandible. British Museum Quarterly **35**, 230–250, pls. LXXXIV–LXXXVI.

Stuart, A.J., 1982, Pleistocene Vertebrates in the British Isles; London: Longman.

Wymer, J.J. (ed), 1977, Gazetteer of Mesolithic Sites in England and Wales; (Council for British Archaeology, Report **20**); Norwich: Geo Books.

Richard Burleigh, (Research Laboratory, The British Museum, London WC1B 3DG, England); current address: The Old George, The Square, Broadwindsor, Beaminster, Dorset DTB 3 QD, England.

Andrew Currant, Department of Palaeontology, British Museum (Natural History), Cromwell Road, London SW7 5BD, England.

Elizabeth Jacobi, 8 Heron Close, Rickmansworth, Hertfordshire WD3 1NF, England.

Roger Jacobi, Department of Classics and Archaeology, University of Lancaster, Lonsdale College, Bailrigg, Lancaster LA1 4YN, England.

JULIET CLUTTON-BROCK AND RICHARD BURLEIGH

The Mandible of a Mesolithic Horse from Seamer Carr, Yorkshire, England

Contents

1. Summary

Illustrations and measurements are given of a complete left mandibular ramus of a wild horse, *Equus ferus,* recently excavated from the Mesolithic site of Seamer Carr, Yorkshire. Radiocarbon analysis of fragments of the right ramus has yielded a date of 9790 ± 180 bp (BM-2350).

2. Material

In August 1984 during excavation of the Mesolithic (Pre-Boreal) levels preserved in peat deposits at Seamer Carr in Yorkshire, the complete left side of the mandible of an adult male horse was retrieved (Site L1, No. 11933/4, Context 5012/5100). All the teeth of the mandible are present, and there were also fragments of the right ramus. The mandible was in very friable condition and had to be consolidated immediately on lifting. This was done while the bone was wet by coating it with polyvinyl acetate emulsion, leaving aside the fragments of the right ramus for radiocarbon dating. Unfortunately, the poor condition of the mandible meant that the teeth could not be completely cleaned. Therefore it has not been possible to measure the occlusal surfaces of the cheekteeth following the procedure of EISENMANN (1981).
Views of the mandible are shown in Plates 1–3 where it can be seen that, although the ramus is complete, the surfaces of the cheekteeth are partly concealed by a mineralized peaty coating. The cheekteeth however, show all the characters that are diagnostic of the true horse. The dimensions of the mandible and cheekteeth are given in Table 1.

Plate 1 Labial view of mandible of *Equus ferus* from Seamer Carr.

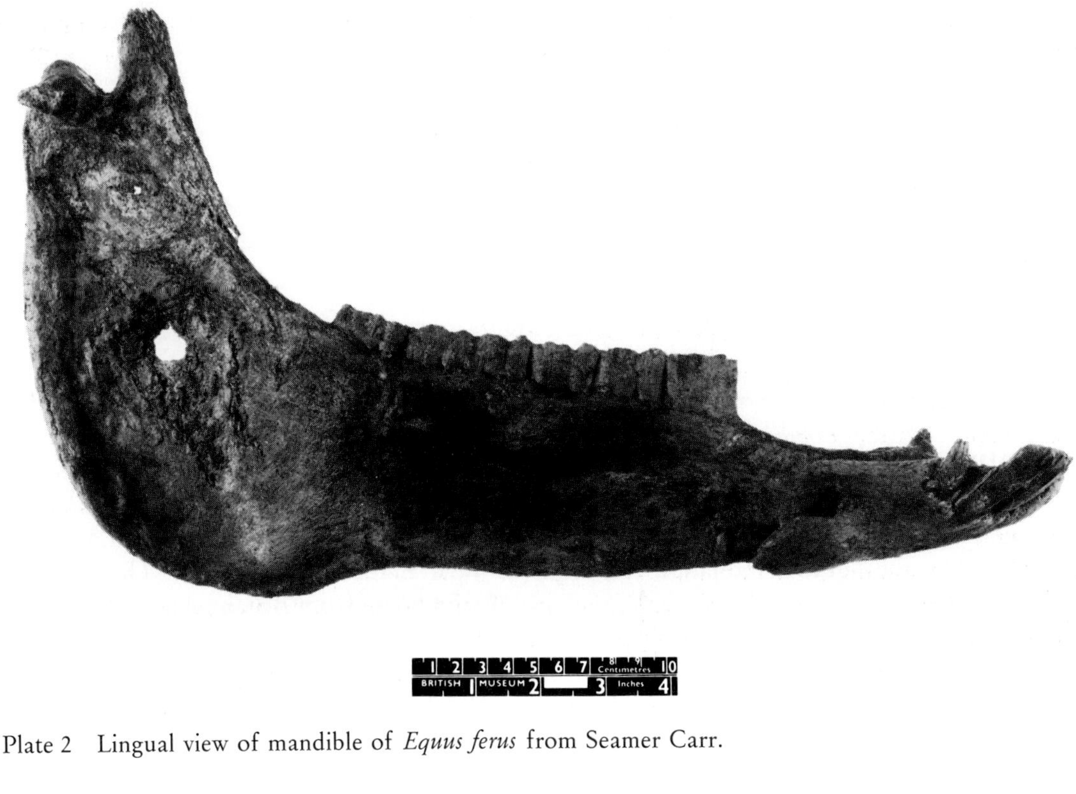

Plate 2 Lingual view of mandible of *Equus ferus* from Seamer Carr.

Plate 3 Occlusal view of the cheekteeth of the mandible of *Equus ferus* from Seamer Carr.

TABLE 1

Dimensions of the horse mandible (in millimeters) measured as in von den Driesch (1976: 52–54, figs. 19 a & b)

1	380	12	25.2/17.7
2	430	13	25.5/16.6
3	130	14	31.5/13.0
4	281	15	90.4
5	294	16	64.3
6	168	17	64.9
6 a	162	18	43.6
7	81.5	19	233.5
7 a	78	20	203
8	85	21	(240)
8 a	85	22 a	102.6
9	28.9/16.3 (L/B)	22 b	76
10	26.0/17.8	22 c	57
11	25.1/16.4		

3. Dating and discussion

Collagen separated from the fragments of the right ramus was dated at the British Museum laboratory by the liquid scintillation technique to 9790 ± 180 bp (BM-2350). This result is comparable with two other dates for the remains of wild horse from Britain. These are 9770 ± 80 (BM-1619) for a metapodial from the Darent River gravels at Sevenoaks in Kent (Harrison/Clutton-Brock/Burleigh 1981), and 10,000 ± 200 (OxA-111) for a decorated mandibular fragment from Kendrick's Cave, Llandudno, North Wales. No other direct dates have been obtained for remains of horse in Britain until the introduction of the domestic horse in the late Neolithic, c. 4000 bp.[1] Our presumption is that the wild horse became extinct or at least very rare in Britain soon after 10,000 bp, probably as a result of hunting by humans and the spread of forests in the Boreal period, a situation that appears to have prevailed throughout western Europe. The mandible described here, together with a number of other bones and teeth of horses from Seamer Carr may be evidence of the hunting element of this process. Finally, it is of interest to note that this is the most completely preserved mandible of wild horse known from this period in Britain and possibly also from western Europe as a whole. The mandible is held in the British Museum (Natural History) and has the registration number ARC 84.5103.

4. Acknowledgements

We should like to thank Tim Schadla-Hall, Director of the excavations at Seamer Carr, for allowing us to publish this report in advance of publication of the excavations, Mrs. P. Long for her help with the conservation of the mandible, and Janet Ambers, Research Laboratory, The British Museum, for the radiocarbon analysis.

5. Literature

Driesch, A. von den, 1976, A Guide to the Measurement of Animal Bones from Archaeological Sites; (Peabody Museum Bulletin 1); Cambridge (MA): Peabody Museum, Harvard University.

Eisenmann, V., 1981, Étude des dents jugales inférieures des *Equus* (Mammalia, Perissodactyla) actuels et fossiles. Palaeovertebrata **10**(3–4), 127–226.

Harrison, D. L./Clutton-Brock, J./Burleigh, R., 1981, Remains of mammals from the Darent river gravels at Sevenoaks Reserve, Kent. Archaeology of Canterbury **97**, 27–52.

Juliet Clutton-Brock, British Museum (Natural History), Cromwell Road, London SW7 5BD, England.

Richard Burleigh, (Research Laboratory, The British Museum, London WC1B 3DG, England); current address: The Old George, the Square, Broadwindsor, Beaminster, Dorset DT8 3QD, England

[1] Since this account was written in 1985 a series of radiocarbon dates has been obtained for remains of wild horse (*Equus ferus*) from Late Glacial levels at Gough's Cave, Cheddar, Somerset. These range from 12,200 ± 250 bp (BM-2185 R) to 12,470 ± 240 bp (BM-2186 R), see Burleigh/Currant/Jacobi/Jacobi (this volume).

JULIET CLUTTON-BROCK AND RICHARD BURLEIGH

The Skull of a Neolithic Horse from Grime's Graves, Norfolk, England

Contents

1. Introduction

During the last (1976) season of the British Museum investigations of flint mining at Grime's Graves, Norfolk, a single horse skull was excavated from within the thrown-up chalk debris that formed the lip of a mine shaft. The skull was in a small pit that appeared to have been intentionally dug, and it overlay a number of chalk blocks in the bottom of the pit, three of which had been decorated with patterns of deeply engraved lines. The skull was lying lengthwise in a horizontal position in the pit and was covered by additional chalk blocks. The method of burial and the associated finds are of obvious interest, but the particular relevance of this find here is that it is the earliest skull of a domestic horse yet found in Britain, coming as it does from an undoubted late Neolithic context and dated as it has been by direct radiocarbon measurement of the bone.[1]

The wild horse was a common animal in the Late Glacial of Europe, but its remains had become scarce by the early Flandrian. A few scattered, and usually badly documented, finds have been recorded from the Mesolithic and Neolithic of Spain, France, Britain, the Nether-

[1] Since this account was written in 1982/83 the skull of a horse has been retrieved from the Neolithic causewayed camp of Etton in Cambridgeshire. The skull appeared to have been intentionally buried in a pit directly underneath a red deer antler pick which signified its Neolithic context (Armour-Chelu 1985, personal communication). The first radiocarbon accelerator date obtained on this skull was clearly anomalous (OxA-1031: 1440 ± 100 bp) so four further samples from the horse teeth and the antler were directly dated by the Oxford Accelerator Unit. Although these dates form a consistent group they fall into the Late Bronze Age and are at least 1000 years too late, according to the archaeological evidence which, if it can be substantiated, means that the horse skull from Etton should be older than the one from Grime's Graves. The radiocarbon dates are: Etton horse tooth 3040 ± 80 bp (OxA-1313) and 3050 ± 80 bp (OxA-1314); Etton red deer antler 3080 ± 80 bp (OxA-1311) and 3020 ± 60 bp (OxA-1312).

lands, and Scandinavia (GRIGSON 1966). In northern Europe there are now some well-stratified remains of horse from the early Neolithic in the Netherlands as described by CLASON in this volume and from the middle Neolithic (TRB) of Denmark (DAVIDSEN 1978). The best known finds of early domestic horse are, however, the few isolated bones and teeth from the Beaker levels of Newgrange in County Meath, Ireland (WIJNGAARDEN-BAKKER 1974). Since the horse is unknown in the fossil fauna of Ireland, it is certain that the Newgrange specimens must be from domestic animals introduced at about 1900 bc.[2] The horse skull from Grime's Graves also belongs to this period, and indeed it appears from the increase in finds of horse from late Neolithic sites in Britain and the Netherlands that the "Beaker people" were responsible for the diffusion of this domestic animal.

2. Osteological Description and Discussion

The skull, although remarkably complete, was in friable and delicate condition when excavated since it had been buried in a chalky, alkaline deposit. To consolidate and preserve the bone, most of the skull was therefore immersed in a dilute emulsion of polyvinyl acetate, the loose fragments being kept separate so that they could be used for dating. As can be seen from Plates 1, 2, and 3, the only parts missing from the skull are the occipital region, the base of the cranium, the right side of the frontal bone, and the ends of the nasal bones. The animal was a female to judge from the absence of alveoli for the canine teeth (Plate 2). All of the upper teeth as well as the mandible are missing and were not found lying loose in the surrounding deposit. If there are no teeth, it is usually impossible to determine with any accuracy the age at which a horse has died. With this skull, however, it is possible to assert that the mare was an extremely aged animal. The alveoli of the cheek teeth are extraordinarily shallow, indicating that the mare was well past the age when the crowns of the teeth could erupt any further. If a horse lives to excessive age, for example, under the protection of a human owner, it can "outlive its teeth", and this appears to have happened in the mare from Grime's Graves. The alveoli of the central incisors had grown together after the teeth were lost, and there is an infilling of secondary bony tissue, just as there is in many of the alveoli of the cheekteeth. It is probable that when the mare died she was over 35 years of age and that she retained only the left and right second and third incisors and the stumps of some of the cheekteeth. Plate 3 presents the palatal view of the skull to show the condition of the alveoli, and Plate 4 provides a comparison with a similar view of the skull of an aged Welsh mountain pony, "Dyoll Starlight". (The teeth from this latter skull are excessively worn, and as they were very loose, they could be temporarily removed for the purpose of this photograph). This renowned pony was a grey stallion that was foaled in 1894 and died in 1929, aged 35 years. He stood 117 cm high at the withers and was said to have had Arab blood. "Dyoll Starlight" at one time belonged to Lady Wentworth, the famous breeder of Arab horses, and he sired a great number of noted Welsh mountain ponies. The dimensions of his skull are given in Table 1 where it can be seen that

[2] The lower case notation is used to indicate uncalibrated radiocarbon dates based on the 5570 year half-life.

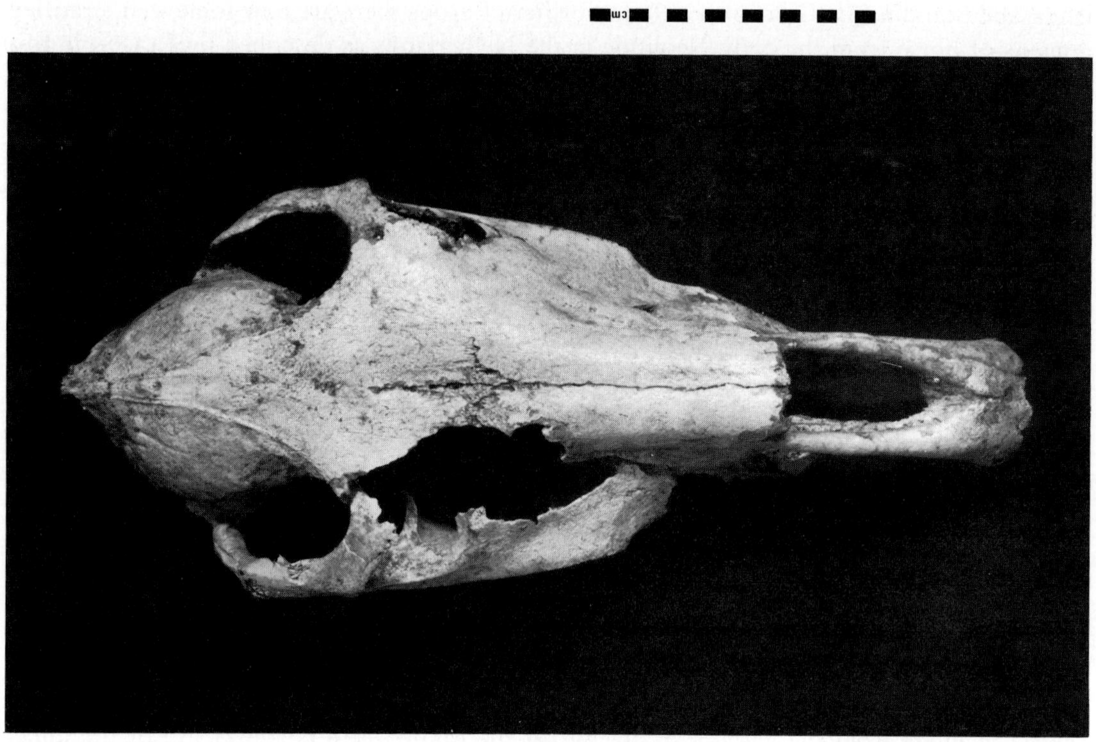

Plate 1 Frontal view of the skull of the mare from Grime's Graves (Photo. British Museum).

Plate 2 Left lateral view of the Grime's Graves skull (Photo. British Museum).

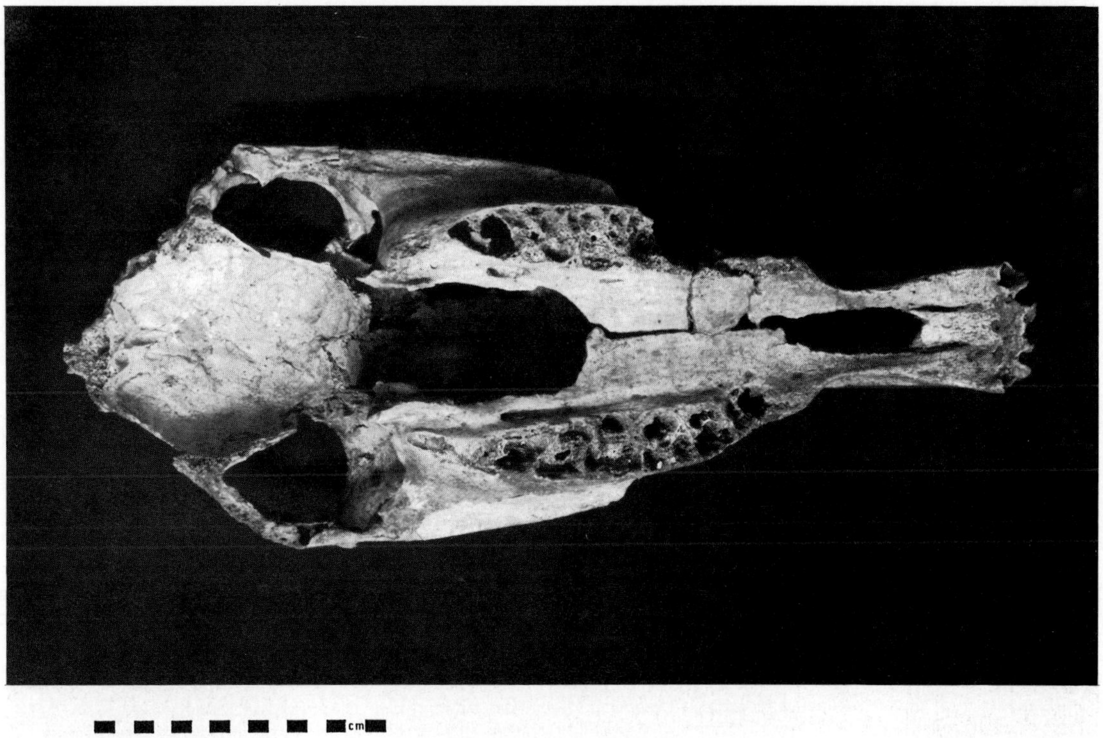

Plate 3 Palatal view of the Grime's Graves skull (Photo. British Museum).

those of the mare from Grime's Graves are generally greater. From these measurements we can guess that the mare stood about 132 cm high at the withers.

The extremely old age at death of the mare from Grime's Graves suggests that she was a highly valued domestic animal and that care was taken to see that she was properly fed. This assumption is supported by the healthy state of the skull bone. Apart from a restricted area of osteoporosis around the left maxillary bone in the region of the premolars, there is no sign of disease in the skull or of abscesses in the alveoli.

Although the mare from Grime's Graves may have been a healthy old animal, the skull does have one peculiarity. This is a marked, saucer-shaped concavity in the frontal bones, between the orbits, that is approximately 100 mm in length and 50 mm in width. At first sight it could be assumed that this hollow had been caused by the pressure of the soil on the bones after burial, but close inspection shows that such an interpretation is unlikely.

A slight depression of this kind in the frontal and nasal bones is often found in ass skulls but it is rare in those of horses or ponies. In skulls of the latter in the collection of the British Museum (Natural History), where a depression is present, it is only very slightly developed. Two exceptions are skulls from Iron Age contexts, one from Gussage All Saints, Dorset (WAINWRIGHT/SPRATLING 1973), that shows a depression even more marked than the one on the skull from Grime's Graves and the other, a skull from Bramdean, Hampshire (PERRY 1972), that has a slight depression. No explanation has been found in the literature for this feature which is not connected with the characteristic, so-called "Roman nose" of the Arab horse. The

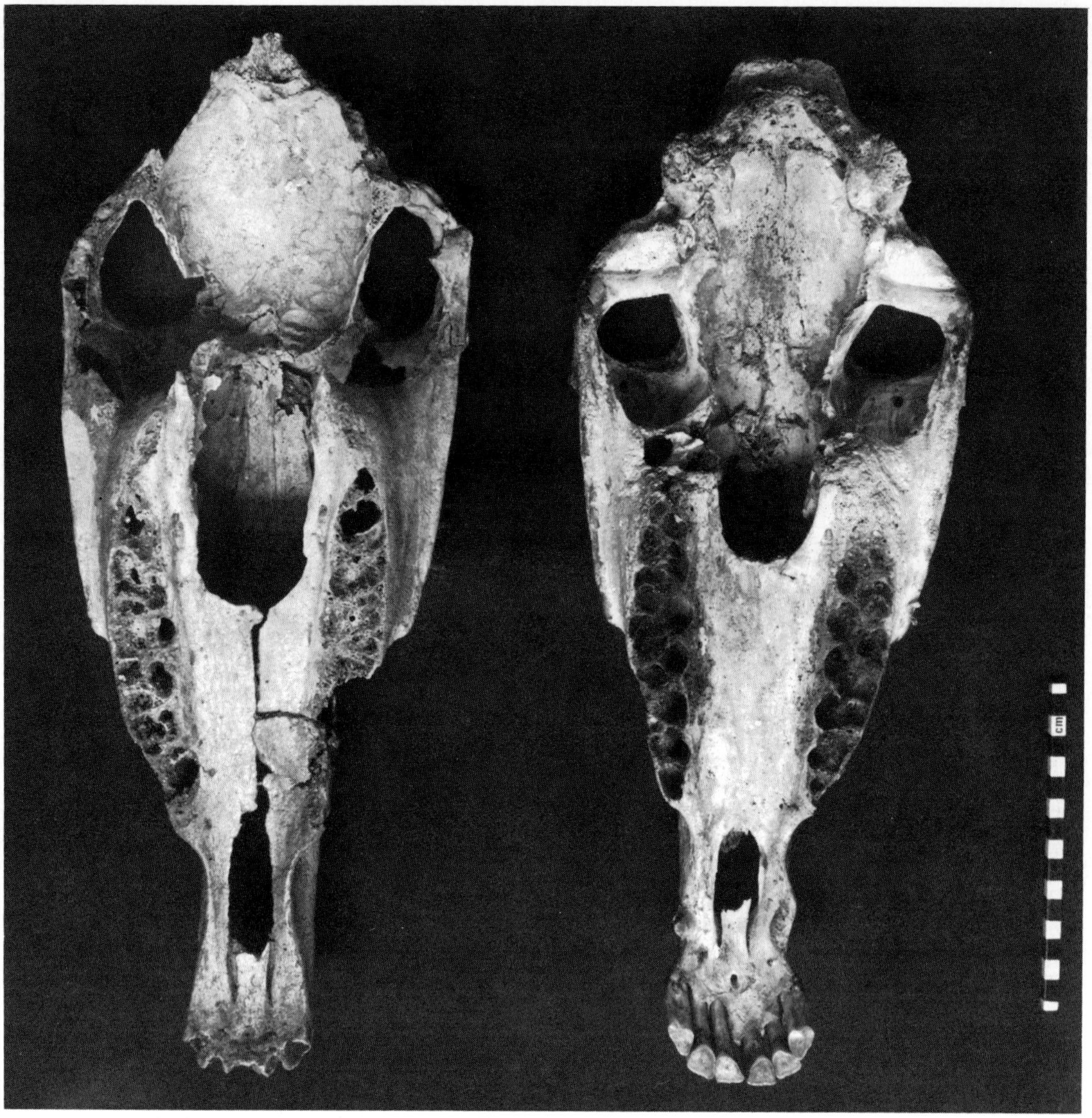

(A) (B)

Plate 4 Palatal views of the Grime's Graves skull (A) and that of "Dyoll Starlight" (B) to show the condition of the alveoli of the cheekteeth (Photo. British Museum).

concavity in the facial region of the Arab is caused by a slight dishing of the nasal bones in their mid-region and a rise at their extremities.

Until the excavation of the mare's skull from Grime's Graves, to our knowledge there was no cranium of a Recent horse in Britain older than Iron Age in date. Therefore, it has not been possible to compare the skull with any British material of contemporary or earlier date. The skull has, however, been compared to those from later contexts that are held in the British

TABLE 1

Dimensions of the horse skull from Grime's Graves together with those of four other prehistoric skulls and one modern Welsh mountain pony. Measurements in millimeters

Measurement no. from VON DEN DRIESCH (1976)	Description of Measurement	Grime's Graves female 79.5017	Bramdean (?)female 76.5113	Bramdean male 76.5112	Bramdean male 76.5111	Gussage All Saints male 75.5691	Clapton male 75.5242	"Dyoll Starlight" male 1937.1.26.3
1	Total length	(485)	476	523	485	-	467	446
13	Anterior orbit to prosthion	281	279	304	-	-	285	255
15	Posterior orbit to prosthion	340	337	363	347	-	328	303
18	Length palate	219	236	261	245	-	(255)	214
21	Length diastema	108	84	98	77	74	83	84
22	Length cheektooth row	152	160	169	168	158	156	138
31	Length orbit	60	57	60	(64)	58	57	51
32	Height orbit	51	50	43	48	50	53	50
38	Greatest breadth cranium	102	99	106	100	103	95	101
41	Greatest breadth across orbits	193	185	201	194	190	187	185
43	Facial breadth across anterior of malar bone	(147)	142	160	-	149	143	134
44	Breadth across infraorbital foramina	70	64	75	-	70	69	65
45	Breadth at posterior edge of alveolus at I^3	61	63	72	67	-	63	59
47	Least breadth diastema	57	51	61	65	-	54	54
48	Greatest breadth across cheektooth alveoli	125	115	125	-	117	113	118

Museum (Natural History), the dimensions of five examples being given in Table 1. Besides the skulls from Gussage All Saints and Bramdean that have already been mentioned (nos. 75.5691 & 76.5112) there are two additional skulls from Bramdean that were measured as well as one from the Lea Valley deposits at Clapton in Essex for which there is no documentation. This last skull was excavated at the beginning of this century together with many other animal bones and artifacts, and although ascribed to the Neolithic period, it is unlikely to be that early in date. The measurements and general appearance of all of these skulls point to a type of pony that must have been common in Britain from the time of its first introduction, perhaps at the peak of flint-mining activity at Grime's Graves.

3. Radiocarbon Dating

The circumstances of the discovery of the skull of the Grime's Graves horse suggested that it could be as early as Neolithic in date and therefore a find of great interest and rarity. Unfortunately this very possibility precluded the sacrifice of any appreciable part of the specimen for radiocarbon dating. Furthermore, the need for consolidation of the skull with PVA was an overriding consideration, and this also rendered it unsuitable for radiocarbon analysis. Some pieces of cranial bone that were too fragmentary for reconstruction were, however, deliberately not treated with PVA. and 23 g of these fragments were demineralized with dilute acid to provide a sample of collagen for radiocarbon measurement. This yielded sufficient material for synthesis of approximately 0.5 g of benzene, roughly one-tenth of the amount normally needed for radiocarbon dating by the liquid scintillation method. Inevitably when so small a sample is measured, the error (standard deviation) of the resulting date tends to be large, but by counting this particular sample over a period of about three weeks (three to four times the period normally required), the error was reduced to an acceptable level. An approximately tenfold extension of the period of counting would have been needed to achieve a significant further reduction of error. The final date obtained was 3740 ± 210 bp or 1790 bc (BM-1546; Burleigh et al., 1982), indicating that the skull of the horse belongs to the late Neolithic period or, at the latest, to the earlier part of the Bronze Age.

4. Conclusion

As discussed above, remains of horses from the prehistoric period are rare in Europe as a whole. Definite finds of domestic horse only begin with the Beaker period, for example the post-cranial bones found at Newgrange (Ireland) in a Beaker horizon dated to 3885 ± 35 bp, 1935 bc (GrN-6342; O'Kelly, 1972), a date close to that for the skull from Grime's Graves.

5. Acknowledgement

We wish to thank Mr. G. de G. Sieveking, Director of the British Museum excavations at Grime's Graves from 1972–1976, for making the horse skull available for study and dating, and for his kind cooperation in allowing us to publish this report in advance of the archaeological publication of the excavations.

6. Literature

Burleigh, R./Ambers, J./Matthews, K., 1982, British Museum natural radiocarbon measurements XV. Radiocarbon **24**(3), 262–290.

Clark, R. M., 1975, A calibration curve for radiocarbon dates. Antiquity **49**, 251–266.

Davidsen, K., 1978, The final TRB culture in Denmark: a settlement study (4. The horse and the TRB culture); (Arkaeologiske Studier Series, Monograph **5**); Copenhagen: Akademisk Forlag, 142–148.

DRIESCH, A. von den, 1976, A Guide to the Measurement of Animal Bones from Archaeological Sites; (Peabody Museum Bulletin 1); Cambridge (MA): Peabody Museum, Harvard University.

GRIGSON, C., 1966, The animal remains from Fussell's Lodge long barrow (including a possible ox-hide burial, with discussion on the presence of the horse in neolithic Britain). Archaeologia (London) **100,** 63–73. (Appendix II in Ashbee, P., The Fussell's Lodge Long Barrow Excavations 1957).

O'KELLY, M.J., 1972, Further radiocarbon dates from Newgrange, Co. Meath, Ireland, Antiquity **46,** 226–227.

PERRY, B.T., 1972, Excavations at Bramdean, Hampshire, 1965 and 1966, and a discussion of similar sites in southern England. Proceedings of the Hampshire Field Club and Archaeological Society **29,** 41–77.

WAINWRIGHT, G./SPRATLING, M., 1973, The Iron Age settlement of Gussage All Saints. Antiquity **47,** 109–130.

WIJNGAARDEN-BAKKER, L.H. VAN., 1974, The animal remains from the Beaker settlement at Newgrange, Co. Meath - first report. Proceedings of the Royal Irish Academy **74** (Section C), 313–383.

JULIET CLUTTON-BROCK, British Museum (Natural History), Cromwell Road, London SW7 5BD, England

RICHARD BURLEIGH, (Research Laboratory, The British Museum London WC1B 3DG, England); current address: The Old George, the Square, Broadwindsor, Beaminster, Dorset DT8 3QD, England

DAVID W. ANTHONY

The Domestication of the Horse

Contents

1. Introduction

The domestication of the horse was an event of revolutionary significance in human history. Nevertheless relatively little research has been conducted concerning the dynamics of the original domestication process. There is general agreement, based on the evidence discovered to date, that the earliest episode of domestication occurred in the Ukrainian steppes, north of the Black Sea (Figure 1), by about 4200–4000 BC[1] (HANCAR 1956; BIBIKOVA 1967, 1969; NOBIS 1971; BÖKÖNYI 1974: 237–252). It is this consensus that allowed PIGGOTT (1983: 87) to state in a recent review of the subject that "The east European evidence for early horse domestication ... is now well known." Yet the Ukrainian evidence is known to western European scholars primarily through very general summaries (TELEGIN 1973a; BÖKÖNYI 1974, 1980) that have presented little of the original raw data and that have made no attempt to analyze the interpretations offered by Soviet palaeozoologists. The primary studies are available only in Russian or Ukrainian and in fact, have been quite superficial, dealing primarily with questions of cranial typology and making use of very small samples (BIBIKOVA 1967, 1969). There has been little systematic examination of how this domestication might have occurred, of why it occurred when and where it did, or even of how we are able to distinguish physically the earliest domesticates from their contemporary wild cousins. Although there has been some speculation on the subject of how the acquisition of domesticated horses might have affected the societies that first acquired them (CLARK 1941; BIBIKOVA 1969; KUZMINA 1974; GIMBUTAS 1963,

[1] All dates are given in true calendar years, recalibrated from radiocarbon dates according to the interlaboratory consensus reported in KLEIN/LERMAN/DAMON/RALPH (1982). A detailed regional chronological table with specific radiocarbon/calendar equivalents appears in ANTHONY (1986).

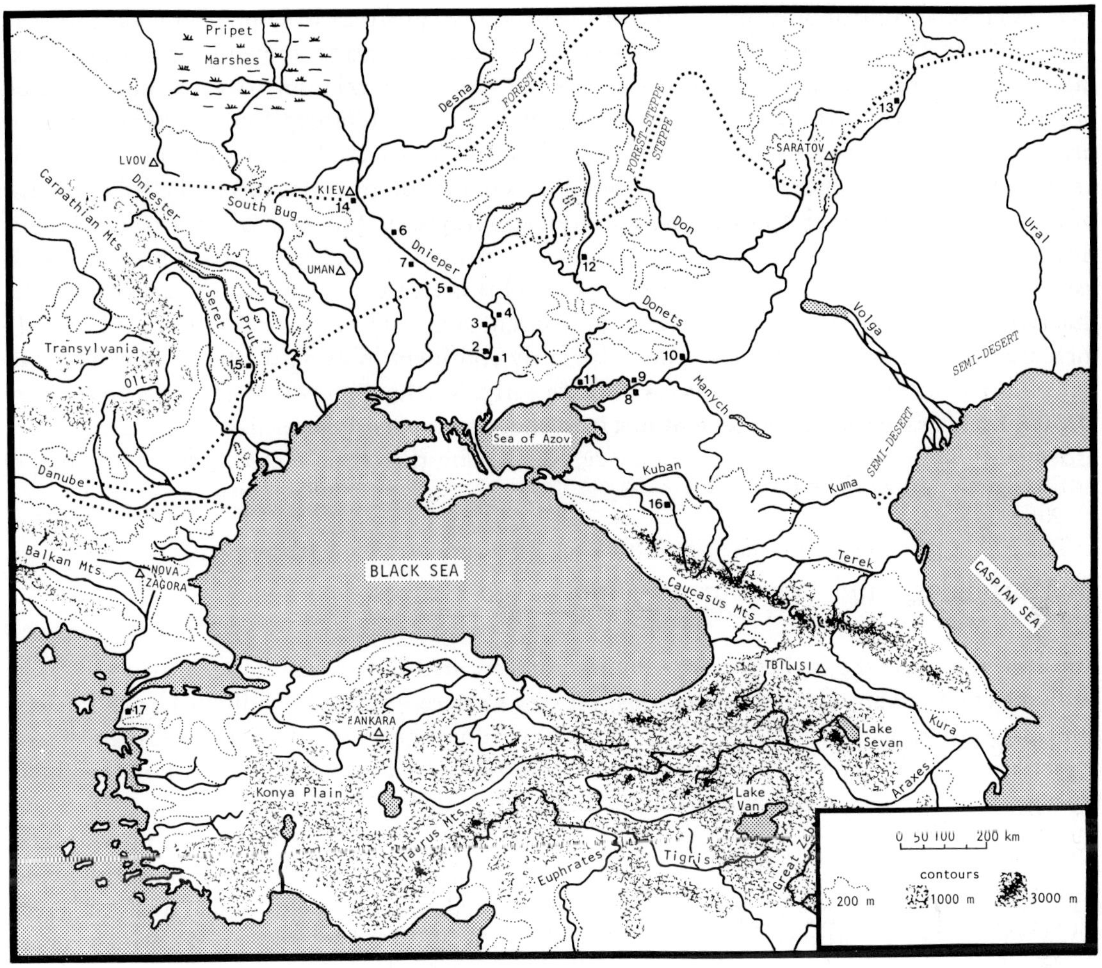

Fig. 1 The Black Sea Region, 5000–2500 BC.

1. Sredni Stog
2. Sobachki
3. Nikol'skoe
4. Igren
5. Dereivka
6. Buz'ki
7. Moliukhor Bugor
8. Rakushechni Yar
9. Samsonovka
10. Konstantinovka
11. Mariupol
12. Aleksandriia
13. Khvalynsk
14. Tripolye
15. Cucuteni
16. Maikop
17. Troy

1970, 1977; Bökönyi 1980), few of these studies are based upon well-founded behavioral or economic models, and they are all skewed by the imposition of the "Kurgan Culture" concept on the North Pontic archaeological data base, a concept I have criticized in detail elsewhere (Anthony 1986). A reappraisal of the entire matter is in order.

I should comment briefly on my use of the term "domestication." A generally accepted definition of domestication involves physical removal of the animal from its wild cousins, and at least occasional control of its breeding, for the benefit or profit of its controller (Jarman/Wilkinson 1972). There are, of course, numerous behavioral gray areas, some of the most perplexing involving specialized herd followers like the Lapps. In relation to early domesticated horses, it is not yet possible to determine on the basis of osteological criteria whether their breeding was being controlled. Nobis (1973) has attempted to differentiate specific breeds among Roman cavalry horses through a detailed examination of variations in the fold patterns of molar enamel, but such studies have not been conducted on the teeth of the earliest domesticates. Indeed, there are no firmly documented, systematic osteological changes of any kind associated with the earliest domesticated horses. However, the appearance of age/sex profiles that strongly suggest a managed herd, in conjunction with a general pattern of greatly increased horse exploitation, allow one to be fairly confident in using the term "tamed." Controlled breeding might well have occurred quite early, for the domesticators of the horse were already familiar with the management of cattle, sheep, and pigs. I therefore use the term "domesticated." Those who wish to do so may replace it with the term "tamed."

2. Morphology and Behavior of Copper Age Horses

The appearance of the earliest North Pontic domesticated horses can be reconstructed from both skeletal remains and representations. Horses were represented vividly in Upper Palaeolithic cave paintings some 10,000 years before their domestication (Baskett 1980), and these depictions agree in many respects with the well-known portrayal on a silver vase in the chieftan's grave at Maikop in the North Caucasus, dated to about 2300 BC (Gimbutas 1956: Plate 14; Gilbert, this volume: Figure 4; Glumac/Anthony in press). The Maikop vase provides the most naturalistic representation of a Pontic-region horse from the two millennia following domestication. It portrays a short-legged, large-bodied, large-headed animal with an erect mane, similar in proportions to modern Przewalski horses. The Dereivka horses, which represent a very early North Pontic domesticated population, stood 126–144 cm high at the withers, based on metacarpal lengths. They had large heads and robust leg bones (Bibikova 1970; and, here, Tables 3–7).

The behavior of the wild horses of the North Pontic Copper Age is a variable that must be considered in any reconstruction of the domestication process. Prehistoric horse behavior was molded by natural selection and by many of the same adaptive pressures that affect the behavior of modern feral horses (Berger 1986: 228). The behavior of modern feral horses in grassland/desert environments therefore can be used as a rough quide to the behavior of prehistoric North Pontic horses. Several studies of modern feral grassland/desert horse populations have been published, and these can be supplemented with historic accounts of true wild horses (Berger 1986, Groves 1974, Bökönyi 1974 a, Mohr 1971, Ryden 1972). For the purposes of this discussion only a few points concerning horse behavior need be stressed.

In the wild, horses reside most often within one of two types of social group: the stallion-harem-and-juvenile-offspring band and the bachelor group of young males. Other types of group exist, but are generally less stable than these two. Group size is determined by many factors, but both harem bands and bachelor groups can contain over 15 individuals, even within the semi-desert refuges now occupied by most free-ranging feral grassland/desert populations. However, 4–7 is a more common group size in these environments.

Both types of group occupy home ranges. The home range of a harem band tends to be smaller and to contain richer resources than that of a bachelor group. Bachelors range across large, relatively resource-poor areas. Bachelors also change their home ranges from year to year, while harem bands return to the same seasonal home ranges with great regularity. On a given day a harem band is likely to travel a shorter distance than a bachelor group (Berger 1986: 56). For a prehistoric hunter, a harem band would be significantly easier to locate because of its smaller home range, shorter movements, and repetitive behavior.

When traveling, harem bands are often led by a lead mare. The stallion often takes a position on the flank or to the rear. A visible predator might be challenged by a stallion, but a pedestrian ambush conducted from behind cover would often result in the death of a mare.

For reasons which remain unfathomable, stallions will occasionally purposefully assemble harems consisting entirely of mares of a specific coloring. One black stallion in the Pryor Mountain range of Nevada fought for and won 11 reddish-brown mares with black manes and tails, in a region where blue grulla and yellow buckskin were prevalent colors (Ryden 1972: 69–70). Stallions therefore might produce their own "breeds", or something approaching a controlled breed, even in the wild.

During the late fall and winter, wild horses move to protected valleys where they can find shelter and browse. Aggressive behavior is much reduced among the stallions, and several bands might be encountered sharing the same loosely defined range. Historic travellers' accounts describe very large herds of wild horses, numbering in the thousands, and divided into "troops" (individual bands?) (Cox [1816] 1957: 244, Falkner [1774] 1935, Smith [1841] in Ridgeway 1905: 31). This might have been a phenomenon of the late fall, when hundreds of bands were simultaneously moving towards winter pastures (Lydekker 1912: 171). Falkner's ([1774] 1935: 39) description of feral horses in the Pampas conveys a vivid impression of the scale that such movements could attain: "... They were in such vast numbers that, during a fortnight [two weeks], they continually surrounded me. Sometimes they passed by me, in thick troops, on full speed, for 2 or 3 hours together." Migrations of such magnitude would have had obvious seasonal significance for the Copper Age European horse hunter.

The age structure of wild horse bands varies significantly from one year to the next, and even between seasons. Under conditions in which predators and competitors have been removed artifically, rates of increase can be very high. A Nevada population benefitting from the removal of both cattle competitors and human predators exhibited an age structure in which 54 percent of the horses were under two years of age and only seven percent were more than 14 years (Berger 1986: 77). In areas where predators are a significant factor, however, the proportion of the population under two years can apparently fall to eight percent or less (Ryden 1978: 212–213). Under prehistoric conditions it is unlikely that the proportion of the population under two years of age rose above about 20 percent except under unusual circumstances. The prehistoric horses of the North Pontic region were subject to predation by wolves, bears, and humans, and these factors might have contributed to a juvenile mortality rate substantially higher than that of many modern feral horse populations.

The sex structure of wild equids is better understood. Among most wild equids, males experience higher mortality than females, presumably due to the costs of inter-male competition, and adult females therefore generally outnumber adult males (Smuts 1976, Berger 1986). In a harem band, of course, females compose the great majority of the social group.

These demographic data have archaeological significance in that the skeletal remains of a hunted population should exhibit age and sex characteristics quite different from those of a managed, domesticated population. Hunters focus their activities on the most predictable, regular aspects of prey behavior, and it is the harem band that exhibits the most regular behavior within the smallest home range. Bachelor groups, in contrast, range across larger areas, move longer distances, and rarely return to the same home range in consecutive years. Hunters who set ambushes along the well-worn, dung-lined trails often used by a wild harem band would obtain a sample consisting almost entirely of females. In a managed population, however, it is the young males that would be culled for slaughter, for they do not contribute to the increase of the herd, and their behavior is disruptive.

A faunal sample exhibiting a relatively high proportion of young adult or mature females, with relatively low percentages of very old or very young animals, would therefore indicate a hunted, wild population. A sample exhibiting a high proportion of young males, with a substantial percentage of animals younger than two years, would strongly indicate a managed, domesticated population. Indeed, it is difficult to reconstruct any realistic hunting strategy that would result in the latter age/sex pattern.

3. Ecological Circumscription and Economic Change

The conversion of horses from an uncontrolled, hunted resource to one that was managed and domesticated required human societies to participate in a new set of tasks, responsibilities, and divisions of labor. These behaviors were applied to an animal that was extremely fast, alert, suspicious, moderately dangerous, and that spent the majority of its time in a steppe environment that was only lightly exploited by river-valley and forest-steppe Copper Age societies. Domestication was therefore unlikely to occur until the existing river-valley or forest-steppe resource base became deficient in some respect. Recent suggestions that horses were first domesticated in Upper Palaeolithic Western Europe (Bahn 1980), the middle Volga forest-steppe region during the local sub-Neolithic (Petrenko 1984), or the Transcaucasian highlands during the Early Bronze Age (S. K. Mezhlumian cited by Munchaev 1975: 387–390) suffer from a failure to establish the existence of antecedent economic conditions that might have encouraged domestication.[2] The North Pontic region, in contrast, presents a social, economic,

[2] The possibility of an earlier North Caspian/Southern Ural center of horse domestication has received increasing attention (Matiushin 1986: 143–44; Mallory 1986: 180–82). Domesticated horses, sheep, and cattle have been claimed for Mullino, Davlekanovo, and other Southern Ural sites in levels said to date to around 6500 BC (5800 bc) (Matiushin 1986). The sheep, aspects of the material culture, and elements of the human population are seen as intrusions from the Neolithic settlements of Turkmenia and the South Caspian. However, cultures of the intervening arid zone of Central Asia exhibit no animal domesticates until the fourth millennium (Dolukhanov 1986: 127–28; Vinogradov 1981), and the earliest ceramics of the Southern Urals exhibit parallels to those of the North Pontic/Caspian region. Moreover, one review has pointed out that there is significant stratigraphic mixing at Mullino and Davlekanovo, throwing doubt on the dating of some material from these sites, specifically including the fauna (Vasiliev/Vybornov/Morgunova 1985: 281).

The concept of a Southern Ural/North Caspian Neolithic independent of and earlier than the European Neolithic is an old one (Danilenko 1969), ultimately deriving from a period in the 1950's when the discovery of the Djeitun culture in Turkmenia attracted "Neolithic origin" theories toward the east

and ecological context in which the domestication of the horse can be viewed as a predictable development. In this section I shall examine the ecological aspects of this context.

The Black Sea lies at the bottom of an enormous geological bowl that is still subsiding, drowning coastal estuaries and some prehistoric sites. On the north, the bottom of the bowl is exposed, forming a vast coastal plain. The northern, northwestern, and northeastern edges of the bowl consist of a discontinuous series of ridges and uplands (the Moldavian, Podolian, Dnieper, and Central Russian uplands, in order from west to east). Prevailing northwesterly storm systems drop most of their moisture on these uplands, leaving the lowland plain a dry and treeless grassland. An intermediate zone of forest-steppe, some 200–300 km wide, intercedes between the closed broadleaf forest of the interior uplands and the dry steppe grasslands of the coastal lowland plains. The lowland steppes widen towards the east and north, joining in that direction with the interior continental steppes of Central Asia.

The boundary between the steppe and the forest-steppe was, in antiquity, clearly marked. (Deforestation in the forest-steppe has now softened the contrast.) Because of the broken nature of its forests, the forest-steppe offered a high percentage of forest-edge environment, which is the most biotically diverse part of any forest community. The animals that prefer forest-edge environments at least seasonally include deer, aurochs, fox, hare, numerous bird species, and, in winter, wild horses. Edible berry and fruit trees/shrubs are also concentrated there. In addition, the loess soils that underlay the Moldavian-Podolian forest-steppe, northwest of the Black Sea, were among the most fertile and readily cultivated soils in Europe, making the region attractive to early farmers as well as to hunter-gatherers.

Adjacent to this resource-rich mosaic environment was the resource-poor environment of the lowland steppes, a region of low biotic diversity, occupied by a very different constellation of plant and animal communities. The subsistence techniques and tactical organizations that were best suited to humans for exploitation of forest-steppe resources (deer, aurochs, pig, nuts, berries) were not transferable to exploitation of steppe resources (horses, antelope, burrowing rodents, and highly localized forest resources).

partially because the true antiquity of the European Neolithic had not yet been discovered. It is now more likely that the techniques of ceramic manufacture and food production were diffused to the Urals from the West, given the presence of a North Pontic Dnieper-Donets Neolithic by 4200–4500 BC. Contact between the societies of the Dnieper and the middle Volga is suggested by the discovery of middle Volga cemeteries at S'ezzhe (similar to Dnieper-Donets) and at Khvalynsk. At the latter site, 200 km north of Saratov on the Volga, a cemetery produced Sredni-Stog-like ceramics, copper spiral bracelets (probably acquired, like similar Sredni Stog items, from the Carpathians), and a horse head-and-hoof „cult deposit" like that at Dereivka.

During this period of Neolithic and Copper Age Dnieper-Volga interaction the Southern Urals might have acquired ceramics and limited food production, particularly if it is true that Mullino II can be equated chronologically with S'ezzhe and Mullino III (Davlekanovo), but elk was the primary food animal at Mullino (80% NISP). Farther east in north-central Kazakhstan, the settlement of Botai has yielded over 10,000 bones, 99% of which are horse (Zaibert 1985). There are no young animals; all are 3–4 years and older. Stature at the withers is 128–152 cm. Their wild/domesticated status is uncertain. The site is probably Eneolithic, or Copper Age, though no copper is present. Its age is estimated typologically and pedologically at about 3500 BC, although radiocarbon determinations of 3530 BP ± 160 (hearth) and 4340 BP ± 120 (bone) suggest a much later period. A horse tooth from Varfolomeyevka, a North Caspian Neolithic site of the Seroglazivska culture, was examined by this writer under a Scanning Electron Microscope (see note 5) and exhibited no bit wear.

Moreover, the transition was relatively sharp. Kishinev, (elevation 150 m), located in the more open, southern fringes of the forest-steppe zone on a tributary of the lower Dniester, receives about 470 mm of rain per year with peaks in June and November. Rainfall agriculture and natural tree growth outside the river valleys are discouraged in areas receiving less than 475–500 mm of rainfall per year. Only 150 km to the southeast of Kishinev, at the coastal city of Odessa (elevation 0–20 m), rainfall averages 390 mm per year, inadequate for the growth of most tree species except in river floodplains where high water tables provide adequate moisture (figures from PLUMMER/HANNE/BRUNER/THUDIUM 1971). Odessa lies in a flat, arid steppe, while Kishinev stands among rolling hills that once supported extensive, if broken, forests. The contrast between these environments strongly influenced cultural evolution in the region and played a major role in the domestication of the horse.

A theoretical context for the domestication process is provided by a variety of R. CARNEIRO's "Circumscription Theory", which describes the relation between increases in societal complexity and population growth within circumscribed environments (CARNEIRO 1967, 1970; BALL 1977). In the sharply contrastive boundary-zone environments of the North Pontic region, where resource-rich biotic communities are juxtaposed to or (in the case of the wooded steppe river valleys) are surrounded by markedly different resource-poor communities, populations dependent upon a given environmental adaptation might become quite circumscribed, making CARNEIRO's paradigm highly appropriate (ANTHONY 1986). The forest-steppe/steppe borderlands and the river valleys constituted attractive, but fragile and limited environments that were likely to have been subject to demographic "pooling" and population pressure at an earlier date than other more homogeneous environments to the north and south. Within such boundary-zone communities these stresses were likely to generate a more rapid pace of innovation than would occur in areas removed from the highly contrastive ecological borders.

The prehistory of the region can be interpreted generally as a continuous, graded series of cultural responses to a chronic condition of adaptive stress along the contrastive boundary between the upland forest-steppe zone (extremely resource-rich) and the lowland steppe zone (extremely resource-poor) north of the Black Sea. These responses included both the initial adoption of Neolithic farming and stockbreeding (economic intensification within the forest-steppe) and the domestication of the horse (expansion of the resource base into the steppe). The domestication of the horse can now be understood not as an isolated event, but as a stage in a larger process of change.

4. Culture and Economy in the North Pontic Region

Neolithic subsistence techniques were introduced into the North Pontic region by Criş-culture immigrants from the lower Danube valley. They entered the forest-steppe uplands of Moldavia (avoiding the steppe lowlands entirely) by c. 5600 BC, bringing with them domesticated cattle, sheep, and pigs; emmer, spelt, and einkorn wheats; barley; ceramic technology; hammered copper ornaments; and new house and settlement forms (TRINGHAM 1971: 96, MARKIEVICH 1974: 12–16). The forest-steppe region that they entered was densely occupied by a thriving, archaeologically well-documented, indigenous Late Mesolithic population (MARKIEVICH 1974, CHERNYSH 1974, MATSKEVOI 1977, TELEGIN 1966, 1982). It was arguably the existence of this population that brought the Criş pioneers to a halt west of the Prut River. They had entered one of the richest farming, fishing, and/or hunting environments in all of Europe but their settlements never extended east of the Prut River drainage.

The Late Mesolithic population that occupied the forest-steppe uplands east of the Prut soon adopted ceramic technology and selected aspects of the Criş Neolithic subsistence economy, giving rise to the first indigenous Neolithic culture of the region: the Bug-Dniester culture (Danilenko 1969, Markievich 1974, Tringham 1971: 47–50). For almost a full millennium (c. 5600–4700 BC) the Bug-Dniester culture acted as a sort of filter through which ceramic technology was diffused to a variety of Late Mesolithic groups living to the north and east, while Neolithic subsistence techniques were adopted only very slowly, if at all, beyond the South Bug Valley.

Between about 4700–4500 BC a complex process of cultural consolidation and integration resulted in the evolution of a widespread, unified cultural tradition in the Moldavian forest-steppe uplands: that of the Cucenteni-Tripolye culture (Marinescu-Bîlcu 1974, Ellis 1984). This event also marked the initial emergence of complex hierarchical societies in the region, a development marked by the appearance of craft specialization, settlement hierarchies, fortified settlements, specialized shrine structures, advanced metallurgy, and very large towns within the Cucuteni-Tripolye area. Settlements of the Cucuteni-Tripolye culture expanded to the north and east, entirely absorbing the old Bug-Dniester territory, and ultimately reached the valley of the Dnieper River. Here again a substantial hunter-gatherer population – represented archaeologically by period I of the Dnieper-Donets culture (Telegin 1968) – was encountered, and Cucuteni-Tripolye expansion halted. At about the same time (c. 4500 BC), the Dnieper-Donets culture adopted many aspects of the Cucuteni-Tripolye economy, signaling the beginning of the fully Neolithic phase (period II) of the Dnieper-Donets culture.

The Bug-Dniester and the Dnieper-Donets cultures were the first, and for millennia remained the only, indigenous North Pontic societies that adopted Neolithic subsistence techniques. Culture groups occupying the closed forest zone to the north and parts of the marine coast of the Black Sea to the south retained economies based largely or entirely upon hunting, fishing, fowling, and gathering until after 3800 BC. It was only in the forest-steppe zone that Neolithic economies were adopted earlier. It is likely that the arrival of immigrant farmers (Criş or Cucuteni-Tripolye) produced the kind of regional population increase that would have endangered the delicate balance of resources and territories in the vicinity of the steppe/forest-steppe boundary zone. A new balance was attained when the indigenous Mesolithic societies living nearest the immigrants adopted Neolithic subsistence techniques themselves, which had the result of intensifying subsistence production and increasing effective environmental carrying capacities within the forest-steppe and the steppe-zone river valleys.

5. The Dnieper-Donets Culture and the Horse

The domestication of the horse occurred within the social context of the Dnieper-Donets (Telegin 1968) and the Sredni Stog (Telegin 1973) cultures. Faunal data from Dnieper-Donets sites are rather thin. There are only four sites with published faunal material, and the largest sample (from Buz'ki) contains 310 identified bones (excluding molluscs). The low count probably results from a combination of relatively brief, low-intensity prehistoric occupations and relatively small modern excavation areas. Nevertheless the samples are adequate to provide a basis for a few generalizations.

In order to place the sub-Neolithic and Neolithic economic patterns of the Dnieper-Donets culture in perspective, we must first step back and briefly examine the Late Mesolithic (c. 7200–5500 BC). Wild horses were an important game animal for some indigenous Late Mesolithic groups in the North Pontic region. At Girzhevo in the northwestern Pontic steppes, the Late Mesolithic levels contained primarily wild horse and aurochs, and at related sites of the Grebenikov technological group, the bones of the wild horse, wild onager (*Equus hemionus*), and saiga antelope attest to the intensive exploitation of steppe resources (Stanko 1966, 1976). Girzhevo was located very near the forest-steppe/steppe boundary in a small river valley (the Kuckurgan) feeding into the Dniester estuary. The Late Mesolithic occupants of this region clearly regarded steppe resources as an integral part of their subsistence economy and had developed effective hunting techniques for exploiting those resources. After the evolution of the Cucuteni-Tripolye culture in the adjacent uplands, however, the hunter-gatherers of the northwest Pontic steppes seem to have been absorbed or to have left the area, for their sites do not appear to have been occupied into the Cucuteni-Tripolye period.

During the Late Mesolithic and the initial sub-Neolithic periods (Surskii, Dnieper-Donets I) in the Dnieper rapids region, wild horses were a resource of only incidental importance, a pattern that contrasts with that of the northwestern Pontic steppes. The faunal data from layer F at Igren 8 (Dobrovolskii 1949) and from Surskii occupations on Shulaev Island (Bodianskii 1949), as well as data from other Surskii sites in the Dnieper rapids region (Bibikova 1969), indicate that wild horses were almost always present, but never surpassed 3 percent of the minimum number of individuals represented. Prior to the wholesale economic changes of the Dnieper-Donets II period, the subsistence economies of the Dnieper rapids region depended on fishing, shellfishing, and fowling and on the hunting of aurochs, red deer, wild pig, beaver, and hare. This animal exploitation pattern appears to reflect a situation of relatively low population pressure and low resource stress. It is significant that steppe resources were only lightly utilized, in spite of the fact that the entire Dnieper rapids region was surrounded by steppes, and the river-valley gallery forest environment was here quite constricted (often less than 2 km wide).

During the Dnieper-Donets I period (ca. 5100–4600 BC) a new "Cro-Magnon" physical type and burial posture, a new lithic industry, and a new set of ceramic traditions appeared in the Dnieper rapids region, apparently as a result of a southward population movement down the Dnieper Valley from the fringes of the forest-steppe zone (Stoliar 1959, Telegin 1957). The faunal sample from layer D1 at Igren 8 (Table 1), representing the remains of this new Dnieper-Donets I population, nevertheless exhibits continuity with preceding Late Mesolithic and earlier sub-Neolithic patterns. Aurochs, beaver, red deer, roe deer, and wild pig dominated the list of hunted animals, indicating a well-forested catchment area rich in wild game. Horses were not hunted at this camp (Telegin 1968: 205).

During the Dnieper-Donets II period (ca. 4600–4100 BC) Neolithic subsistence techniques appear to have been widely adopted in the region. Domesticated animals are present at all three sites with published faunal data (Table 1) and are attested at two more coeval sites on the lower Don (Rakushechni Yar, Samsonova: Gei 1979). Cultivated barley (*Hordeum* sp.) is known from seed impressions at Vita Litovskaia (Telegin 1977). Flint sickle blades (some of which exhibit silica polish), disc-shaped hand-held grinding stones, and stone mortars were used for the first time in the preparation of cereal foods. On the basis of both the published

TABLE 1

DNIEPER DONETS ANIMAL EXPLOITATION PATTERNS
(Birds, turtles, shellfish, and dogs excluded)

	PHASE SITE	I IGREN 8				IIa SOBACHKI				IIb SREDNI STOG 1				II BUZ'KI				PERIOD II, AVERAGE			
		Ind.	% Ind.	Meat Weight	% M.W.	Ind.	% Ind.	Meat Weight	% M.W.	Ind.	% Ind.	Meat Weight	% M.W.	Ind.	% Ind.	Meat Weight	% M.W.	Ind.	% Ind.	Meat Weight	% M.W.
	Cattle, domestic	+?	?	?	?	5	11.9	1750	43.3	2	18.1	700	60.1	3	11.5	1050	53.4	3.3	12.6	1155	48.3
DOMI-	Ovicaprids, dom.					8	19.0	100	2.5	4	36.4	50	4.2	1	3.8	12.5	0.6	4.3	16.4	53.75	2.2
NANT	Pig, domestic					3	7.1	45	1.1	1	9.0	15	1.2	1	3.8	15	0.7	1.7	6.5	25.5	1.0
FOOD	**HORSE**					4	**9.5**	**1200**	**29.7**	**1**	**9.0**	**300**	**25.7**					**1.7**	**6.5**	**510**	**21.4**
ANIMALS	Red deer	3	14.2	285	10.2	3	7.1	285	7.0	1	9.0	95	18.1	3	11.5	285	14.5	2.3	8.8	218.5	9.1
	Roe deer	1	4.7	10.5	0.3									4	15.3	42	2.1	1.3	4.9	13.7	0.6
	Pig, wild	1	4.7	53.75	1.9	1	2.3	53.75	1.3					4	15.3	215	10.9	1.7	6.5	91.4	3.8
	Elk													1	3.8	300	15.2	0.3	1.1	100	4.1
	Aurochs	6	28.5	2400	85.9	1	2.3	400	9.9									0.3	1.1	133.3	5.6
	Onager					1	2.3	175	4.3									0.3	1.1	58.3	2.4
	Fish					12	28.6	24	0.5	2	18.7	4	0.3	1	3.8	2	0.1	5	19.0	10	0.4
	% of site fauna represented above		52.3		98.3		90.1		99.6		99.6		99.6		68.8		97.5		84.5		98.9
	Beaver	4	19.0	30	1.0									5	19.2	37.5	1.9	1.7	6.5	12.75	0.5
	Fox	2	9.5	9	0.3	1	2.3	4.5	0.1					1	3.8	4.5	0.2				
FUR-	Otter																	0.7	2.6	3.15	0.1
BEARING	Marten																				
ANIMALS	Wildcat					1	2.3	1.0	N									0.3	1.1	0.3	N
AND	Wolf																				
PESTS	Hare	4	19.0	4	0.1	2	4.7	2.0	N					2	7.7	2.0	0.1	1.3	4.9	1.3	N
	Wild/Domestic		% Ind.		% M.W.		% Ind.		% M.W.		% Ind.		% M.W.		% Ind.		% M.W.		% Ind.		% M.W.
	Animal Resources	Wild	100		100	Wild	62.0		53.1	Wild	36.5		34.5	Wild	80.9		45.3	Wild	64.5		48.3
	By Site	Dom.	?		?	Dom.	38.0		46.9	Dom.	63.5		65.5	Dom.	19.1		54.7	Dom.	35.5		51.5

Note: Ind. = Individuals ≈ MNI; Dom. = Domestic; M.W. = Meat Weight as calculated from MNI

faunal data and the character of the environment, Telegin (1968: 210) has suggested that stockbreeding (and presumably agriculture) would have been much more important in the steppe-zone river valleys of the Dnieper-Azov region than in the moister, more forested regions to the north where traditional forest game would have been more abundant. The adoption of stockbreeding and agriculture in the steppe-zone river valleys, however, would have accelerated deforestation through the establishment of garden plots and pastures and would have encouraged population growth as the labor requirements and increased sedentism associated with cereal cultivation and stockbreeding made it more advantageous to produce larger families (Blanton 1975, Handwerker 1983). Both of these developments would have placed strains on the surviving river-valley forest reserves and on the populations dependent on them.

In the Dnieper rapids region, these strains are evident in an apparent shift in animal exploitation patterns during period II of the Dnieper-Donets culture. Forest-adapted game (beaver, roe deer, wild pig) are virtually absent in the relevant samples (from Sobachki and Sredni Stog I). Wild horses appear to have become the dominant game animal, averaging over 9 percent of the individuals represented in and over 25 percent of the meat weight calculated for both samples. This change, although only tenuously documented, suggests a significant economic re-orientation when compared to the forest-oriented hunting patterns displayed consistently in all earlier samples from the region. Wild horses, previously almost ignored by river-valley hunters, now appear to have become a consistently sought quarry in a context where domesticated stock supplied most of the meat diet. Thus, it seems likely that hunting parties still remained within the familiar territory of the river valleys, picking off horses as they came down for water or shelter. Deep steppe animals – importantly saiga antelope – are not represented in the published samples. The apparent rise in the importance of horses, therefore, might be seen as an adjustment related to a reduction in forest-adapted game populations rather than as an indication of a conscious penetration of the steppes by period II hunters. The sample from Buz'ki, a settlement located in the moister forest-steppe zone 300 km north of the Dnieper rapids, reinforces this impression. There, forest game animals (roe deer, wild pig, beaver) were still abundant at this camp. Domesticated animals (cattle, pigs, ovicaprids) were kept but supplied a significantly smaller portion of the meat diet. Not a single horse bone was identified from the site.

Higher human population densities in the steppe river valleys might also be indicated by a series of cultural innovations that appeared only within the steppe and forest-steppe fringe groups of the Dnieper-Donets II culture. These innovations included a new mortuary ritual involving the secondary burial of skeletal remains in vertical layers within long trenches or pits. Single cemeteries such as Mariupol and Nikol'skoe now contained up to 161 individuals, many more than earlier recorded cemetetries, implying longer usage and perhaps increased sedentism. In addition, a few burials were accompanied by small gold rings; beads made of carnelian, jet, and copper; and drilled porphyry mace-heads. All of these objects were imported, and their concentration with only a few individuals implies the existence of at least moderate ranking. Increased sedentism and the development of ranking might both be associated with increased population densities..

In sum, it was in the constricted environment of the steppe-zone river valleys that deforestation and rising population densities placed the most serious strains on traditional subsistence practices, and brought the hitherto untapped resources of the surrounding steppes to the attention of a rapidly changing society.

6. The Sredni Stog Culture and the Horse

The Sredni Stog culture represents the society within which domesticated horses were first regularly exploited, according to the weight of current evidence (TELEGIN 1973, 1973a, 1986; BÖKÖNYI 1974; KUZMINA 1974; PIGGOTT 1983). The early phases of the Sredni Stog culture date to approximately 4200–3900 BC, and the latest phase extended to about 3500–3400 BC (TELEGIN 1973: 124–130). Diagnostic assemblages are documented at sites located between the Ingul River on the west and the middle Volga on the east, within the steppe zone and the fringes of the forest-steppe. Faunal collections have been published for five sites; these data are summarized in Table 2. Horses account for an average of 13.8 percent of the individuals represented at and 48.6 percent of the meat weight calculated for these sites. This is more than twice the average frequency documented in Dnieper-Donets II samples (Table 1). Even at Moliukhor Bugor, a well-watered forest-steppe site at the northern periphery of the Sredni Stog region and a site in which the presence of wild pigs, elk, bear, and wildcats attests to the forested nature of the site catchment area, horses still account for 23.2 percent of the meat weight calculated for the sample.

Many aspects of Sredni Stog material culture exhibit continuity with preceding Dnieper-Donets II patterns (lithics, settlement locations, cemetery locations, some aspects of ceramics). The earliest known Sredni Stog sites are located in the steppe-zone portion of the Dnieper Valley, in the heart of a Dnieper-Donets II settlement region. There is, then, substantial evidence for a local evolution of the Sredni Stog culture (ANTHONY 1985: 318–331). Other aspects of material culture, however, differ from Dnieper-Donets II patterns. The differences might reflect a variety of changes related to deforestation in the river valleys, the domestication of the horse, and the expansion of the resource base into the steppe environment.

Sredni Stog dwellings were large (ca. 12 × 5 m), three times larger than most Dnieper-Donets II dwellings, and contained two to three open hearths. Larger households have the potential for greater flexibility in dealing with very diverse or scattered economic resources that require simultaneous labor (WILK/RATHJE 1982: 632). The larger Sredni Stog residential units therefore might well have evolved in response to an expansion of the resource base into the steppe, and a consequent need to manage more dispersed resources. Sredni Stog cemetery populations, on the other hand, were quite small (three burials at Mariupol, compared to 124 Dnieper-Donets II burials at the same cemetery). This might reflect somewhat less stable settlements and a more mobile residential pattern, while the continued usage of the same cemetery location suggests an element of continuity. The new Sredni Stog mortuary ritual, involving primary single burial, might have been related to changes in household size, subsistence pattern, and the intensity of cemetery usage.

If horses were fairly rapidly converted to mounts, as seems likely, then many additional social and economic changes might have occurred, as I and others have described elsewhere (SHERRATT 1983, ANTHONY 1985, 1986). The evidence that Sredni Stog horses were ridden will be described separately.

The Sredni Stog site with the largest published faunal sample is the phase IIa site of Dereivka, probably occupied some time between about 4000 and 3700 BC. The Dereivka settlement was located on a low, semi-forested river terrace on the west bank of the Dnieper in the steppe/forest-steppe borderlands. Rounded bluffs rose steeply behind the site to a height some 200 feet above the valley floor. Excavations conducted sporadically between 1959 and 1967 revealed an area of 2200 sq.m, constituting most of a horizontal block approximately

TABLE 2

SREDNI STOG ANIMAL EXPLOITATION PATTERNS
(Birds, turtles, shellfish, and dogs excluded)

PHASE SITE	Ib SREDNI STOG 2				IIa DEREIVKA				IIa ALEKSANDRIIA				IIa MOLIUKHOR BUGOR				IIb KONSTANTINOVKA				AVERAGE			
	Ind.	% Ind.	Meat Weight	% M.W.	Ind.	% Ind.	Meat Weight	% M.W.	Ind.	% Ind.	Meat Weight	% M.W.	Ind.	% Ind.	Meat Weight	% M.W.	Ind.	% Ind.	Meat Weight	% M.W.	Ind.	% Ind.	Meat Weight	% M.W.
Cattle, domestic	2	11.7	700	45.8	18	8.3	6300	24.1	5	20	1750	42.9	3	9.0	1050	27.0	8	9.7	2800	40.0	7.2	9.7	2520	29.6
Ovicaprids, dom.	8	47.0	100	6.5	16	7.4	200	0.7	3	12	37.5	0.9					17	20.7	212.5	3.0	8.8	11.8	110	1.3
Pig, domestic	1	5.8	15	0.9	9	4.1	135	0.5					2	6.0	30	0.7	4	4.9	60	0.8	3.2	4.3	48	0.6
HORSE	**2**	**11.7**	**600**	**39.3**	**52**	**24.2**	**15600**	**59.7**	**6**	**24**	**1800**	**44.1**	**3**	**9.0**	**900**	**23.2**	**6**	**7.3**	**1800**	**25.7**	**13.8**	**18.5**	**4140**	**48.6**
Red deer	1	5.8	95	6.2	18	8.3	1710	6.5	2	8	190	4.6	5	15.1	475	12.2	12	14.6	1140	16.3	7.6	10.2	722	8.5
Roe deer					12	5.5	126	0.4	1	4	10.5	0.2	4	12.1	42	1.0					3.4	4.6	35.7	0.4
Pig, wild					11	5.1	591	2.3	5	20	269	6.6	6	18.1	322.5	8.3	7	8.5	376	5.3	5.8	7.8	311.75	3.7
Elk					3	1.3	900	3.4					3	9.0	900	23.2					1.2	1.6	360	4.2
Bear					2	0.9	300	1.1					1	3.0	150	3.8					0.6	0.8	90	1.0
Onager																	3	3.6	525	7.5	0.6	0.8	105	1.2
Fish					37	17.2	74	0.3									8	9.7	16	0.2	9.0	12.1	18	0.2
% of site fauna represented above		82.3		98.7		82.3		99.0		88		99.3		81.3		99.4		79		98.8		82.2		99.3
Beaver	2	11.7	15	0.9	15	6.9	112.5	0.4									3	3.6	22.5	0.3	4.0	5.4	30	0.3
Fox					7	3.2	315	0.1	3	12	13.5	0.5					3	3.6	13.5	0.2	2.6	3.5	11.7	0.1
Otter	1	5.8	1.5	0.1	2	0.9	3	N					1	3.0	1.5	N	1	1.2	1.5	N	1.0	1.3	1.5	N
Marten													1	3.0	0.5	N	3	3.6	1.5	N	0.8	1.1	0.4	N
Wildcat					4	1.8	4	N					3	9.0	3	N	1	1.2	1.0	N	1.6	2.1	1.6	N
Wolf					2	0.9	14	N					1	3.0	7	0.1	3	3.6	21	0.3	1.2	1.6	8.4	N
Hare					7	3.2	7	N									3	3.6	3	N	2.0	2.7	2.0	N
Wild/Domestic		% Ind.		% M.W.		% Ind.		% M.W.		% Ind.		% M.W.		% Ind.		% M.W.		% Ind.		% M.W.		% Ind.		% M.W.
Animal Resources	Wild	23.8		7.5	Wild	56.0		15.0	Wild	44.0		12.1	Wild	76.0		49.1	Wild	57.4		30.5	Wild	55.7		19.9
By Site	Dom.	76.2		92.5	Dom.	44.0		85.0	Dom.	56.0		87.9	Dom.	24.0		50.9	Dom.	42.6		69.5	Dom.	44.3		80.1

Note: Ind. = Individuals ≈ MNI; Dom. = Domestic; M.W. = Meat Weight as calculated from MNI

60 × 40 m in size (TELEGIN/ZHILAEVA 1964; TELEGIN 1968: 89–90, 1973: 29–59, 1986). A small associated cemetery containing 12 inhumations lies less than 20 m west of the settlement. An earlier Dnieper-Donets IIb ossuary-type cemetery and traces of a settlement lie farther to the west. The settlement area exposed and investigated at Dereivka is the largest exposure obtained for any excavated Sredni Stog settlement.[3]

Animal bones, mollusc shells, and broken or discarded tools formed a rich cultural deposit some 10 cm thick over much of the occupation area at Dereivka. Among the 3938 identified animal bones recovered from the site, 2555 (61 percent) came from horses, representing at least 52 individual animals and over 15,000 kg of usable meat or more than half of the meat estimated to have been consumed by the site occupants (see Table 2) (TELEGIN 1973: 132–133). The wide variety of artifact types and the sheer quantity of animal bones revealed at Dereivka make it highly unlikely that the site was a specialized horse-hunting camp. Cattle (18 individuals) and ovicaprids (16 individuals) were also consumed and a variety of game animals were hunted including red deer, roe deer, wild pig, elk, bear, beaver, otter, and wildcat. With portions of three preserved dwellings revealed, the Dereivka site actually exhibits a more substantial and permanent character than any other excavated Sredni Stog settlement. It also contained what might have been a ritual deposit of a horse head (with hide attached) and a series of pierced antler objects that have been interpreted as bit cheekpieces (TELEGIN 1973: 137, TELEGIN 1971).

The ritually deposited horse skull, that of a stallion 7–8 years old, was used as the basis for BIBIKOVA's metric analysis which purported to demonstrate that the animal was domesticated (BIBIKOVA 1967, 1969). Although her methodology and presentation of the data can be criticized (see below), her conclusion was probably correct. The predominance of males among the sexable skull fragments (15 of 17), the high number of juveniles (over 23 percent), and the complete absence of old individuals make for a population profile that strongly suggests a managed, domesticated herd (BIBIKOVA 1969: 64–65). Domestication is also suggested by the simple fact that no known earlier North Pontic community relied to such an extent on the exploitation of horses.

The economy of the Dereivka settlement probably included cereal cultivation, which is suggested by the recovery from the site of six polished schist mortars, three flat, ovoid grinding stones, and numerous silica-polished flint sickle blades (TELEGIN 1973: 69, 139). Net weights, bone fish hooks, and freshwater shellfish (*Unio* sp. and *Anadonta* sp.) attest to the exploitation of riverine resources.

The site's occupants cultivated grain and kept horses, cattle, sheep and a few domesticated pigs. Their hunting parties were able to obtain elk, bear, roe deer, and beaver. Fox, wildcat, and otter were perhaps trapped for their fur with the necessary skinning and butchering being performed at the settlement. Horses supplied most of the meat in the diet. Red deer were eaten as often as cattle and more often than sheep. Fish and shellfish were abundantly exploited. All of this implies a site catchment area rich in timber, forest resources, riverine resources, and pastures.

The ritual horse-head deposit at Dereivka was discovered in the eastern part of the site, over 12 m away from the nearest known dwelling structure but very near a large open-air hearth

[3] I regret that I have been unable to examine TELEGIN's 1986 English summary of the Dereivka excavations prior to submission of this paper. All descriptions contained herein are based upon his earlier site reports.

(Hearth VI) and a small storage or refuse pit containing numerous ceramic sherds (Pit 8). The hearth was surrounded by a dense scatter of ceramic sherds, animal bones, and (particularly on the east and south) mollusc shells. North of the hearth, between it and the storage pit, there was a row of stones which TELEGIN (1973: 44) interpreted as an intentional divider or low wall. Some 2 m northwest of the hearth and less than 1 m southwest of the pit was the "cultic" horse deposit.

TELEGIN does not describe or illustrate a burial pit. The deposit seems to have rested on the prehistoric ground surface. The horse skull, with mandible attached, was oriented towards Pit 8. On either side of the skull were the metacarpals and phalanges, laid out in anatomical order. This signifies that the deposit originally consisted of the horse head with the hide and front hoofs attached. The skull was that of a stallion 7–8 years old which was, according to BIBIKOVA's reconstruction (1969: 64) based on the length of the metacarpals, 144 cm high at the withers. This made him the largest horse documented at the site. The head-and-hoofs deposit was associated with the remains of two dogs. TELEGIN (1973: 45) says that the dogs lay "... partially beneath the horse skull and a little to the side." The superimposition of the horse over the dogs does not seem to have been sharply defined, such as one would encounter in a stratified burial deposit or in a pit. The dogs were largely anatomically complete with all of the vertebrae of one being present and in anatomical order.

Near the dog bones, in the narrow space between the anterior end of the horse skull and Pit 8, were 2 ceramic figurine fragments and a curved "cheek-piece" made of an antler tine. In this area there was also another perforated antler tine which TELEGIN (1973: 47) has classed as a mattock but which might represent a second cheekpiece. Two additional figurine fragments were found nearby, to the south of the deposit. One figurine represented a pig, and the others appear to have been fragments of cone-headed anthropomorphic figures with straight joined legs, perhaps modelled after similar late Tripolye female figurines. TELEGIN (1973: 45) considers the hearth, the pit, the row of stones, the horse and dogs, and the figurines to represent the remains of a single "cultic complex."

The deposit near Hearth VI and Pit 8 was the only such deposit that has been clearly documented at the site. However, BIBIKOVA (1969: 63) notes that only 18 complete horse metacarpals were found at Dereivka, and that all of these were found in small groups of 2 or 3 near large skull fragments. Furthermore, 17 of the 18 were *left* metacarpals, implying that only the left leg was retained. Unfortunately neither she nor TELEGIN indicates where in the settlement these additional assemblages were found. This does suggest that other such head-and-hoof deposits or offerings might have once been mounted around the settlement.

Numerous Late Copper/Early Bronze Age head-and-hoof animal offerings have been documented in the region, but the animals involved are sheep or cattle in most cases, all other examples are later, and the context is almost always mortuary (PIGGOTT 1962, SMIRNOV 1961, MERPERT 1974: 103–105). Horse skulls and metacarpals do occur in another Sredni Stog-related context, but in a mortuary setting, in graves at S'ezzhee and Khvalynsk on the middle Volga (VASILIEV/MATEEVA 1976) as well as in later Yamna graves in the same region (MERPERT 1974: 104). The Dereivka head-and-hoof assemblage is the earliest and the only one of its kind recognized in a settlement context. It might have consisted of multiple head-and-hoof groupings.[4] The performance of a horse-centered ritual, the association of the horse with

[4] Conversations with Dr. TELEGIN in 1989 established that there is little possibility of other horse head-

domesticated dogs, and the association with the antler "cheek-pieces" are all highly suggestive of a domesticated status, whatever the precise nature of the ritual might have been.

The horse remains of Dereivka reveal several important trends. Age-sex profiles suggest a managed population. The overwhelming reliance on horse exploitation is, by itself, highly significant, for it documents a reorientation of subsistence strategies towards steppe resources, namely horses. This would have produced an immediate widening of the resource base in the steppe-zone river valleys, and it constituted a critical first step in the opening of the steppe environment to efficient human exploitation. This aspect of the Dereivka subsistence system is confirmed and repeated at every other Sredni Stog site with published faunal data. Moreover, the evidence from Dereivka, from Moliukhor Bugor, and to a lesser extent from Konstantinovka establishes that horses were now, for the first time, relied upon as a significant food resource even in environments where traditional riverine and gallery forest resources where abundantly available. Specialized hunting of wild horses in such environments is an unlikely explanation for their dietary importance, as other more traditional game was available and was in fact still actively exploited, drawing hunters' energies into forest and riverine environments that hosted few wild horse herds.

7. The Evidence for Riding[5]

I have argued elsewhere that the initial penetration and occupation of the deep steppes by large, stable human societies was made possible by new forms of transport (ANTHONY 1986). Traditionally, intensive sheep herding has been interpreted as the key that opened the steppes (SHILOV 1975, MERPERT 1982: 327, KHAZANOV 1984). The Yamna, or pitgrave horizon, c. 3500–2400 BC, is recognized as representing the earliest systematic occupation of the

and-hoof groupings having been present at Dereivka. The mandible-metacarpal groupings hinted at in BIBIKOVA (1969) might have been butchering or other types of assemblages.

[5] Riding is now firmly documented at Dereivka. The "cult stallion" at Dereivka exhibits clear and unambiguous microscopic bit wear on his lower second premolars. Bitting at this date (ca. 4000 BC, well before the acquisition of wheeled vehicles) indicates riding. Bit wear has been studied and defined by the author and D. BROWN, on the basis of modern metal bit-worn specimens and control specimens from feral horse populations in Nevada and Maryland. The pilot study (ANTHONY/BROWN in press) identified Kaftari-phase (Ur III) bit wear on an equid from Malyan, Iran, the earliest documented usage of a bit in the Near East. In 1989 we studied a series of horse teeth in Kiev, including those from Dereivka, at the invitation of the Institute of Zoology and the Academy of Sciences of the Ukrainian S. S. R. Paleolithic, Mesolithic, and Neolithic horse teeth, including teeth from Cucuteni-Tripolye and Seroglazivka (North Caspian Neolithic) contexts, do not exhibit bit wear, nor do the other horse teeth from Dereivka. However, the "cult stallion" exhibits very clear bit wear. Research continues on the question of the type of bit that caused the Dereivka wear pattern, and on other aspects of bit wear. We are grateful to Dr. Prof. Natalya BELAN and Dr. Prof. Dimitri TELEGIN for their advice and encouragement in Kiev, and to the Ukrainian Academy of Sciences for its generosity. The bit wear project has been supported by the American Philosophical Society, the Wenner-Gren Foundation, and Hartwick College Trustee Research Grants, with cooperation from the University of Pennsylvania, Cornell University, and West Chester University. The formal report on the Dereivka results is in preparation (12/89).

steppes. It is clear that sheep were numerically predominant in the animal herds of the earliest eastern Yamna (pitgrave) horizon societies, but a glance at Tables 1 and 2 will reveal that sheep had been an important North Pontic resource for millennia prior to the emergence of the Yamna horizon. It was only after sheep herding was combined with new forms of transport (horses and ox-carts) that the steppe environment was opened to efficient human exploitation. The reason for this lies in the distribution of steppe resources.

The steppes contained many localized areas that were rich in resources, but some of these shifted locations frequently (animal herds, good pastures), and most, including the river valleys, were separated by wide expanses of resource-poor territory. Scouting was required to locate the more mobile resources, and transport was required to carry any stored resources from one rich location to the next. The means of transport directly determined the amount of material that could be carried and thereby placed a limit on the accumulation of any surplus. Pedestrian societies were not able to transport significant quantities of surplus or stored resources, including such basic commodities as tents. They, also, were not able to carry more than a few of the very old, very young, or infirm, nor were they capable of the kind of rapid, long distance scouting that riding provided. Horse transport removed all of these limitations during the Sredni Stog period, providing the basis for a limited penetration of the steppes. The acquisition of wheeled vehicles, importantly the ox-cart, provided the final impetus (high-volume transport), permitting the systematic, regular exploitation of the steppes during the Yamna period, after about 3500 BC. A fully loaded ox travels more slowly than a man walking, however, so the scouting and transport capabilities of the horse (low-volume, high-speed transport) remained essential. It was the ox-cart and the horse, combined with intensive sheep herding, that opened the steppes.

The early evidence for horse riding can be divided into four categories: cheekpieces, representations of horses showing apparent harness trappings, new cultural patterns, and inherent probability. Let us examine the last category first.

Within the Sredni Stog culture horses were maintained as a food resource for over 600 years. Mounting a horse requires very little imagination and no new technology. A fairly docile mare might be ridden successfully by any boy with a modicum of determination. Riding a horse was a qualitatively different experience from riding a cow, which was the only other domesticated North Pontic animal big enough to mount. Native Americans, who were in most cases self-taught riders, realized the advantages of riding and learned to ride expertly within decades after first acquiring horses. Their horses were no larger than Sredni Stog horses (12–14 hands, or 120–140 cm at the withers: Ewers 1955: 33). We might conservatively allow the Sredni Stog culture several centuries for a similar period of experimentation. It seems extremely unlikely that horses were kept for over six centuries without ever being ridden. The question is not, in fact, whether a Sredni Stog horse was ever ridden (this seems almost certain) but rather whether riding became an accepted innovation in transport. Here we might note as an analogy that no Native American society on either American continent exhibited any significant resistance to this innovation.

Riding is also suggested by the appearance of objects thought to be cheekpieces within Sredni Stog material culture. A variety of antler, wood, and metal objects have been advanced as putative early cheekpieces (Danilenko/Shmagli 1972, Munchaev 1975: 390, Cherniakov/Shmagli 1983), but none are as convincing as the Sredni Stog specimens. Sredni Stog cheekpieces were made on the end of an antler tine, pierced with one or rarely two holes. The perforations have saucer-shaped worn areas around them, apparently from the movement of cords. They occasionally occur in pairs, as in Grave 18 at Aleksandriia (TELEGIN 1973: 137, 1959), and perhaps at Dereivka. At the latter site at least one, perhaps two, such object(s) occurred in association with the stallion skull in the ritual deposit near Pit 8. Similar objects occur slightly later in pairs in rich late Funnel Beaker culture graves in the DDR (LICHARDUS 1980). Later still, during the Late Yamna (pitgrave) period (c. 2800–2400 BC), similar objects appeared in Afanasievo tumulus graves in the Central Asian Altai Mountains near the headwaters of the Yenisei (KOZHIN 1970). By this time artifacts of this type were appearing in Southeast Europe in mortuary contexts that establish their direct association with horses (HÜTTEL 1977, 1981). The earliest Sredni Stog specimens were perforated and manufactured in much the same manner as these more familiar Early Bronze Age Carpathian examples. They were deposited in pairs, were cord-worn at the perforation, and occur at Dereivka in association with the skull of a domesticated horse. Their identification as cheekpieces is as firm as any such identification can be.

Some sort of early horse harnessing system is also suggested by the appearance of stone horse-head effigy maces, or "sceptres", made of exotic stones, in a variety of circum-Pontic and middle Volga archaeological contexts during the late Sredni Stog and succeeding Early Yamna periods (ANTHONY 1986, GIMBUTAS 1977, DANILENKO/SHMAGLI 1972). These objects appear to portray, in relief, halter straps attached to equine heads. The portrayal of horses on stone mace heads that were presumably symbols of authority and prestige was in itself indicative of a special relationship between horses and power. The earliest Sredni Stog examples may date to about 3800 BC.

Finally there is the argument that the development of horses as mounts should have caused predictable changes in North Pontic society and economy. Just such changes are, in fact, documented archaeologically in the appearance of the Sredni Stog culture and the Yamna horizon (ANTHONY 1986). These changes included increased household size, individual-centered mortuary rituals, increased social differentiation, intensified warfare, territorial expansion, and increased exchange in exotic prestige goods.

The evidence for riding is circumstantial, but it is cumulatively convincing. Riding provides a unified explanation for a great many observed changes that must otherwise be interpreted piecemeal. Until a better interpretive model becomes available, it seems safe to accept the hypothesis that Sredni Stog people rode horses.

8. The Osteological Evidence for Domestication[6]

V. I. BIBIKOVA (1967) conducted the analysis that is usually referred to as providing firm osteological evidence for the domesticated status of Sredni Stog horses. In her analysis she selected one skull and mandible from a collection of 2555 identifiable horse bones excavated at Dereivka (TELEGIN 1973: 28–59). The skull was that of a stallion, 7–8 years old (specimen 44–1192, Institute of Zoology AN). It had been damaged by soil pressure, so that the left side

[6] The osteometric analyses of BIBIKOVA were considerably enhanced by those of NOBIS (1971), who conducted his own osteometric analyses of the Dereivka material. More recently, the entire sample of Dereivka horse bones has been reanalyzed by Marsha LEVINE of Cambridge University, who visited Kiev in 1988.

of the cranial box, the nasal-frontal contact area, and parts of the premaxillary and maxillary bones were missing. Nevertheless, Bibikova felt that reconstruction was successful and that measurements could be taken with accuracy. She compared the specimen with skulls from known domesticated horses from Yamna (pitgrave) burials, c. 2500 BC; from Scythian burials, c. 500 BC; from modern Kirgiz herds; and with skulls from modern Przewalski and tarpan horses. She also made some less formal comparisons to fossil Upper Paleolithic and Middle Paleolithic populations (see, here, Table 3 and Bibikova 1967).

The results of the cranial comparison were not conclusive, although Bibikova (1967, 1969) felt that they strongly indicated that the specimen was domesticated. The comparison involved 20 cranial measurements, as shown in Table 3. For seven of these dimensions the Dereivka skull measured larger than any comparative specimen consulted (including the categories of overall and basal length, skull height, and width of the nasals). The Dereivka skull was, in fact, pro-

TABLE 3

Comparative measurements of horse skulls
(from Bibikova 1967)

Measurements, mm*	Dereivka skull	Domesticated: Bronze Age tumuli, 6 ind.		Domesticated: Modern Kirgiz 2 ind.	Wild: Tarpan, Tavrik ind.	Wild: Tarpan, Kherson ind.	Wild: Przewalskii, 5 ind.	
		range	mean				range	mean
Basal length	500	462–472	466.2	476; 488	468	ca. 470	474–495	485
Overall length	546	520–525	523	520; 533	520	–	535–545	539
Prosthion to posterior edge of palate	ca. 270	–	–	–	255	ca. 260	252–269	263
Posterior edge of palate to basion	ca. 225	215–222	217	220; 230	213	210	219–225	223
Prosthion to most posterior edge of eye orbit-frontal eye line	390	362–372	368.5	372; 370	363	370	375–391	383
Acrocranion to most anterior edge of eye orbit-posterior eye line	198	192–200	197	195; 208	198	195	196–204	199
Anatomical facial axis	ca. 386	370–370	370	364; 370	356	363	367–381	376
Anatomical cranial vault axis	182	175–181	179	180; 190	175	171	175–188	181
Length of diastema	108	83–107	97.2	102; 104	–	–	–	–
Length of tooth row, by alveolar cavities	170	149–175	163	167; 161	171	153	182–193	185.5
Width between outer edges of incisors	76	66– 73	70	65; 72	67	71	70– 73	72
Width at narrowest point of diastema	53	45.5–52	50	42; 46	–	–	–	–
Muzzle width at anterior point of the facial ridge	ca. 160	157–161	158.4	169; 162	157	158	162–165	164
Width on a horizontal plane between the anterior edges of eye orbits	ca. 155	153–160	156.2	–; 162	183	181	186–192	189
Maximum width of skull at temporals	ca. 215	200–208	204.2	211; 220	202	205	203–212	207
Maximum width of skull at occiput	ca. 120	113–122	121.6	122; 121	104	97	109–117	112.5
Minimum width of occiput	90	82– 90	85	81; 89	77	82	86– 93	89
Height of skull at lowest point	ca. 136	116–128	124.4	120; 126	122	122	125–132	128
Maximum length of nasal bones	ca. 220	206–224	210	212; 232	239	–	253–262	258
Maximum width of nasal bones	ca. 120	112–117	116	–; –	108	–	108–113	–

* *Editors' note: Measurement descriptions are direct translation of* Bibikova*'s language by* Anthony *and are generally more accurate than those found in* Telegin *(1986: 138) in which there are a number of errors.*

bably atypically large even within its own population, for the stallion's stature (as indicated by the metacarpals that were found directly associated with the skull) was estimated at 144 cm, making it the tallest single animal documented at Dereivka. The sample utilized in the comparison therefore consisted of a single atypical animal, and it is thus impossible to draw valid conclusions concerning the population from which it came. Bibikova felt, however, that several traits were strongly suggestive of a domesticated status. These were a large, but lightly built cranium; a very long nasal portion or snout; a short tooth row; and a wide occiput (BIBIKOVA 1967: 113).

Bibikova's analyses also included age and sex classifications (1969) and measurements of the metacarpals and phalanges of the Dereivka horse population (1970). The age and sex data have not been presented in complete detail, but they are nonetheless interesting. Of the 17 sexable large cranial and mandibular fragments identified at Dereivka, 15 were males. Moreover, 23 percent of the ageable bone fragments examined by BIBIKOVA came from juveniles, and there were absolutely no old individuals in the sample. The great majority were young adults or juveniles, especially juvenile males. It is this aspect of the Dereivka data that is particularly suggestive of domestication, for juvenile males are precisely the individuals one would first select for slaughter in a managed herd, while they are not the individuals one would be most likely to obtain through the predation of wild populations.

BIBIKOVA's descriptions of the metacarpal and phalanx dimensions of the Dereivka horses are presented in Tables 4–7. These data are from a 1970 study that attempted to determine whether the Dereivka horses were more similar to Przewalski horses or tarpans. The samples used were inadequate for such a purpose, but the measurements are valuable for other comparisons. BÖKÖNYI has observed (1974: 238) that the extreme variability in the dimensions of these bones is suggestive of a domesticated population, and it does appear that the Dereivka horses are more variable in this respect than the sample of modern wild Przewalskis.

The horses of Dereivka present the sex and age characteristics of a domesticated population culled for slaughter. Those who have directly examined the bones (BIBIKOVA 1967, BÖKÖNYI 1974: 238) find them to conform in several traits to the proportions of domesticated horses, although these observations are far from definitive. The fact that, in the site sample as a whole, horses represented 24 percent of the individuals represented, 60 percent of the calculated meat weight, and 71 percent of the domesticates (if counted as domesticates) is by itself indicative of, at the very least, a specialized exploitation pattern.

TABLE 4

Comparative measurements of Metacarpals
(from BIBIKOVA 1970)

Measurements, mm	Dereivka, N=21		Tarpan, N=1	Przewalski, N=8	
	range	mean		range	mean
Length	206.0–234.5	221.05	207.5	216.0–235.0	225.0
Breadth of proximal end	47.0– 56.0	52.86	47.0	47.5– 52.5	50.0
Breadth of distal end	47.0– 55.5	51.18	47.0	45.0– 49.0	47.0
Breadth of diaphysis	30.5– 39.0	35.81	33.0	30.0– 33.0	32.0

TABLE 5

Comparative measurements of first Phalanges
(from Bibikova 1970)

Measurements, mm	Dereivka, N=25/43*		Tarpan,	Przewalski, N=8	
	range	mean	N=1	range	mean
Length	79.0–92.5	86.2	74.5	79.5–88.0	83.0
	76.5–91.0	83.8	71.0	74.0–84.0	79.0
Breadth of	51.0–62.5	55.9	52.7	48.0–53.5	50.0
proximal end	51.0–62.5	57.2	53.7	49.5–54.0	52.0
Breadth of distal end	42.0–49.0	45.3	42.0	41.0–44.0	42.5
	41.0–48.0	44.9	41.5	40.0–43.0	41.5
Breadth of diaphysis	34.0–39.0	36.6	33.8	32.0–35.0	34.0
	32.0–40.0	36.2	34.0	30.6–35.0	33.0

* Numerator = number of anterior specimens
Denominator = number of posterior specimens

TABLE 6

Comparative measurements of second Phalanges
(from Bibikova 1970)

Measurements, mm	Dereivka, N=23/29*		Tarpan,	Przewalski, N=8	
	range	mean	N=1	range	mean
Length	44.5–52.0	47.9	37.5	42.0–47.0	44.2
	47.0–53.0	50.2	38.5	43.0–48.0	45.5
Breadth of	48.5–58.5	54.6	52.3	47.0–52.5	49.0
proximal end	48.5–59.0	53.6	52.5	47.5–53.0	49.6
Breadth of distal end	46.0–53.5	50.6	49.3	48.0–52.0	49.0
	45.0–51.0	48.4	46.7	45.0–48.0	46.0
Breadth of diaphysis	42.5–50.0	46.7	47.3	42.0–45.0	43.0
	42.0–48.5	44.5	43.3	44.0–42.5	41.0

* Numerator = number of anterior specimens
Denominator = number of posterior specimens

TABLE 7

Comparative measurements of third Phalanges
(from Bibkova 1970)

Measurements, mm	Dereivka, N=26/27* range	Dereivka, N=26/27* mean	Tarpan, N=1	Przewalski, N=8 range	Przewalski, N=8 mean
Length to anterior edge	51.0–61.0 / 51.0–63.5	55.9 / 57.3	52.5 / 54.3	48.0–54.0 / 51.0–59.0	51.0 / 55.0
Height	38.0–45.0 / 40.0–48.0	41.8 / 43.1	36.8 / 38.3	/ 39.0–41.0	40.0 / 40.0
Width	75.0–91.0 / 68.0–87.0	82.7 / 77.5	72.5 / 70.0	71.5–78.0 / 69.0–73.0	74.0 / 71.0

* Numerator = number of anterior specimens
Denominator = number of posterior specimens

9. The Domestication Process

Domestication required, as a first step, the capture of healthy, live animals. The capture and control of a healthy wild horse was not a particularly easy task for a man on foot and could have been accomplished in only a limited number of ways. Juveniles or colts might theoretically have been somewhat easier to capture than adults, but for the protective instincts of stallions. A mare will hang back to defend a straggling colt, and the stallion often will hang back to protect his mare. If the stallion perceives that the entire band is being exposed to danger because of the stragglers, he may leap forward and kill the colt by breaking its neck, then drive the mare back into the retreating band (Ryden 1978: 143). Such tactics would have added to the substantial difficulties involved in capturing a live animal. It is highly unlikely that any human group would have set about purposefully to capture and domesticate wild horses. The domestication process must have occurred as a side-effect associated with some other exploitative pattern.

The only technique that could have resulted in the capture and then in the unpremeditated maintenance of live animals was the cooperative drive. Not coincidentally, the drive was also the optimal hunting technique for the systematic pursuit of wild horses. Cooperative drives were likely to have been employed first because they offered the opportunity for a very substantial "kill" and also because they were the technique best suited to take advantage of wild horses' natural shyness, suspiciousness, and tendency to "spook." Once a horse band or an allied group of bands had been driven into the prepared trap or corral, the hunters might well have found themselves in control of many more animals than could immediately be consumed or could even be killed and processed (dried, smoked, etc.) for storage. The best way to store the extra animals would have been to leave them alive in the trap and perhaps even feed them for a period of time. Hunters who were experienced at herding and raising cattle, sheep, and pigs would very soon have realized the possibilities offered by their new captives. A horse kept in a corral was, after all, a much easier and more convenient source of food than a horse running free on the steppes.

The domestication of the horse was therefore almost certainly the result of the employment of cooperative drive hunting techniques by a community that was already familiar with herding and stockbreeding. Domestication might actually have occurred very soon after cooperative drives were first employed.

Cattle herders would have made the transition to horse keepers most easily, for cattle are also large herbivores, and certain aspects of cattle behavior are paralleled among horses. For example, the cows in a cattle herd, like the mares in a horse herd, arrange themselves in a distinct social hierarchy headed by a lead cow (hence the common name "Bossy") whose parallel among horses is the lead mare. The lead cow and the lead mare choose the daily route and lead the herd, which follows almost blindly. The clever cowherd, like the bull or the stallion, knows that if he controls that single cow or mare he controls the movement of the entire herd. Such parallels in behavior might have made the transition to tending and controlling horses much easier for a community already familiar with cattle herding. The Dnieper-Azov (or Nikolskoe-Mariupol) group of the Dnieper-Donets II culture, a highly distinctive steppe river valley sub-group described above, was just such a society. Cattle were the single most important source of meat in their diet, accounting for an average of 51.7 percent of the estimated amount of meat consumed and 15 percent of the individual animals slaughtered in the faunal samples from Sredni Stog (Level 1) and Sobachki (see Table 1).

The maintenance of newly captured, or later of newly bred, horses should not have required substantial resources, an important consideration for the resource-poor peripheral communities likely to have first experimented with systematic horse hunting. Steppe horses probably consumed approximately 30 pounds of vegetation daily, mostly in the form of grasses, herbs and shrubs (Hafez/Williams/Wierzbowski 1962). If they were removed from the restrictive environment of a corral and were instead restrained only with leg hobbles they could easily have foraged for their own food. Although they would have naturally tended to migrate to wooded valley bottoms during the winter, horses did not need extensive forest resources and could survive without them. In 1816, the North American explorer Ross Cox described the life of Indian horses in the American northwest (Cox 1957 [1816]: 244):

> In summer they have no shelter from the heat,
> in winter no retreat from the cold, and their
> only provender throughout the year is the wild
> loose grass of the prairies, which in the
> latter season is generally covered with
> snow ...

The cold-weather adaptability of horses might have played a role in their development as a managed food resource. Horses can break through iced-over water holes using their tough hoofs; their high gait allows them to move through snows that would paralyze sheep; and they find forage in the winter by scraping away the snow with their hoofs. Both cattle and sheep use their noses to push through snow, and in deep or prolonged snows their noses will become so raw and bloody that they will starve unless given fodder (Ryden 1978: 163–164). In snowy conditions, horses are therefore more likely to survive than cattle or sheep, and they require less winter care.

Given these behaviors, the climatic change known as the "Piora oscillation" might have played an important role in promoting horse domestication. The Piora oscillation was a brief but well-documented, and therefore probably severe, climatic cooling episode which occurred sometime during the 4200–3800 BC time-range and which lasted perhaps a century (FRENZEL 1967: 103–106). During this period the summer growing season was shortened and winter frosts lasted significantly longer, with the combined effect causing a drop of 2° C in the mean annual temperature. More recent studies support the occurence of the oscillation (OSBORNE 1975, JANOSSY/KORDOS 1976). During this event horses would have been better adapted to the increased cold than either cattle or sheep. The Piora oscillation might well have played a contributing role in the acceptance of the horse as a new domesticate by those stronger communities that otherwise might not have invested their time and energy in such a new venture (e.g., the occupants of Dereivka). By the time more moderate climatic conditions returned, the horse already might have been established as an integral part of the resource base, a part that required less maintenance than cattle or sheep during the critical winter months.

10. Conclusion

The domestication of the horse was one outcome of an interplay of social, economic, ecological, and climatic factors that can be documented for the steppe-zone river valleys of the North Pontic region. Population growth, economic intensification, and an elaboration of status positions were accompanied by deforestation within the circumscribed environments of the steppe river valleys. The resultant stresses established the necessary preconditions for an expansion of the resource base through the increased utilization of steppe resources, namely horses. The initial use of horses was as a managed, cold-adapted source of food, but there is persuasive circumstantial evidence for riding by the late Sredni Stog period, about 3800–3500 BC. After about 3500 BC, when wheeled vehicles were acquired in the North Pontic steppes, the combination of horses, ox-carts, and sheep herding made possible the intensive exploitation and occupation of the vast Central Asian grassland environment. Large, stable human societies moved rapidly into the steppes, resulting in the widespread appearance of the Yamna, or pit-grave horizon.

Beyond the steppes, the exploitation of horses became widespread in the Carpathian Basin during the Corded Ware period, the equivalent of the Early Yamna. Horse remains are widespread in the Early Bronze Age Kura-Araxes sites of the Caucasus at about the same time (3500–3000 BC). Prior to this period in the Caucasus, horses were attested in only a few sites in the lower Kura River valley. Horses were not widely adopted in the Near East until a full millennium later (c. 2300 BC), but by 2800–2700 BC Bell Beaker societies in Europe were probably engaged in an active commerce in domesticated horses (SHERRATT 1983), and the societies of the North Pontic steppes were drawn into a larger European world perhaps partially through such exchanges.

The use of domesticated horses resulted in great cultural transformations both within the North Pontic region (ANTHONY 1986) and in the world beyond (SHERRATT 1983). The origin and beginning of these changes has been traced to the North Pontic region through an analysis that places the domestication process within a clear ecological, economic, and social context. It is only within such a broad context that the domestication of the horse or of any other animal can be understood.

11. Acknowledgements

The research that led to this paper began with the encouragement of Dr. Dexter Perkins, Jr. and Patricia Daly, to whose spirits I send the warmest regards. It was continued and brought to fruition through the guidance of Prof. Bernard Wailes and Prof. Ward Goodenough at the University of Pennsylvania to whom I acknowledge an enormous debt. Finally, this paper was substantially affected in its final form by the helpful comments of Dr. Richard Meadow and several anonymous but incisive reviewers. All errors and omissions are, of course, my own.

12. Literature

Anthony, D., 1985, The Social and Economic Implications of the Domestication of the Horse; Dissertation, University of Pennsylvania, Philadelphia.

Anthony, D., 1986, The "Kurgan Culture," Indo-European origins, and the domestication of the horse: a reconsideration. Current Anthropology **27**(4), 291–313.

Anthony, D./Brown, D., in press, Looking a gift horse in the mouth: bit wear and the origins of riding. In: Crabtree, P./Campana, D./Ryan, K., eds., Animal Domestication and Its Cultural Context: Essays in Honor of Dexter Perkins and Pat Daly; Philadelphia: University of Pennsylvania M.A.S.C.A. Publications.

Bahn, P.G., 1980, Crib-biting: tethered horses in the Palaeolithic? World Archaeology **12**(2), 212–217.

Ball, J.W., 1977, The rise of the northern Maya chiefdoms: a socioprocessual analysis. In: Adams, R.E.W., ed., The Origins of Maya Civilization; Albuquerque: University of New Mexico Press, 101–132.

Baskett, J., 1980, The Horse in Art; Boston: New York Graphic Society.

Belanovskaia, T.D., 1977, Poselenie Rakushechni Yar. Arkheologicheskie Otkrytiia (goda 1976), 87.

Berger, J., 1986, Wild Horses of the Great Basin: Social Competition and Population Size; Chicago: University of Chicago Press.

Bibikova, V.I., 1967, K izucheniu drevneishkh domashnikh loshadei Vostochnoi Evropy (Studies of ancient domestic horses in Eastern Europe). Bulleten Moskovskogo Obshchestva Ispitatelei Prirody, Otdel Biologicheskii **72**, 106–118.

Bibikova, V.I., 1969, Do istorii domestikacii konia na Pivdennom Skhode Evropy. Arkheologiia **22**, 55–67.

Bibikova, V.I., 1970, K izucheniu drevneishkh domashnikh loshadei Vostochnoi Evropy (Studies of ancient domestic horses in Eastern Europe) **2**. Bulleten Moskovskogo Obshchestva Ispitatelei Prirody, Otdel Biologicheskii **75**, 118–125.

Blanton, R.E., 1975, The cybernetic analysis of human population growth. American Antiquity **40**(2), 116–126. (Memoir **30**).

Bodianskii, O.V., 1949, Neolitichna stoianka na Ostrovi Shulaevomu. Arkheologichni Pam'iatki URSR (Kiev) **2**, 253–263.

Bökönyi, S., 1974, History of Domestic Animals in Central and Eastern Europe; Budapest: Akadémii Kiadó.

Bökönyi, S., 1974a, The Przewalsky Horse; London: Souvenir Press.

Bökönyi, S., 1980, The importance of horse domestication in economy and transport. In: Sørbom, P., ed., Transport Technology and Social Change; (Tekniska Museet Symposia **2**); Stockholm, 15–21.

Carneiro, R.L., 1967, On the relationship between size of population and complexity of social organization. Southwestern Journal of Anthropology **23**, 234–243.

CARNEIRO, R. L., 1970, A theory of the origin of the state. Science **169,** 733–738.

CHERNIAKOV, I. T./SHMAGLI, N. M., 1983, Derevyani psalii yamnoi kultury. Arkheologiia (Kiev) **42,** 10–16.

CHERNYSH, A. P., 1974, Mezolitichne naselennia podnistrov'ia. Starodavne Naselennia Prikarpattia I Volini; Kiev: Naukova Dumka, 75–88.

CLARK, G., 1941, Horses and battle-axes. Antiquity **15**(57), 50–70.

COX, R., 1957 (1816), The Columbia River. STEWART, E./STEWART, J., eds., Norman: University of Oklahoma Press.

DANILENKO, V. M., 1969, Neolit Ukraini; Kiev: Naukova Dumka.

DANILENKO, V. M./SHMAGLI, M. M., 1972, Pro odin povorotnii moment v istorii eneolitichnogo naselennia Pivdennoi Evropi. Arkheologiia (Kiev) **6,** 3–20.

DOBROVOLSKII, A. V., 1949, Vosma Igrenska neolitichna stoianka. Arkheologichni Pamiatki URSR (Kiev) **2,** 243–251.

DOLUKHANOV, P. M., 1986, Foragers and farmers in west Central Asia. In: ZVELEBIL, M., ed., Hunters in Transition: Mesolithic Societies of Temperate Central Asia and Their Transition to Farming; Cambridge: Cambridge University Press, 121–132.

ELLIS, L., 1984, The Cucuteni-Tripolye Culture; (BAR International Series **217**); Oxford: British Archaeological Reports.

EWERS, J. C., 1955, The Horse in Blackfoot Indian Culture; (Bureau of American Ethnology, Bulletin **159**); Washington, D. C.: Smithsonian Institution.

FALKNER, T., 1935 (1774), A Description of Patagonia and the Adjoining Parts of South America; Chicago: Armann and Armann Co.

FRENZEL, B., 1967, Climatic change in the Atlantic/sub-Boreal transition on the Northern Hemisphere: botanical evidence. In: Proceedings of the International Symposium on World Climate 8000–0 BC; London: Royal Meteorological Society, 99–123.

GEI, A. N., 1979, Samsonovskoe mnogosloinoe poselenie na Donu. Sovietskaia Arkheologiia **3,** 119–131.

GIMBUTAS, M., 1956, The Prehistory of Eastern Europe, Part I; (American School of Prehistoric Research, Bulletin **20**); Cambridge, Mass.: Peabody Museum, Harvard University.

GIMBUTAS, M., 1963, The Indo-Europeans: archeological problems. American Anthropologist **65,** 815–836.

GIMBUTAS, M., 1970, Proto-Indo-European culture: The Kurgan Culture. In: CARDONA, B./HOENIGSWALD, H. M./SENN, A., eds., Indo-European and the Indo-Europeans; Philadelphia: University of Pennsylvania Press, 155–197.

GIMBUTAS, M., 1977, The first wave of Eurasian steppe pastoralists into Copper Age Europe. The Journal of Indo-European Studies **5**(4), 277–337.

GLUMAC, P./ANTHONY, D., in press, Culture and environment in the prehistoric Caucasus, Neolithic to Early Bronze Age. In: EHRICH, R., ed., Chronologies in Old World Archaeology; Chicago: University of Chicago Press, (3rd edition).

GROVES, C. P., 1974, Horses, Asses, and Zebras in the Wild; London/Hollywood (FL): David and Charles/Curtis Books.

HAFEZ, E. S./WILLIAMS, M./WIERZBOWSKI, S., 1962, The behavior of horses. In: HAFEZ, E. S., ed., Behavior of Domestic Animals; Baltimore: Williams and Wilkins Co., 370–396.

HANCAR, F., 1956, Das Pferd in Prähistorischer Zeit und Früher Historischer Zeit. In: Beiträge zur Kulturgeschichte und Linguistik **11**; Wien: Verlag Herold.

HANDWERKER, W. P., 1983, The first demographic transition: an analysis of subsistence choices and reproductive consequences. American Anthropologist **85**(1), 5–27.

HÜTTEL, H., 1977, Altbronzezeitliche Pferdetrensen. Jahresberichte Institut für Vorgeschichte Universität Frankfurt a. M., 65–86.

HÜTTEL, H., 1981, Bronzezeitliche Trensen in Mittel- und Osteuropa. Prähistorische Bronzefunde **16**(2).

JANOSSY, D./KORDOS, L., 1976, Pleistocene-Holocene mollusc and vertebrate fauna of two caves in Hungary. Annales Historico-Naturales Musei Nationalis Hungariei (Budapest) **68**, 5–29.

JARMAN, M.R./WILKINSON, P.F., 1972, Criteria of animal domestication. In: HIGGS, E.S., ed., Papers in Economic Prehistory; Cambridge: Cambridge University Press, 83–96.

KHAZANOV, A.M., 1984, Nomads and the Outside World; Cambridge: Cambridge University Press.

KLEIN, J./LERMAN, J.C./DAMON, P.E./RALPH, E.K., 1982, Calibration of radiocarbon dates: tables based on the consensus data of the workshop on Calibrating the Radiocarbon Time Scale. Radiocarbon **24**(2), 103–150.

KOZHIN, P.M., 1970, O psaliakh iz afanas'evskikh mogil. Sovietskaia Arkheologiia **4**, 189–193.

KUZMINA, E.E., 1974, Kolesnyi transport i problema etnicheskoi i sotsial'noi istorii drevnego naseleniia Iuzhnorusskikh stepei. Vestnik Drevnei Istorii **4**, 68–87.

LICHARDUS, J., 1980, Zur Funktion der Geweihspitzen des Typus Ostorf: Überlegungen zu einer vorbronzezeitlichen Pferdeschirrung. Germania **58**, 1–24.

LYDEKKER, R., 1912, The Horse and Its Relations; London: George Allen and Co.

MALLORY, J.P., 1986, Editor's addendum. In: TELEGIN, D., Dereivka: A Settlement an Cemetery of Copper Age Horse Keepers on the Middle Dnieper; (BAR International Series **287**); Oxford: British Archaeological Reports, 180–182.

MARINESCU-BÎLCU, S., 1974, Cultura Precucuteni Pe Teritoriul Romaniei; Bucharest: Academia de Ştiinţe Sociale şi Politice, Institutul de Arkheologie.

MARKIEVICH, V.I., 1974, Bugo-Dnestrovskaia Kultura na Territorii Moldavii; Kishinev: Akademiia Nauk.

MATIUSHIN, G., 1986, The Mesolithic and Neolithic in the southern Urals and Central Asia. In: ZVELEBIL, M., ed., Hunters in Transition: Mesolithic Societies of Temperate Eurasia and Their Transition to Farming; Cambridge: Cambridge University Press, 133–150.

MATSKEVOI, L.G., 1977, Nekotorye osobennosti mezolita predkarpat'ia. In: Paleoekologiia Drevnego Cheloveka: 10th Congress of INQUA; Moscow: Akademiia Nauk, 123–129.

MERPERT, N.I., 1974, Drevneishie Skotovody Volzhsko-Ural'skogo Mezhdurech'ia; Moscow: Akademiia Nauk.

MERPERT, N.I., 1982, Eneolit Iuga SSSR i Evraziiskie stepi. In: Eneolit SSSR; Moscow: Akademiia Nauk, 322–331.

MOHR, E., 1971, The Asiatic Wild Horse; London: Allen and Co.

MUNCHAEV, R.M., 1975, Kavkaz na Zare Bronzovogo Veka; Moskow: Akademiia Nauk.

NOBIS, G., 1971, Vom Wildpferd zum Hauspferd; (Fundamenta Reihe B., Band **6**); Köln, Wien: Böhlau-Verlag.

NOBIS, G., 1973, Zur Frage römerzeitlicher Hauspferde in Zentraleuropa. Zeitschrift für Säugetierkunde **38**, 224–252.

OSBORNE, P.J., 1975, Evidence from the insects of climatic variation during the Flandrian period. World Archaeology **8**(2), 150–158.

PETRENKO, A.G., 1984, Drevnee i srednevekovoe zhivotnovodstvo srednego Povolzh'ia i Predural'ia; Moscow: Academiia Nauk.

PIGGOTT, S., 1962, Heads and hoofs. Antiquity **36**, 110–118.

PIGGOTT, S., 1983, The Earliest Wheeled Transport; Ithaca: Cornell University Press.

PLUMMER JR. T.R./HANNE, W.G./BRUNER, E.F./THUDIUM, C.C., 1971, Landscape Atlas of the USSR; West Point (NY): U.S. Military Academy.

RIDGEWAY, W., 1905, Origin and Influence of the Thoroughbred Horse; Cambridge: Cambridge University Press.

RYDEN, H., 1972, Mustangs: A Return to the Wild; New York: Penguin Books.

RYDEN, H., 1978, Americas' Last Wild Horses; New York: E.P. Dutton.

SHERRATT, A., 1983, The secondary exploitation of animals in the Old World. World Archaeology **15**(1), 90–104.

SHILOV, V. P., 1975, Modeli skotovodcheskikh khoziaistv stepnykh oblastei Evrazii v epokhu eneolita i rannego bronzovogo veka. Sovietskaia Arkheologiia **1**, 5–16.

SMIRNOV, K. F., 1961, Arkheologicheskie dannye o drevnikh vsadniakh Povolzhosko-Ural'skikh stepei. Sovietskaia Arkheoligiia **3**, 1–14.

SMITH, C. H., 1841, Horses; (The Naturalist's Library **12**); Edinburgh: W. H. Lizars.

SMUTS, G. L., 1976, Population characteristics of Burchell's zebra (*Equus burchelli antiquorum*) in the Kruger National Park. South African Journal of Wildlife Research **6**, 99–112.

STANKO, V. N., 1966, Mezoliticheskaia stoianka Girzhevo v Odesskoi oblast. Sovietskaia Arkheologiia **2**, 96–103.

STANKO, V. N., 1976, Periodizatsiia pamiatnikov mezolita severnogo prichernomor'ia. Materialy po Arkheologii Severnogo Prichernomor'ia **8**, 15–21.

STOLIAR, A. D., 1959, Pervyi Vasil'evskii mezoliticheskii mogilnik. Arkheologicheskii Sbornik Gosudarstvennyi Ermitazh. Epoka Kamnia **1**, 78–165.

TELEGIN, D. IA., 1957, Tretii Vasil'evskii mogil'nik. Kratkie Soobshcheniia Instituta Arkheologii (Kiev) **7**, 9–12.

TELEGIN, D. IA., 1966, Mezolit Levoberezhnoi Ukrainy i ego mesto v slozhenii Dnepro-Donetskoi neoliticheskoi kul'tury. Materiali i Issledovaniia po Arkheologii SSR (Moscow-Leningrad) **126**.

TELEGIN, D. IA., 1968, Dnipro-Donets'ka Kul'tura; Kiev: Naukova Dumka.

TELEGIN, D. IA., 1973, Seredno-Stogivska Kul'tura Epokhi Midi; Kiev: Naukova Dumka.

TELEGIN, D. IA., 1973 a, Über einen der ältesten Pferdezuchtherde in Europe. In: Actes du VIIIe Congrès International des Sciences Préhistoriques et Protohistoriques **2**; Beograd, 324–327.

TELEGIN, D. IA., 1977, Pro stanovlennia skotarstva i zemlerobstva na pivdennomu zakhodi Evropeiskoi chastini SRSR. Arkheologiia (Kiev) **21**, 17–26.

TELEGIN, D. IA., 1982, Pidsumki ta zavdannia vivchennia dobi mezolitu i neolitu na Ukraini. Arkheologiia (Kiev) 40, 34–48.

TELEGIN, D. IA., 1986, Dereivka: A Settlement and Cemetery of Copper Age Horse-Keepers on the Middle Dnieper; (BAR International Series **287**); Oxford: British Archaeological Reports.

TELEGIN, D. IA./ZHILAEVA, S. I., 1964, Dereivs'kii neolitichnii mogil'nik. Arkheologiia (Kiev) **16**, 44–59.

TRINGHAM, R., 1971, Hunters, Fishers, and Farmers of Eastern Europe. 6000–3000 BC; London: Hutchinson.

VASILIEV, I. B./MATEEVA, G. I., 1976, Poselenie i mogilnik u s. S'ezzhee. In: Ocherki Istorii i Kultury Povolzhya; Kuibyshev, 76–96.

VASILIEV, I. B./VYBORNOV, A. A./MORGUNOVA, N. L., 1985, Review of G. N. MATIUSHIN, Eneolit Iuzhnogo Urala, 1982. Sovietskaia Arkheologiia **2**, 280–289.

VINOGRADOV, A. V., 1981, Drevnie okhotniki i rybolovy Sredni Aziatskogo mezhdurechya; (Trudy Khorezmskoi Arkheologo-Etnograficheskoi Ekspeditsii **13**); Moscow: Akademiia Nauk.

WILK, R. R./RATHJE, W. L., 1982, Household Archaeology. American Behavioral Scientist **25**(6).

ZAIBERT, V. F., 1985, Poselenie Botai i zadachi issledovanii eneolita severnogo Kazakhstana. In: Eneolit i Bronzovyi Vek Uralo-Irtyshskogo Mezhdurech'ia; Cheliabinsk: Cheliabinskii Gosudarstvennyi Universitet, 3–17.

DAVID ANTHONY, Department of Anthropology, Hartwick College, Oneonta, NY, 13820, USA.

JÉRÔME DIVE AND VÉRA EISENMANN

Identification and Discrimination of First Phalanges from Pleistocene and Modern *Equus,* Wild and Domestic

Contents

1. Summary

Three aspects of the first phalanges of fossil and modern *Equus* were studied including 1) anteroposterior dimorphism and discrimination, 2) interspecific differences and identification, 3) tentative morphofunctional interpretation.

The analyzed sample comprises more than 500 first phalanx specimens (measured by V.E.). Fossil species include one Villafranchian *E.stenonis* (Spain: La Puebla de Valverde) and four caballine forms ranging from the Middle to the Upper Pleistocene (France: La Caune de l'Arago, Lunel-Viel, Jaurens, and Solutré). The modern sample includes the three extant species of zebras, wild and domestic asses, various forms of half-asses, and the wild horse *E.przewalskii.*

Biometrical data were elaborated by J.D., using various statistical methods (correspondence analysis, principal components analysis, ascending hierarchic classifications with the chi-square or the Euclidian metrics, and discriminant analysis).

In most cases, the discrimination between anterior and posterior phalanges is possible, both "at first glance" and by biometrical techniques. Anterior phalanges are more developed in their proximal part than are the posterior phalanges. This is expressed biometrically by the longer supra-tuberosital segments and the longer *trigonum phalangis* of the anterior phalanges. In a few cases, evidence for a moderate amount of sexual dimorphism was found, large phalanges being more frequently male.

In *E.stenonis,* the anterior phalanges have a morphology rather similar to that of the posterior phalanges of modern *Equus*: long infra-tuberosital segments and short *trigonum phalangis.* The posterior phalanges of *E.stenonis* have the same characters but they are even more marked. The phalanges of modern *Equus* exhibit few interspecific differences other than in size and gracility. But using these characters, one can hope to achieve a correct specific identification provided that the sample is good enough.

In several cases, the characters distinguishing anterior and posterior phalanges were found to also distinguish taxonomic groups: fossil versus modern forms, wild versus domesticated animals, one subspecies versus another. An hypothesis is proposed relating these characters to the distribution of weight on the front and the hind limbs. The morphofunctional mechanism, however, is still unclear and the hypothesis has yet to be tested.

2. Introduction

In many fossil samples, first phalanges are well preserved and numerous. Even when samples are poor, a few first phalanges are usually present. Accordingly they deserve some attention in order to be used to best advantage.

Phalanges differ in size and in shape not only from one species to another but also from the front limb to the hind limb. In some cases, posterior phalanges belonging to one species can be mistaken for the anterior of another species. There are therefore two problems, often connected, these being to distinguish fore and hind phalanges of a given species and to distinguish fore and hind phalanges of different species.

From previous work devoted to the first of these problems has emerged a pragmatic way of sorting "at first glance" anterior from posterior first phalanges. Such techniques are the better for being tested, and that was our initial aim, this constituting the first part of this paper. In addition, whenever the material was suitable, we looked for the presence of sexual dimorphism. As the study progressed, we came to analyze more specimens from more species and began to look into the second problem. Thus, interspecific differences make up the second part of this paper. In carrying out our analyses there emerged some unexpected facts, possibly related to evolutionary trends. These, however, need more time and work to be completely elaborated and so presently they will only be introduced in a short third part of this paper.

Our sample includes material from all modern *Equus* species (although inequally represented) and a few fossil forms. The total amounts to more than 500 specimens. Owing to the fact that our methodology calls for 14 or 15 measurements for each bone, the very bulk of the data has required statistical treatment including not only calculation of means and standard deviations but also multivariate analyses (Benzécri et al. 1973). Comments by both archaeologists and biostatisticians have led us to try to explicate, at least a little, the more elaborate statistical methods we have used. We have thus devoted a few paragraphs to that in the "Methods" section. We, however, do not feel that it is necessary to again return to the matter of ratio diagrams (Simpson 1941) nor to the biometrical method (Fig. 1) explained in previously published papers (Eisenmann 1986, Eisenmann/de Giuli 1974). Individual measurements which may be useful for archaeozoologists and paleontologists can be found in the appendix.

Computations were carried out at the "Centre Informatique du Muséum National d'Histoire Naturelle" and at the "Centre Inter-Régional de Calcul Electronique" with funds from U.A. 184 (CNRS). Multivariate analyses have been performed with A.D.D.A.D. software (JAMBU/ LEBEAUX 1983).

3. Methods

Multivariate analyses are convenient whenever it is necessary to synthesize data provided by a large number of specimens described by numerous measurements (variables). The data are treated so as to form a cloud of points in multidimensional space, there being as many points and as many dimensions as there are specimens and variables. The cloud is projected on appropriate planes that are defined by factorial axes. Factorial axes and planes are chosen to show as much as possible of the information ("variance" or "inertia") contained within the cloud. In practice, the first two or three factorial axes express most of the useful information. The interpretation takes into account apparent relations between specimens and variables. Roughly speaking, when specimens appear grouped on a plane, it means that they are similar; when variables are grouped, it means they are correlated; when some specimens are grouped with some variables, it means that these variables best "characterize" those specimens.
The appearance of the relations between specimens and variables depends upon the mathematical formulae ("metrics") by which distances between points are calculated within the multidimensional cloud. Two metrics are commonly used to calculate, respectively, the "Euclidian" and "chi-square" distances. If x_{ij} is the value for the jth variable of the ith specimen in a sample of "n" specimens:

the mean $m_j = 1/n \sum_i x_{ij}$;

the standard deviation $\sigma_j = \left(1/n \sum_i (m_j - x_{ij})^2\right)^{\frac{1}{2}}$;

$x_{i.} = \sum_j x_{ij}$;

$x_{.j} = \sum_i x_{ij}$;

the Euclidian distance $d^2_{i,i'} = \sum_j \left((x_{ij} - x_{i'j})/\sigma_j\right)^2$;

the chi-square distance $d^2_{i,i'} = \sum_j 1/x_{.j}(x_{ij}/x_{i.} - x_{i'j}/x_{i'.})^2$.

Multivariate analyses can be based on either of these metrics but will give somewhat different results. Principal components analysis uses Euclidian distance and takes into account, first, the absolute size of the specimens. Correspondence analysis uses the chi-square distance and is concerned not with size but with relative proportions (shape); with this method factorial axes are the same for variables and specimens (R and Q method) which is of great interest for interpretation.
In addition to the general ways in which specimens are related to one another, it may be interesting to find out if separate groups appear. That defines the goal of "automatic" classifications. The first step is to calculate the distances between the members of each possible pair of

specimens. Afterward, the ascending hierarchic classifications (used in this study) proceed by aggregating the most similar specimens (where the distance is the smallest) into one class that is thereafter considered as a single specimen and then aggregated again with the next most similar specimen or class. Naturally, the distances between specimens can be calculated either by the Euclidian or the chi-square metrics. There are, moreover, different ways of calculating distances between a single specimen and a class composed of several specimens, as well as between one class and another. We used the criterion of the second order central moment of a partition (close to Ward's method, JAMBU/LEBEAUX 1983).

When factorial analyses show that differences exist between two groups of specimens (for instance, anterior and posterior phalanges), discriminant analysis helps to more precisely explain those differences. This type of analysis considers two "specimens": the average anterior phalanx and the average posterior phalanx and a double set of variables: means of the 14 variables measured on the anterior sample and means of the 14 variables measured on the posterior sample. Once again, distances between the specimens and the variables can be calculated in different ways; here we have used the chi-square metric. Naturally, the variables that discriminate best will be those that take the most different values in the two average specimens. A final point should be noted. Discriminant analysis makes use of the whole existing body of material, even incomplete specimens, because it employs means to represent theoretical "average" specimens. In the other factorial analyses, the specimens are real and therefore must be complete in order to be used.

4. Material

4.1. Primitive *Equus*

Equus stenonis cf. *vireti*: The site of La Puebla de Valverde, situated in northeastern Spain in the Teruel basin, has yielded a rich Middle Villafranchian fauna (HEINTZ 1978). A primitive *Equus* is represented by numerous skeletal and cranial remains (EISENMANN 1979, 1980, 1981). The phalanx sample, already studied by EISENMANN/DE GIULI (1974), comprises 85 specimens, 41 of which were considered as anterior and 44 as posterior. Only half of them are well enough preserved to permit the complete set of measurements to be taken so that the factorial analyses were carried out on 19 supposed-anterior and 21 supposed-posterior phalanges.

4.2. Zebras

Equus grevyi: After discarding 4 juvenile, 2 abnormal, and 2 incomplete specimens, the sample comprises 26 anterior and 25 posterior phalanges from 26 skeletons, The origin of this material is Kenya (8), Ethiopia (2), zoological parks (12), and unknown (4). There are 11 males, 13 females, and 2 individuals of unknown sex.

Equus burchelli boehmi: This sample is basically the one studied by EISENMANN/DE GIULI (1974) but several young specimens were removed and a few others added. All are from Kenya. There are 19 anterior and 19 posterior phalanges belonging to 10 males, 8 females, and 1 individual of unknown sex.

Equus zebra: The whole sample comprises 50 phalanges but all measurements are available for only 21 anterior and 23 posterior specimens. The true mountain zebra, *Equus zebra zebra,* is probably represented by 7 skeletons while 16 probably belong to the subspecies *E.zebra hartmannae.* There are 13 females and 10 males.

4.3. Asses

These are quite a problem. Samples are small, identification and origin are sometimes uncertain and sex unknown. For example, one supposedly Nubian Wild Ass (Munich 1952-9) is very small. Another (BM 1904-6-12-1) looks very much like a large domestic donkey (the White Egyptian Ass, AC 1983-634). As it is, our samples are of four sorts:

1. subfossil domestic donkeys from Ikrit, Israel, dating to the Middle Bronze Age (Davis, *in litt.*): 15 phalanges sorted out as 10 anterior and 5 posterior. They are not perfectly preserved so that analyses were run, with 8 variabls only, on 9 "anterior" and 4 "posterior" phalanges;
2. modern supposedly domestic donkeys: 13 skeletons (6 female, 4 male, 3 of unknown sex) with 13 anterior and 12 posterior phalanges;
3. supposed "Somali wild asses": 5 skeletons but only 4 anterior and 4 posterior phalanges are able to be analyzed;
4. supposed "Nubian wild asses": 3 anterior and 3 posterior phalanges.

4.4. Half-Asses

E.hemionus hemippus (Syria): 3 skeletons but only 2 anterior and 2 posterior phalanges with the complete set of measurements;
E.hemionus onager (Iran): 15 skeletons but only 14 anterior and 14 posterior phalanges with the complete set of measurements; 10 males, 3 females, and 1 individual of unknown sex;
E.hemionus kulan (Turkmenia): 6 skeletons but complete sets of measurements for only 3 anterior and 3 posterior phalanges;
E.hemionus khur (India): 3 skeletons with the complete sets of measurements;
E.hemionus hemionus or *luteus* (Mongolia): 6 skeletons but with complete sets of measurements for only 5 of them;
E.kiang (Tibet): 6 skeletons with complete sets of measurements;
E.hemionus or *E.kiang* (origin unknown): 8 skeletons with the complete sets of measurements.

4.5. Horses

4.5.1. Fossil material

La Caune de l'Arago: This is a cave situated near Tautavel (Pyrénées-Orientales, France). It has yielded pre-Neandertal human remains as well as a rich lithic industry and fauna (H. de Lumley/M.-A. de Lumley 1971, H. de Lumley 1979, Renault-Miskowsky 1980). There was some controversy about whether it is Mindel (Crégut 1980 a, Guérin 1980) or Riss (Chaline 1971, 1981) in age, although now an age of 400–500 kya is generally accepted (Bouchez *et al.* 1984).

In the upper levels were found numerous remains of a caballine horse (MOIGNE 1983), first referred to a subspecies of *E.mosbachensis* (CRÉGUT 1980b) and later to a different, short-muzzled, horse resembling *E.chosaricus* (EISENMANN *et al.* 1985). The horse remains include 34 adult, more or less well-preserved, first phalanges that were sorted out as 13 anterior and 21 posterior. A very small number of phalanges (3 supposedly anterior and 9 supposedly posterior) were complete. We were therefore obliged to work with a restricted set of variables and in the end, analyses were run on a sample of 29 (11 supposedly anterior and 18 supposedly posterior) phalanges with 8 measurements only. Even so, 13 values (among the $29 \times 8 = 232$) were missing and had to be replaced by the corresponding means.

Lunel-Viel: These caves in Hérault, France, have yielded a rich and well-preserved fauna as well as lithic and bone industries. They are believed to have been filled during the Mindel-Riss Interglacial (M.-F. BONIFAY/E. BONIFAY 1965). The horse remains have been ascribed to a new subspecies of the Mindelian Mosbach species: *E.mosbachensis palustris* (M.-F. BONIFAY 1980), but it has recently been pointed out (EISENMANN *et al.* 1985) that the skull does not have the long muzzle characteristic of *E.mosbachensis* and looks more like that of the medium-muzzled *E.przewalskii.* Among the horse fossils were 34 phalanges, 23 of which are complete enough to permit multivariate analyses; 10 were considered as anterior, 13 as posterior.

Jaurens: This site, situated south of Brive, Corrèze, France, contained a rich Upper Pleistocene vertebrate fauna dated to about 30,000 years B.P. The horse remains were studied by MOURER-CHAUVIRÉ (1980) who ascribed them to *E.caballus* cf. *gallicus,* a form intermediate between the older *E.caballus germanicus* and the somewhat younger *E.caballus gallicus* from Solutré (PRAT 1968). These remains included 63 first phalanges, 51 of which were suitable for multivariate analyses; 25 have been sorted out by us "at first glance" as anterior and 26 as posterior.

Solutré: The well known Upper Pleistocene site of Solutré, Saône-et-Loire, France (COMBIER 1955), has yielded a very rich collection of horse bones. All of it was supposed to come from the Gravetian breccia dated to about 25,000 years B.P. and was described as *E.caballus gallicus* by PRAT (1968). Recent excavations have shown that older (about 30,000 years old) and younger (about 12,000 years old) horses were also present at Solutré (COMBIER/THÉVENOT 1976). The phalanges studied in the present paper belong to old collections stored in Lyon and form a quite homogeneous sample of 33 bones; they probably come from the Gravetian breccia (COMBIER, personal communication); 30 are complete and were sorted out at "first glance" as 17 anterior and 13 posterior.

4.5.2. *E.przewalskii*

Although relatively rich (23 skeletons, 14 of which are male and 9 female), the sample is far from satisfactory. Nearly all animals were reared in zoological gardens. Different breeding lines (VOLF 1960–1985) are represented, some of which are known not to be pure. As a result, this sample is one of the least homogeneous, and the heterogeneity is even present among those animals that are supposed to be "true" Przewalski horses.

5. Discrimination between anterior and posterior phalanges

5.1. Introduction

PRAT (1957) was probably the first author to devote an entire paper to the practical problem of discriminating between anterior and posterior phalanges of *Equus*. Although the material used is not clearly specified, PRAT was dealing with the domestic horse (*E.caballus*). He stated that first anterior phalanges are not always longer than posterior ones but that the latter are less broad in the middle and relatively deeper proximally. PRAT also pointed to some qualitative differences in the shape of the proximal end, it being more asymmetric in first posterior phalanges, and moreover suggested differences in the second and third phalanges. Nearly 20 years later, EISENMANN/DE GIULI (1974) approached the same problem using other species (*E.burchelli boehmi,* a plains zebra and *E.stenonis* cf. *vireti,* a Villafranchian form) and restricting themselves to the first phalanges for which they introduced a system of 13 measurements per bone. They found that, on anterior phalanges, the *trigonum phalangis* (CAMP/SMITH 1942) is longer and the supra-articular tuberosities are situated closer to the disal end. They confirmed the relatively greater proximal depth noted by PRAT for the posterior phalanges.
In 1981, a Canadian student, GENIA KEDNEY, applied the EISENMANN/DE GIULI biometrical system to a sample of 15 *Equus caballus* and used discriminant function analyses (probably with a Euclidian metric) to distinguish the 30 anterior and 30 posterior phalanges. Judging from the unpublished manuscript, her results are in agreement with what had been pointed by PRAT (1957) and EISENMANN/DE GIULI (1974) with two differences. In her caballine sample, the supra-articular tuberosities of the anterior phalanges are not situated closer to the distal end but farther away from the proximal end and the supra-articular breadth is wider on anterior phalanges. We think it useful to return to the same question, using the same system of measurements (with an additional 14th measurement: the distal supra-articular breadth and, in some cases, a 15th: the distal articular depth), a richer and more varied body of material, and more sophisticated methods: multifactorial correspondence and principal components analyses.
We will begin by comparing a "first glance" sorting of fossil phalanges (La Caune de l'Arago) with the results of multivariate analyses on the same sample. We then use the same methods on a modern sample of *E.grevyi* for which the position of the phalanges are known for sure.

5.2. La Caune de l'Arago

5.2.1. Introduction

Some 34 adult, more or less well-preserved first phalanges were sorted out as 13 anterior and 21 posterior. The main criterion for the sorting was the more "waisted" aspect of the posterior phalanges. In the modern Burchell's zebra studied by EISENMANN/DE GIULI (1974) and many other equids, posterior phalanges seem narrower at the "waist" (smaller breadth) in contrast to the proximal and distal breadths (the latter, at the supra-articular tuberosities). Part of the impression may be caused by the relative shortness of the bone. Moreover, on these posterior phalanges, the supra-articular tuberosities are not only more developed but also situated more proximally, enhancing the contrast between these "hips" and the "waist". After confronting

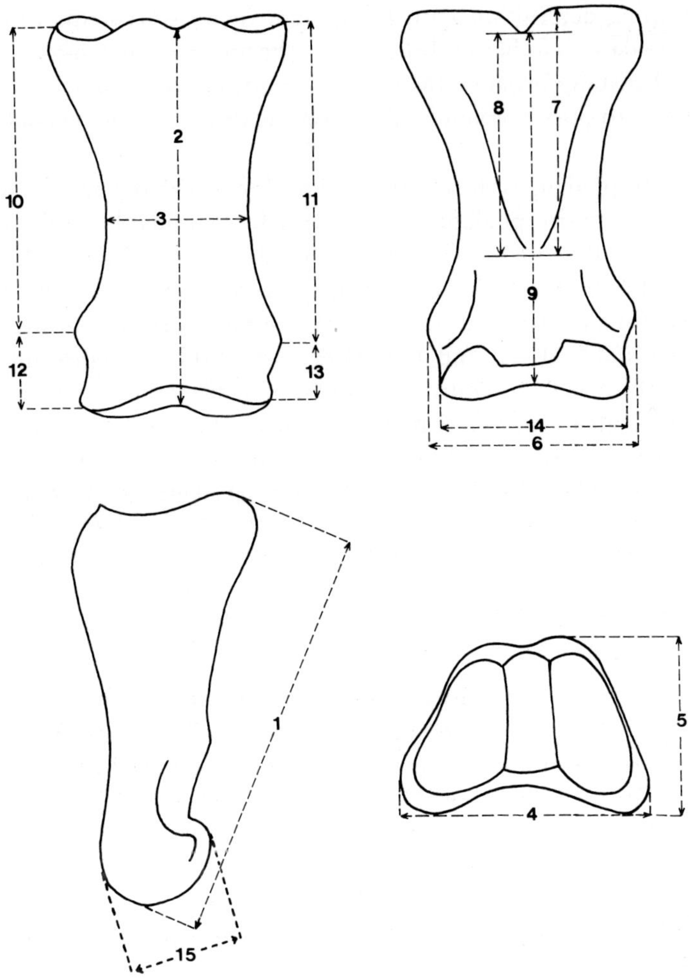

Fig. 1 System of measurements for the first phalanges (Ph 1):

1. Greatest length
2. Anterior length
3. Smallest breadth
4. Proximal breadth
5. Proximal depth
6. Distal supra-articular breadth
7. Greatest length of *trigonum phalangis*
8. Smallest length of *trigonum phalangis*
9. Posterior length
10. Medial supra-tuberosital length
11. Lateral supra-tuberosital length
12. Medial infra-tuberosital length
13. Lateral infra-tuberosital length
14. Distal articular breadth
15. Distal articular depth. This measurement has been added at the suggestion of an international symposium in New York, 1981. We have not yet enough data to evaluate its significance.

this anthropomorphic description with the system of measurements of Figure 1, one could expect posterior phalanges to be: 1) shorter than the anterior phalanges (Measures 1,2, and 9); 2) proximally and distally broader (Measures 4 and 6); 3) possibly narrower in the middle (Measure 3); and 4) to posses smaller values for Measures 10 and 11 and/or larger values for Measures 12 and 13.

After measuring the phalanges, histograms were drawn for the whole sample (pooling supposed anterior and posterior phalanges). The only histograms that seem clearly unimodal are those for Measures 6, 13, and 14. Most seem bimodal (2, 3, 4, 5, 9, 10, 12) or unclear. Coefficients of variation (v = 100 × standard deviation/mean) are usually less than 4 although they are notably larger for the proximal depth (Measure 5) and for the measures related to supra-articular tuberosities (10, 11, 12, and 13). After sorting the collection into anterior and posterior elements, samples were, of course, reduced and histograms as well as coefficients of variation are more difficult to interpret. It is noteworthy, however, that the coefficients of variation drop markedly for precisely those measures for which they were anormally high: 5, 10, 11, 12, and 13. This should mean that the dimorphism between anterior and posterior phalanges is expressed by these measurements.

5.2.2. The analyses

The next step was to see if the "first glance" sorting would be confirmed by multivariate analyses and for what characters. Let us already note that, in general, it was confirmed and that most characters were the same as the ones used in our "first glance" identification. Due to imperfect preservation, several measurements were missing for several phalanges. Accordingly, the analyses were run for 8 measures (2, 3, 4, 5, 7, 10, 12, 14) of 29 phalanges.

Correspondence analyses (Figure 2): The first axis, F(actor) 1, accounts for 67 percent of the total inertia (variance) and opposes variables 5 and 12 to variable 10. On the side of variables 5 (proximal depth) and 12 (internal infra-tuberosital length) are placed phalanges that have been identified as posterior. On the side of variable 10 (internal supratuberosital length) are the phalanges identified as anterior. It can be said that (supposed) anterior phalanges are more

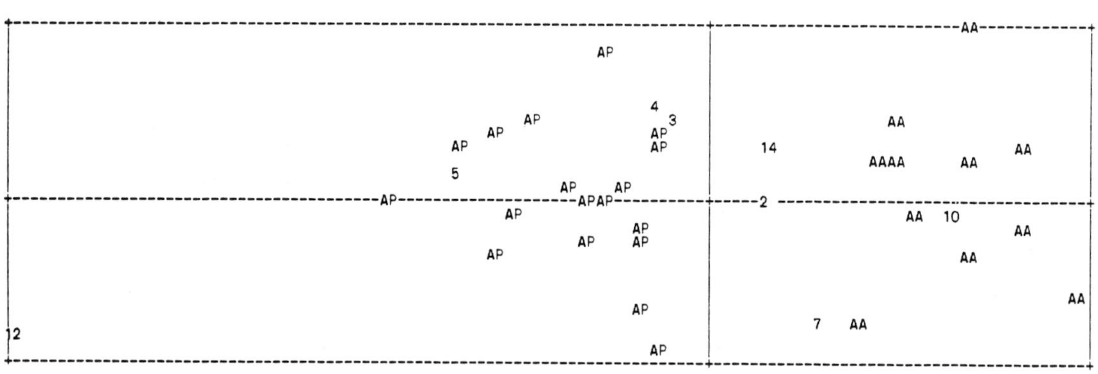

Fig. 2 Correspondence analysis for 8 measures on 29 first phalanges belonging to the fossil horse from La Caune de l'Arago. First axis horizontal, second axis vertical. AA = supposed anterior phalanges; AP = supposed posterior phalanges. The measurements 2, 3, 4, etc. are illustrated on Fig. 1.

developed proximal to their tuberosities whereas the (supposed) posterior are more developed distal to their tuberosities as well as being proximally deeper. The second axis (F2) accounts for 11 percent of the inertia and opposes lengths to breadths, namely the length of the *trigonum phalangis* (7) and the infratuberosital length (12) to the least (3) and proximal (4) breadths. Thus (supposed) anterior phalanges are rather long and (supposed) posterior are rather wide.

Principal components analysis: F1 is a size-axis largely defined by lengths (because in the set of measurements used there are 5 lengths and only 3 breadths). F2 and F3 suggest pretty much the same things as F1 and F2 of the correspondence analysis.

Ascending hierarchic classification (using chi-square metric): Nearly 58 percent of the total inertia is explained by the (supposed) anteroposterior dimorphism. The whole sample is separated into two discrete groups: one of (supposed) anterior and one of (supposed) posterior phalanges.

Discriminant analysis: The distinction between the average (supposed) anterior and average (supposed) posterior phalanges is naturally based on the already discussed characters and variables. Infra-tuberosital lengths (12 and 13) and proximal depth (5) characterize the "posterior" phalanges while "anterior" phalanges are characterized by supra-tuberosital lengths (10 and 11).

5.2.3. Conclusion

Among a sample of a Middle Pleistocene horse phalanges, a clear dichotomy can be made using "first-glance" criteria. Multivariate analyses confirm this dichotomy and identify which measurements actually discriminate. Most of these are related to the anthropomorphic model of contrasting "waist" and "hips". The proximal depth, not used during the first-glance sorting but noted by PRAT (1957) appears equally valid. Supra-articular breadth, noted by KEDNEY, does not appear to discriminate.

Considering that the characters involved in dichotomizing the fossil sample are more or less the same as those found useful in distinguishing anterior and posterior phalanges of modern horses (PRAT, KEDNEY) and Burchell's zebras (EISENMANN/DE GIULI), we have grounds to believe that the dichotomy in the fossil material is the result of body position and not something related to sex, age, or some other factor. It seemed desirable, however, to test this assumption on other samples of *Equus* to see if the same characters were always involved. We tried first *Equus przewalskii* but the sample appeared to be not homogeneous at all, probably because it included different breeding lines (VOLF 1960–1985) and more or less pathological specimens resulting from a long life in captivity. Although *E.przewalskii* will be discussed later, it is the best sample of a modern wild *Equus* we have, namely that of *E.grevyi,* that will be used as the principal example and discussed in some detail next.

5.3. *Equus grevyi*

5.3.1. First set of analyses

The first set of analyses was run with all of the 14 variables for 26 anterior and 25 posterior phalanges.

Correspondence analysis (Figure 3): F1 accounts for 62 percent of the total inertia and opposes infra-tuberosital lengths (12 and 13) characterizing posterior phalanges to the length

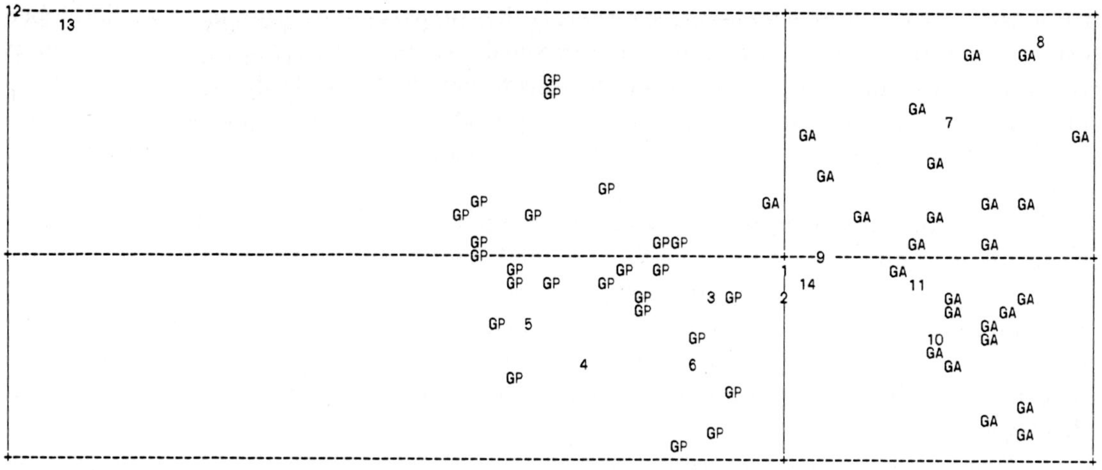

Fig. 3 Correspondence analysis for 14 measures on 51 first phalanges belonging to the modern *Equus grevyi.* First axis horizontal, second axis vertical. GA = anterior phalanges; GP = posterior phalanges. See Fig. 1 for the system of measurements.

of the *trigonum phalangis* (8) characterizing anterior phalanges. F2 (19 percent of the inertia) roughly opposes lengths (7, 8, 12, 13) and breadth (4). No sexual dimorphism appears on any of the axes.

Principal components analysis (Figure 4): On the plane defined by the first two axes (57 percent and 20 percent of inertia), anterior and posterior phalanges are clearly separated by a diagonal. Anterior are longer (1, 2, 9) and have longer supra-tuberosital segments (10, 11). Distal articular breadth (14) is strongly correlated with the lengths so that variable appears also on the side of the anterior phalanges. Posterior phalanges have longer infra-tuberosital segments (12, 13) and are proximally broader (4) and deeper (5). A probable element of sexual dimorphism appears on this plane; in each group of anterior and posterior phalanges, most of the larger specimens belong to male and most of the smaller to females.

Hierarchic ascending classification: Both Euclidian and chi-square based classifications clearly separate anterior and posterior phalanges. Moreover, using the Euclidian metric, males and females tend to cluster in different sub-classes: for posterior phalanges, one is composed of 9 females and 1 male whereas the other includes 4 females and 10 males; for anterior phalanges, one sub-class comprises 8 females and 1 male, the other 6 females and 10 males. No sexual dimorphism appears in the chi-square based classification.

Discriminant analysis: In accordance with the previous observations, posterior phalanges are characterized by long infra-tuberosital segments (12, 13) and large proximal ends (4, 5). Anterior phalanges are characterized by the length of the *trigonum phalangis* (8).

5.3.2. Second set of analyses

Another set of analyses was run with only 8 variables, namely those that had to be used for the fossil Arago sample: measures 2, 3, 4, 5, 7, 10, 12, and 14.

Correspondence analysis: F1 (67 percent of the inertia) correctly discriminates anterior and posterior phalanges. We find again that posterior phalanges are proximally more voluminous

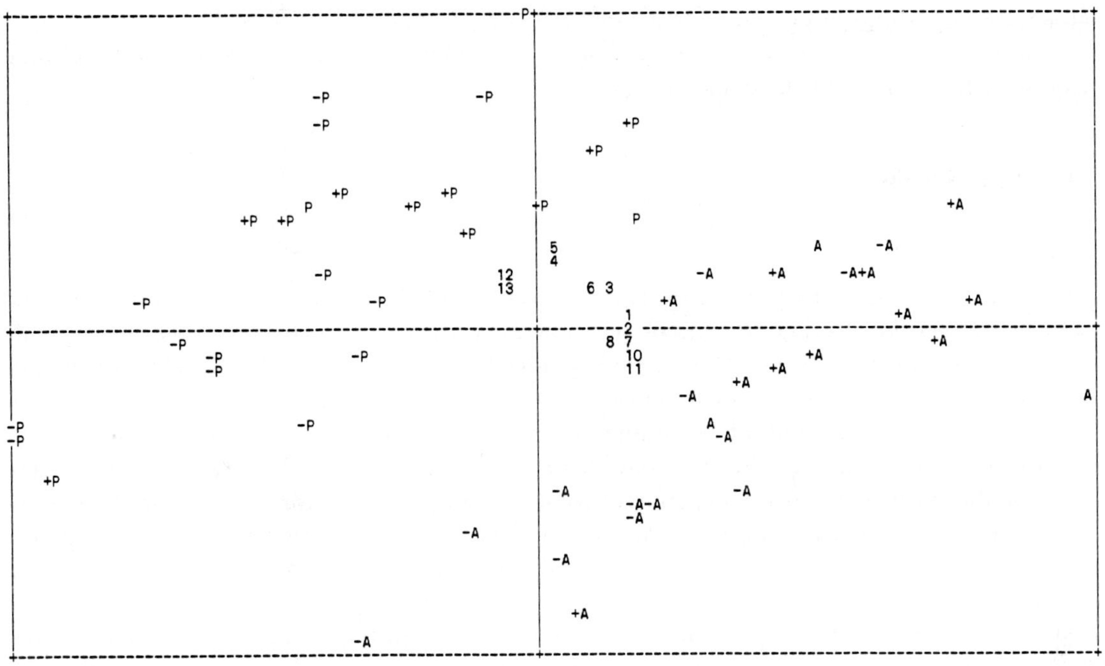

Fig. 4 Principal components analysis for 14 measures on 51 first phalanges belonging to the modern *Equus grevyi.* First axis horizontal, second axis vertical. A = anterior phalanges; P = posterior phalanges; + = males; – = females. Measures 9 and 14 were not printed by the computer because they are too close to other measures. Measure 9 is very close to measures 1 and 2; measure 14 is very close to measures 3 and 6.

(4, 5) and have a longer infra-tuberosital segment (12) while anterior phalanges have a long *trigonum phalangis* (7) and supra-tuberosital segment (10). F2 (10% of the inertia) opposes lengths (7, 12) and breadth (4).

Principal components analysis: The first plane shows the same pattern as in the analysis with 14 variables, namely good separation between anterior and posterior phalanges, the latter having longer infratuberosital segments and more developed proximal ends. Sexual dimorphism is also expressed in the same way.

Ascending hierarchic classifications: Both classifications (based on the chi-square and the Euclidian metrics) perfectly separate anterior and posterior phalanges but no clear separation by sex appears.

5.3.3. Conclusion

Anterior and posterior *E.grevyi* first phalanges can be perfectly separated by multivariate analyses based on chi-square or Euclidian metrics. Among the discriminating characters appear the same ones that provided the dichotomy in the fossil sample previously discussed: supra-and infra-tuberosital lengths, length of the *trigonum phalangis* and proximal depth. In some analyses, greater proximal breadth is an additional character for the posterior phalanx. Male phalanges are not always larger than female but this sexual dimorphism does show up in some of the analyses. Analyses run with a reduced set of variables (8 instead of 14) provide the same results as those run with the complete set.

5.4. Other Samples

5.4.1. General pattern

The same multivariate analyses as for *E.grevyi* were applied to 9 other populations of zebras, asses, half-asses, horses, and *E.stenonis* (i.e., correspondence and principal components analyses, ascending hierarchic classifications using both the chi-square and the Euclidian metrics). All were run both with the complete set of measures (14 or 15) and with the restricted set (the 8 measures used in the study of La Caune de l'Arago). The results are roughly identical.

In correspondence analysis, the first axis (60–84 percent of the total inertia) opposes infra-tuberosital lengths (12, 13) and sometimes proximal breadth and depth (4, 5) to supra-tuberosital lengths (10, 11) and/or lengths of the *trigonum phalangis* (7, 8) – that are also "supra-tuberosital." The second axis (6.5–19 percent of the inertia) opposes lengths to breadths and depths.

In principal components analysis appear the same trends with the difference that total lengths (1, 2, 9) are added to supra-tuberosital in their opposition to infra-tuberosital lengths and proximal breadth and depth.

In both kinds of analyses, the first axis usually separates quite well anterior and posterior phalanges. Principal components analyses may also show a sexual dimorphism, phalanges belonging to males being in general larger than phalanges belonging to females.

In the ascending hierarchic classifications, the separation between anterior and posterior phalanges is usually good with the chi-square metric, poor with the Euclidian metric; the latter may show some sexual dimorphism.

In the discriminant analyses, the measures characteristic of posterior phalanges are 12, 13, and sometimes 4 and 5. Anterior phalanges are characterized by measures 8, 10, and sometimes 7 and 11.

To summarize, sexual dimorphism, when present, is size related. The differences between anterior and posterior phalanges are not so much related to the size as to shape. They lie principally in the relative development of the supra- and infra-tuberosital segments. Proximal breadth and depth and minimal breadth are more rarely involved. The ratio diagrams of Figure 5 illustrate the anteroposterior dimorphism in *E.hemionus onager.* The average anterior phalanx is used as reference (as in the other ratio diagrams of this paper). The average posterior phalanx differs from the anterior by its shorter *trigonum phalangis* (7) and supra-tuberosital length (10). There is nearly no overlap of the two ranges of variation for these measores.

5.4.2. Particular points

E.stenonis cf. *vireti*: The first glance sorting is confirmed by most analyses. Out of the 40 phalanges, one supposedly anterior is placed with the posterior, but only in analyses using the Euclidian metric with the reduced set of variables. In the ascending hierarchic classification with the chi-square metric, most of the partition (72–74 percent) is explained by the (supposed) anterior-posterior dimorphism.

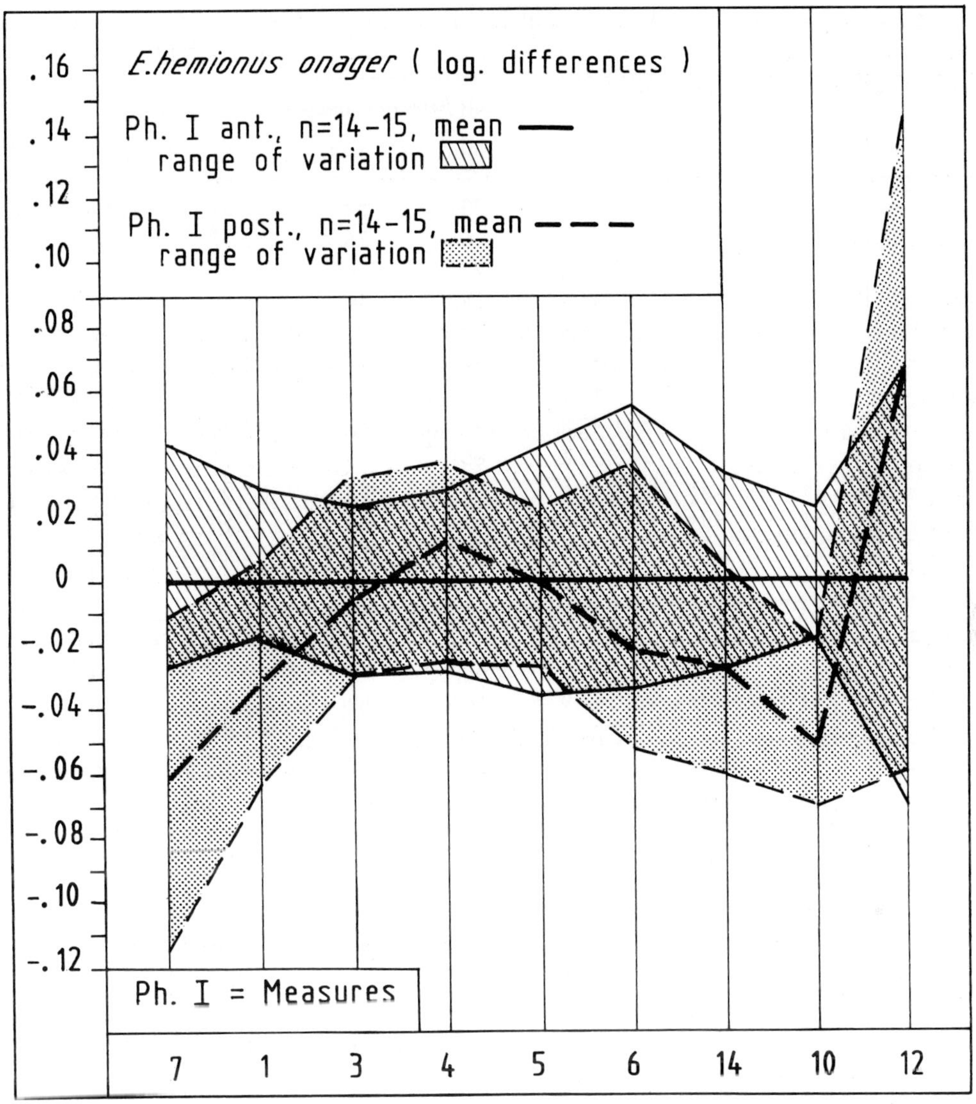

Fig. 5 Ratio diagrams of the means and ranges of variation (between the minimal and maximal observed values) for the anterior and posterior first phalanges of *Equus hemionus onager.* See Fig. 1 for the system of measurements, and Table 1 for the measurements.

E.burchelli boehmi: One of the 38 phalanges is constantly misplaced. This certainly anterior phalanx has a "posterior" morphology because of its well-developed infra-tuberosital lengths. The posterior of the same individual exhibits the same peculiarity which makes it appear as "superposterior" on the analyses. Slight sexual dimorphism appears in the hierarchic classification with the Euclidian metric on 14 variables. One class contains 14 males and 4 females, the other 6 males and 12 females.

E.zebra: Out of 44 phalanges, 3 posterior and 2 anterior are occasionally misplaced. The principal components analysis on 8 variables shows an interesting point. The first axis opposes the

TABLE 1

	Equus hemionus onager							
	mes.	n	x	min.	max.	s	v	log.
Ph. I Ant	1	15	76.5	73.3	82	2.27	2.96	1.884
	2	15	69.8	66	76	2.57	3.68	1.844
	3	15	24.6	23	26	1.02	4.17	1.390
	4	15	41.1	38.5	44	1.68	4.07	1.614
	5	15	30.8	28.5	34	1.53	4.97	1.489
	6	15	36.7	34	41.8	1.98	5.40	1.564
	7	14	48.1	45	53	2.54	5.29	1.682
	8	15	42.8	39	47	2.50	5.83	1.632
	9	14	69.1	66	75	2.21	3.20	1.840
	10	14	58.5	56	62	2.06	3.52	1.767
	11	14	58.4	55	63	2.34	4.00	1.767
	12	14	10.3	8.8	12	0.94	9.09	1.014
	13	14	10.2	8.9	11	0.65	6.42	1.007
	14	15	35.6	33.3	38.5	1.53	4.29	1.551
Ph. I Post	1	15	71.2	66.5	78	2.58	3.62	1.853
	2	15	64.9	60.7	69	2.06	3.18	1.812
	3	15	24.3	23	26.5	1.01	4.15	1.386
	4	15	42.4	39	45	1.85	4.36	1.627
	5	15	30.9	29	32.5	1.19	3.84	1.490
	6	15	35	32.6	40	2.05	5.85	1.544
	7	14	41.8	37	47	2.48	5.92	1.621
	8	15	36.5	32	41	2.42	6.62	1.562
	9	14	63.5	59.8	69	2.10	3.31	1.803
	10	14	52.2	50	56	1.90	3.64	1.717
	11	14	52.6	50	57	2.01	3.82	1.721
	12	14	12	9	14.5	1.44	12.00	1.080
	13	13	11.2	8	13	1.46	13.10	1.049
	14	14	33.3	31	36	1.53	4.60	1.523

Tables 1–5 Measurements in millimeters for anterior and posterior first phalanges.
mes. = dimension (measure); the numbers 1 to 15 refer to the system of measurements defined in Figure 1;
n = number of specimens studied;
x = mean;
min. = minimum value observed;
max. = maximum value observed;
s = standard deviation;
v = coefficient of variation;
log. = decimal logarithm.

phalanges with well-developed supra-tuberosital segments (mostly anterior) to phalanges with well-developed distal ends (mostly posterior). In both sub-classes, *E.zebra zebra* individuals have more distally developed phalanges ("more posterior") than do *E.zebra hartmannae.* The second axis opposes robust to less robust bones. Phalanges belonging to males are robust with the exception of one. The skull of this individual was not available so that the male sex marked on the label could not be confirmed by the presence of canines. Phalanges belonging to females may be robust or not. The sub-specific dimorphism superimposed on the antero-posterior dimorphism is probably responsible for the relatively high number of misplaced phalanges.

Asses: In the analyses run with the 14 variables on the wild and domestic modern asses, the sample appears as a continuum between the anterior and posterior "poles". In consequence, the hierarchic ascending classifications fail to discriminate properly between anterior and posterior phalanges. Other analyses were run with the reduced set of variables in order to include the subfossil Ikrit sample. The results are even worse because Ikrit phalanges are shifted in the posterior direction. Thus, several anterior Ikrit phalanges are placed on the posterior side of the axis. In accordance, in the hierarchic classification with the chi-square metric, one group is purely anterior while the other contains posterior phalanges as well as several anterior (mostly from Ikrit).

E.hemionus onager: Out of 27 or 28 phalanges, 2 are occasionally misplaced in the analyses using the chi-square metric. The anteroposterior dimorphism seems weaker than in the previously studied species. In the hierarchic ascending classifications the two groups fuse into one much earlier than in other species (at 40–50 percent instead of 60–80 percent of the total inertia). The great number of males renders the sample improper for research on sexual dimorphism.

The horse from Lunel-Viel: One supposedly anterior phalanx (LVI-10-16589), and one supposedly posterior (LVI-9-2921) phalanx (out of 23) are "misplaced."

The horse from Jaurens: The first glance sorting was relatively difficult: 10 phalanges were only doubtfully identified as anterior or posterior. Thus we were not surprised to find 6 of the phalanges "misplaced" (or should we say that their identifications were corrected?) by the analyses. If these analyses are correct, FSL 302–212 and 255 should be posterior, FSL 302–205, 211, 216, and 332 should be anterior, and our sample contains 27 fore and 24 hind phalanges instead of 25 fore and 26 hind.

The horse from Solutré: The Solutré sample like the one from Jaurens contains several (4) phalanges doubtfully identified as anterior or posterior. Two of them and a third one are "misplaced" by the analyses. If these are correct, SOL"2" and 144 should be anterior, SOL 141 should be posterior, and our sample should contain 18 anterior and 12 posterior phalanges instead of 17 and 13.

E.przewalskii: The sample (46 phalanges) looks quite heterogeneous with 6 phalanges being usually "misplaced". In some cases (LD 359, AMNH 32686, 32696, AC 1929–35) the phalanges seem shifted in the posterior direction (anterior are classed with posterior whereas some posterior tend to occupy an extreme "hyperposterior" position on the cloud). In other cases, posterior phalanges look like anterior (e.g., MA 1964–107, BA 10877).

Following the suggestion of O. Ryder (personal communication) we looked for possible differences between what are believed to be "pure" *E.przewalskii* and animals belonging to mixed or doubtful breeds. Toward that end, various analyses were run separately on anterior and posterior phalanges. In the correspondence analysis for the anterior phalanges, the first axis opposes, as usual, infra-tuberosital to supra-tuberosital lengths, i.e., "posterior" to "anterior" morphologies. On the "posterior" side, we find 5 "pure" *E.przewalskii* and 4 doubtful whereas on the "anterior" side are 4 pure and 10 doubtful. The analysis for the posterior phalanges is more striking. *All* supposed pure *E.przewalskii* are on the "posterior" side of F1 together with 3 doubtfully pure Przewalski horses. Two of these latter (AC 1935-486 and 1941-322) were also placed on the "posterior" side in the correspondence analysis for the anterior phalanges. Thus it seems that the heterogeneity of our sample is explained by the fact that pure Przewalski horses usually have more developed infra-tuberosital segments both on anterior and on posterior phalanges than do horses belonging to mixed breeding lines.

5.5. Conclusions

In most cases, first phalanges of all species of *Equus* can be discriminated at first glance by looking at the relative position of the supra-articular tuberosities (located more distally in anterior first phalanges). Additional criteria, less reliable, are the length of the *trigonum phalangis* and the greatest length (both larger in anterior phalanges) and, occasionally, the distal articular breadth (larger in anterior phalanges) and the proximal depth (larger in posterior phalanges). In some taxa, the anteroposterior dimorphism is more obvious than in others. For example, zebra phalanges usually contrast more than do those of horses, the latter being at times more difficult to differentiate. Discrimination is easier within homogeneous samples.

In most cases, first glance identification is confirmed by multivariate analyses. Contradictions appear usually only when first glance identification was questionable. The discriminant characters are the same as those used in first glance identifications. When populations are homogeneous, discrimination can be nearly perfect. This is true both for modern (Grévy's and Burchell's zebras) and for fossil samples (La Puebla de Valverde, Arago). When populations are heterogeneous, discrimination is not so good, e.g., 13 percent misplaced phalanges among 46 specimens of *E.przewalskii.* The reason for this is that the shape differences that express anteroposetrior dimorphism within an homogeneous sample may also express other features. For example, *E.zebra zebra* individuals generally have a more "posterior" morphology than do *E.zebra hartmannae* specimens. The same seems to be true for "pure" *E.przewalskii* in relation to mixed breeding lines.

The best statistical discriminations are achieved by correspondence analysis and ascending hierarchic classification based on the chi-square metric. In the principal components analysis appear, at the same time, shape-related anteroposterior dimorphism and size differences that may or may not be related to it. Size and shape are also intricately bound in classifications based on the Euclidian metric. The agreement between analyses using the complete set of measures (14) and the restricted set (8) validates the results obtained when only the latter could be used.

In practice, if an homogeneous sample of 15-20 phalanges has to be sorted as anterior and posterior, one can use either the first glance and/or the biometrical techniques. If the measures recommended in this paper are used, multivariate analyses will differentiate anterior and posterior phalanges with an error of about 1-3 percent. If, however, one isolated phalanx has to be identified as anterior or posterior, one can analyse its measures together with those of a sample including anterior and posterior phalanges belonging to the same (supposed) taxon as the isolated phalanx. A severe mistake in choosing the comparative taxon would become evi-

dent through the analyses. Within the same species, without problems due to pathology or domestication, the risk of error can be evaluated as being 4–6 percent. In bad cases, it can increase up to 15 percent, but there, the method indicates its own limits: overlappings of anterior and posterior clouds and low percentage of inertia explained by anteroposterior dimorphism. For one problematical isolated phalanx, one can also compare the corresponding ratio diagram with those published in this paper.
Naturally, anteroposterior discrimination is closely bound to taxonomic identification that has been (artificially) isolated for discussion in the next part of this study.

6. Interspecific differences

6.1. Introduction

The more extensive studies of equid first phalanges of which we are aware are those of CAMP/SMITH (1942), GROMOVA (1949), and SONDAAR (1968). These authors, however, were more interested in functional evolution (from the primitive tridactyl forms to the modern monodactyl *Equus*) than in specific differences within the genus *Equus.*
We have approached the question of interspecific differences using two kinds of methods. The first includes multivariate analyses that permit us to obtain a general view of the entire genus *Equus* or of its subgroups. The second involves the use of ratio diagrams in order to make more detailed comparisons and to better illustrate some points. Since we were no longer interested in anteroposterior dimorphism, we were able, at times, to use analyses run only on one kind of phalanges (anterior or posterior). Posterior phalanges were frequently chosen because they often reveal greatest differences from one species to another.

6.2. General view of the genus *Equus*

When all phalanges are pooled together in correspondence analyses, the first axis expresses slenderness by opposing measure 3 (and, to a lesser degree, measures 4 and 5) to measures 1, 2, 9, 10, and 11. Slender species are half-asses, asses and Grevy's zebras. Robust species include most of the horses and *E.stenonis* cf. *vireti.* The second axis opposes infra-tuberosital lengths (12 and 13) to lengths of the *trigonum phalangis* (7 and 8). Posterior phalanges are on the first side, anterior on the other. Both anterior and posterior phalanges of *E.stenonis* cf. *vireti* are entirely on the "posterior" side of F2. Figure 6 shows the same analysis run using average dimensions for posterior phalanges of the best represented populations. (If all bones were represented there would be too much overlapping for clarity in the figure. These overlappings do exist, however, even if they are not apparent on Fig. 6.) Average anterior phalanges are entered as supplementary elements which means that they are placed on the figure but do not affect the relative position of posterior phalanges. On Figure 6, the points representing average anterior phalanges are united by straight (interrupted) lines to the points representing the average posterior phalanges of the same species. These lines are roughly parallel, which means that the same differences separate anterior and posterior phalanges of all the studied species.

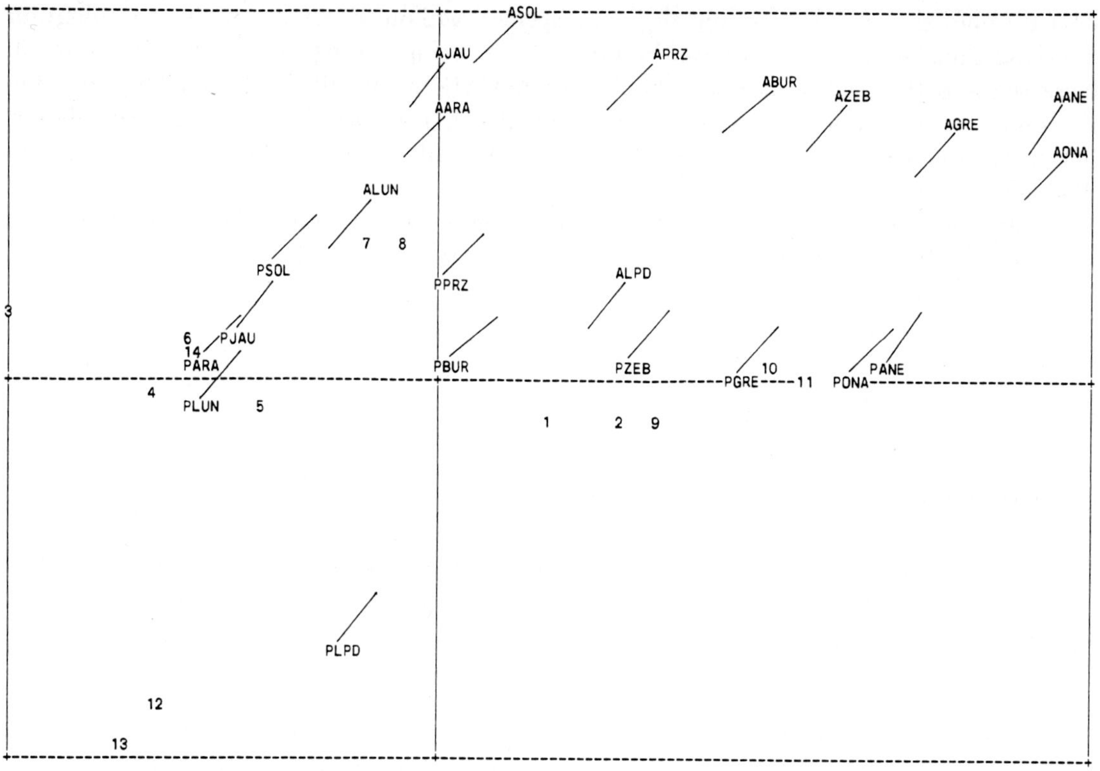

Fig. 6 Correspondence analysis for 14 measures on average posterior (P) phalanges belonging to 11 populations of modern and fossil *Equus.* Average anterior phalanges (A) are introduced as supplementary data. First axis horizontal, second axis vertical.
ANE = domestic ass; ARA = fossil horse from La Caune de l'Arago; BUR = Kenyan Burchell's zebra; GRE = Grevy's zebra; JAU = fossil horse from Jaurens; LPD = *E.stenonis* cf. *vireti* from La Puebla de Valverde; LUN = fossil horse from Lunel-Viel; ONA = *E.hemionus onager*; PRZ = *E.przewalskii*; SOL = fossil horse from Solutré; ZEB = *E.zebra.*

Figures 7 and 8 compare the ratio diagrams of a horse, an *E.stenonis,* and the Kenyan Burchell's zebra. The horse and Burchell's zebra have very similar phalanges, especially anterior ones. The phalanges of *E.stenonis* differ by having a shorter *trigonum phalangis* (measure 7) and a longer infra-tuberosital segment (measure 12). Species with slenderer phalanges are compared in Figures 9 and 10. Anterior and posterior phalanges of domestic asses are very similar to the ones of *E.hemionus onager* (Fig. 9; Figs. 5 and 10). *Equus grevyi* phalanges are clearly more robust. Kiang phalanges are characterized by their relatively narrow distal ends (measure 14) and long supra-tuberosital segments (measures 10). The three species differ from *E.stenonis* by measures 7 and 12.

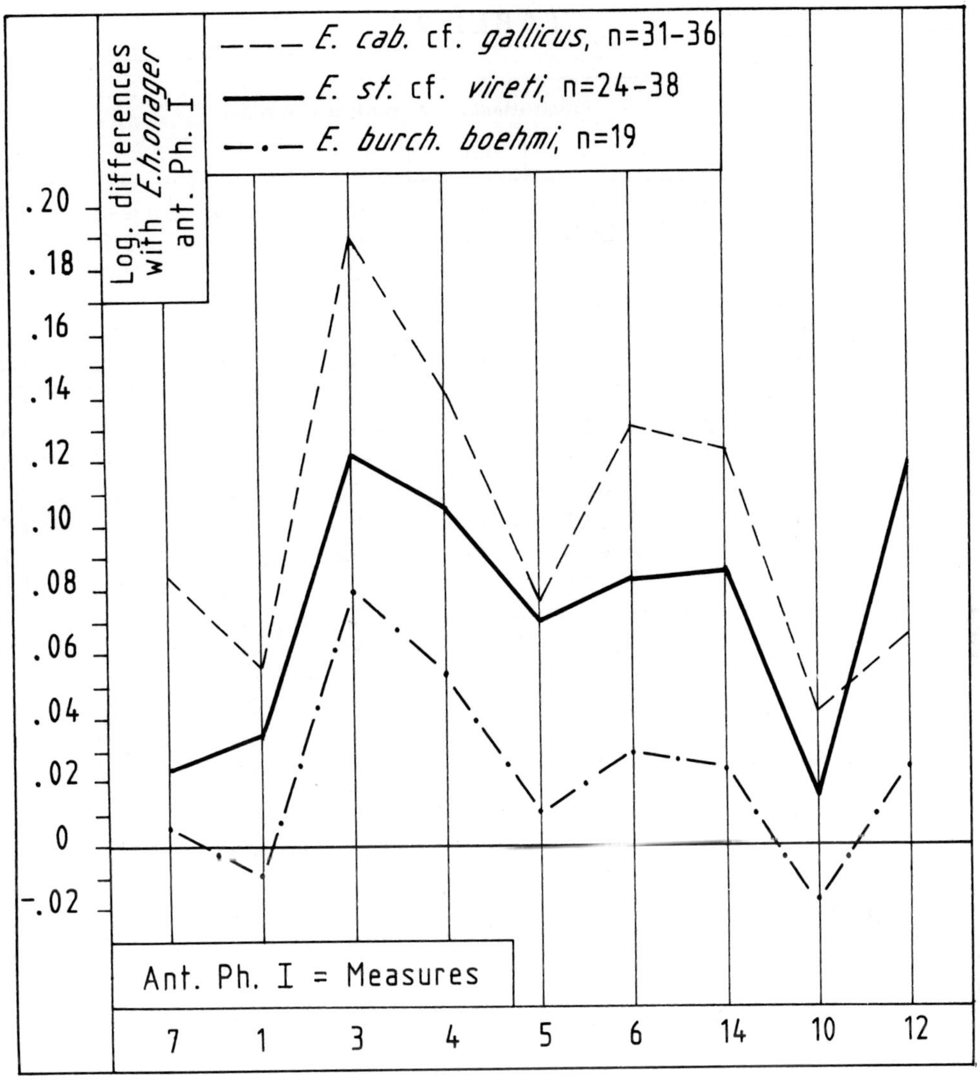

Fig. 7 Ratio diagrams for the average values of robust anterior first phalanges belonging to *E. caballus* cf. *gallicus* from Jaurens, *E. stenonis* cf. *vireti* from La Puebla de Valverde, and *E. burchelli boehmi* from Kenya. n = number of specimens. See Fig. 1 for the system of measurements and Table 2 for the measurements (EISENMANN/DE GIULI 1974: Tables III and IV).

TABLE 2

	Equus caballus cf. *gallicus* (Jaurens)							
	mes.	n	x	min.	max.	s	v	log.
Ph. I Ant	1	27	85.7	82	92	2.24	2.61	1.932
	2	27	76.2	73	84	2.67	3.51	1.882
	3	27	37.5	35	42.5	1.89	5.04	1.574
	4	27	57.6	54	65	2.61	4.54	1.760
	5	27	37.8	35	43	1.84	4.86	1.577
	6	27	49.3	46.5	54.3	1.79	3.64	1.692
	7	27	58.4	54	64	2.40	4.12	1.766
	8	27	50.1	45	55	2.53	5.05	1.699
	9	27	74.8	71	81	2.57	3.44	1.873
	10	27	62.6	59	67	2.07	3.30	1.796
	11	27	62.3	57	68	2.32	3.72	1.794
	12	27	12.7	11	15.5	1.20	9.45	1.103
	13	27	12.5	10	15	1.29	10.32	1.097
	14	27	47	45	50	1.52	3.23	1.672
	15	27	25.7	24	29	1.16	4.52	1.410
Ph. I Post	1	24	83.2	80	89	2.28	2.74	1.920
	2	24	74.4	72	80	2.03	2.73	1.871
	3	24	36.7	34	41	1.66	4.52	1.564
	4	24	58.8	55	65	2.32	3.95	1.769
	5	24	40.8	38	45	1.51	3.70	1.610
	6	24	47.7	45.2	53	2.08	4.35	1.678
	7	24	53.6	50	58	1.91	3.56	1.729
	8	24	45.8	42	51	2.29	5.00	1.660
	9	24	71.6	68	76	2.10	2.93	1.855
	10	24	56	52	61	2.28	4.08	1.748
	11	24	56.6	53	60	1.83	3.23	1.752
	12	24	16.7	15	20	1.32	7.94	1.222
	13	24	15	13	18	1.02	6.82	1.176
	14	24	45.7	43	49	1.48	3.25	1.660
	15	24	26	24	28.2	1.04	4.00	1.415

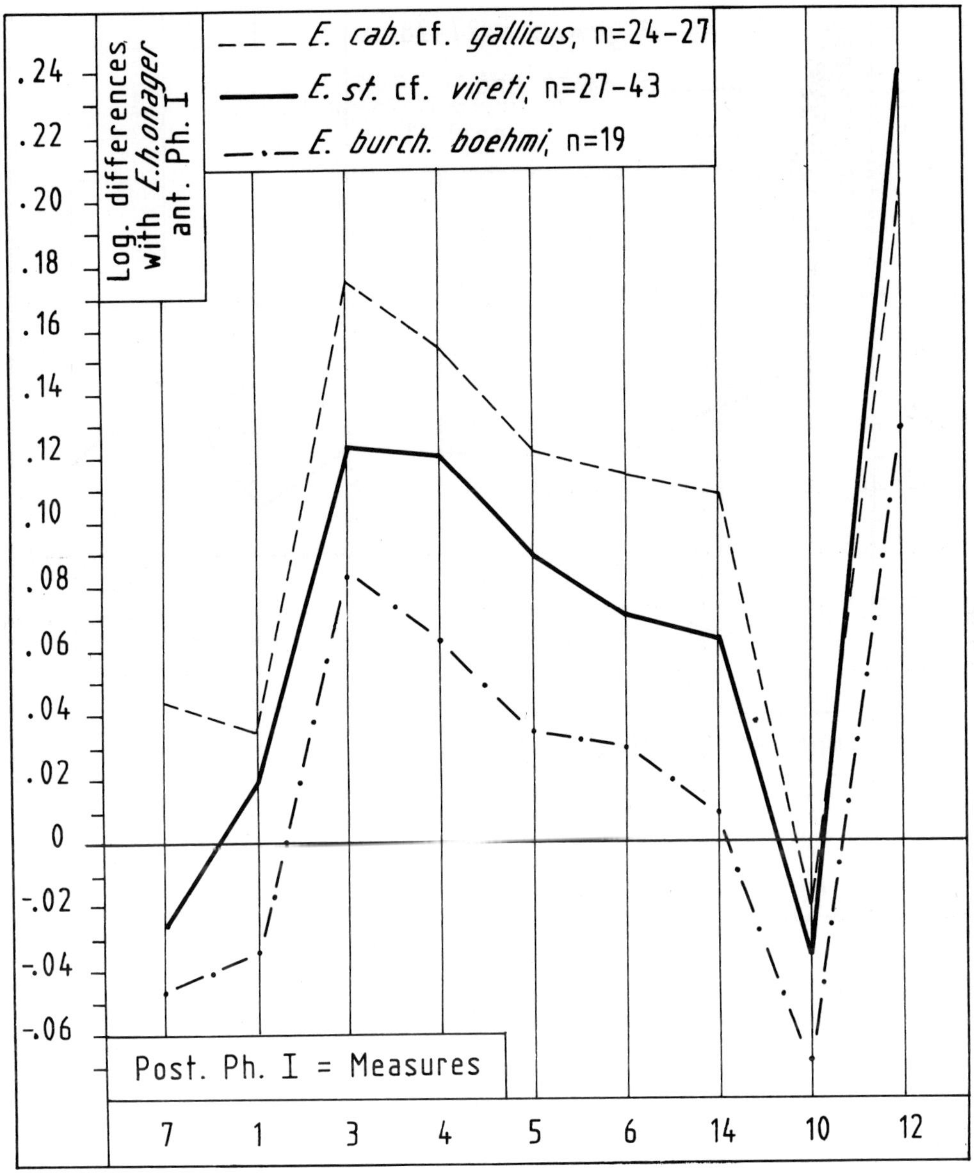

Fig. 8 Ratio diagrams for the average values of robust posterior first phalanges. Same species and abbreviations as in Fig. 7.

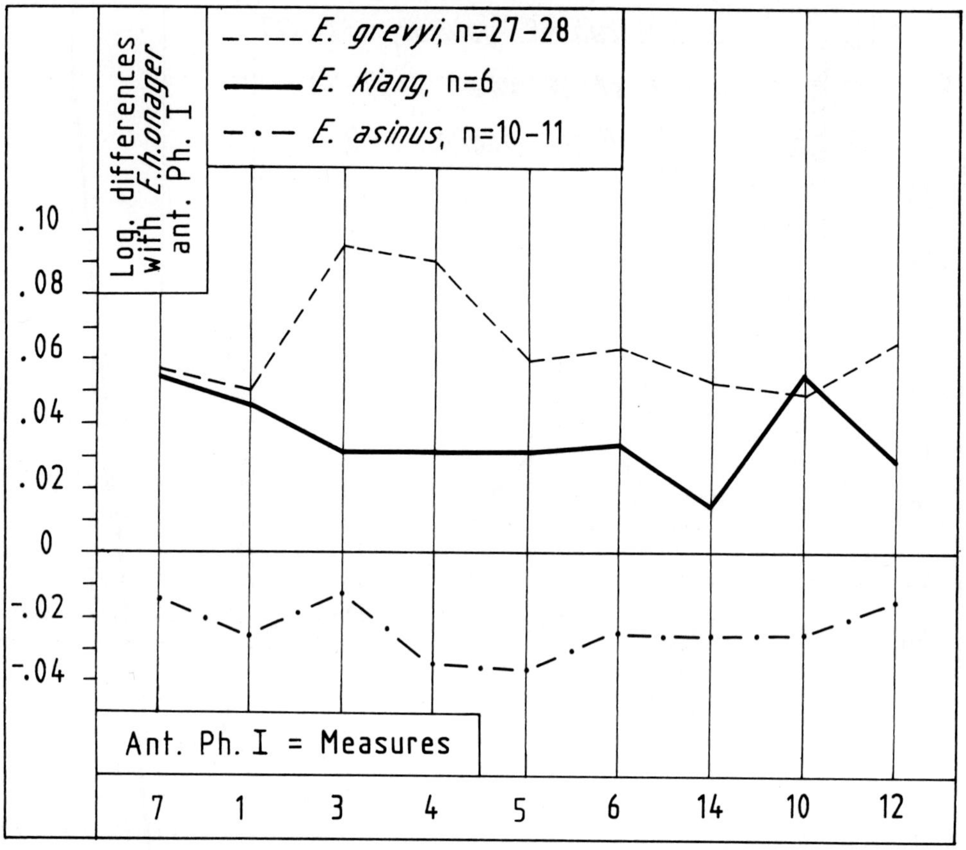

Fig. 9 Ratio diagrams for the average values of slender anterior first phalanges. Same abbreviations as in Fig. 7. See also Tables 3, 4, and 5.

TABLE 3

	Equus grevyi							
	mes.	n	x	min.	max.	s	v	log.
Ph. I Ant	1	28	86.2	80	91	2.69	3.12	1.935
	2	28	79.6	74	85.5	2.69	3.38	1.900
	3	28	30.7	28.5	33	1.28	4.16	1.488
	4	28	50.7	48	54	1.81	3.56	1.705
	5	28	35.4	33	37	1.26	3.56	1.550
	6	28	42.4	40	45	1.26	3.33	1.627
	7	28	54.9	50	61	3.01	5.47	1.740
	8	28	49.1	41	56	3.37	6.86	1.691
	9	28	77.5	70	82	2.80	3.61	1.889
	10	28	65.6	62	72	2.36	3.60	1.817
	11	28	66.7	63	73	2.17	3.26	1.824
	12	28	12.1	9.7	15	1.52	12.63	1.081
	13	28	11.3	9	14	1.34	11.94	1.052
	14	27	40.2	37.5	43	1.25	3.11	1.604
Ph. I Post	1	27	81.1	75	87	2.93	3.61	1.909
	2	27	74.8	69	81	2.75	3.68	1.874
	3	27	29.6	27.5	31.5	1.13	3.84	1.471
	4	27	51.7	48.5	55	1.96	3.78	1.714
	5	27	36.7	34.5	39	1.25	3.40	1.564
	6	27	41.7	39	44.5	1.57	3.76	1.620
	7	27	49.1	44	55	2.51	5.11	1.691
	8	27	42.3	36	48	2.49	5.89	1.626
	9	27	71.9	65	78	2.99	4.16	1.857
	10	27	59.3	55	65.5	3.02	5.09	1.773
	11	27	60.6	54.5	66.5	2.91	4.81	1.783
	12	27	14.5	12	17	1.29	8.85	1.162
	13	27	13.1	10.5	15.5	1.47	11.21	1.118
	14	26	37.6	34.5	40	1.31	3.48	1.575

TABLE 4

	Equus asinus							
	mes.	n	x	min.	max.	s	v	log.
Ph. I Ant	1	11	72.2	64	82	5.45	7.55	1.859
	2	11	65.5	57	76	5.23	7.98	1.816
	3	11	23.9	22	26	1.36	5.68	1.378
	4	11	38	35	41.1	1.77	4.66	1.580
	5	11	28.3	26.1	32.5	1.66	5.87	1.452
	6	11	34.5	31	38.6	1.93	5.59	1.538
	7	11	46.5	41	52	3.39	7.28	1.668
	8	10	41.6	37	47.4	3.51	8.43	1.619
	9	10	64.2	57	74.5	5.37	8.36	1.808
	10	11	55.2	47	65	5.05	9.14	1.742
	11	11	56.1	50	64.8	4.46	7.94	1.749
	12	11	9.9	9	12	0.91	9.11	0.998
	13	11	9.2	8	11	0.94	10.27	0.963
	14	11	33.6	31	36.2	1.62	4.83	1.526
Ph. I Post	1	10	67.5	61	76	4.82	7.14	1.829
	2	10	61.7	56	70	4.55	7.38	1.790
	3	10	23.2	22	25.6	1.28	5.53	1.366
	4	10	40	35	43.2	2.28	5.85	1.591
	5	10	28.4	27	32.3	1.65	5.81	1.453
	6	10	32.9	29	38.2	2.27	6.90	1.517
	7	10	40.8	34	46	3.84	9.41	1.611
	8	10	35.9	29	41.5	3.93	10.94	1.555
	9	10	60.5	53.5	70.1	5.05	8.35	1.782
	10	10	49.9	44	56	4.64	9.31	1.698
	11	10	50.3	45	58	4.26	8.48	1.701
	12	10	12	11	14	0.85	7.08	1.079
	13	10	10.7	10	13	1.03	9.65	1.029
	14	10	31.2	29	35	1.75	5.63	1.494

TABLE 5

	Equus kiang							
	mes.	n	x	min.	max.	s	v	log.
Ph. I Ant	1	6	85.5	83	87.5	1.48	1.73	1.932
	2	6	78.8	78	81	1.17	1.49	1.897
	3	6	29.3	25.5	27	0.59	2.23	1.421
	4	6	44.2	42	46.1	1.50	3.39	1.645
	5	6	33.1	32.2	34	0.60	1.81	1.520
	6	6	39.6	37.5	41	1.22	3.07	1.598
	7	6	54.8	52	58	1.94	3.54	1.739
	8	6	50.1	48	53	1.62	3.24	1.700
	9	6	78.2	77	81	1.51	1.93	1.893
	10	6	66.2	64	70	2.14	3.23	1.821
	11	6	66.8	66	69	1.17	1.75	1.825
	12	6	11	10	12	0.91	8.24	1.043
	13	6	9.9	9	11	1.02	10.29	0.996
	14	6	36.9	36	38	0.90	2.43	1.567
Ph. I Post	1	6	78.6	77	82	2.03	2.59	1.895
	2	6	72.8	71	76.5	2.66	3.65	1.862
	3	6	25.9	25.5	26.5	0.39	1.52	1.414
	4	6	45.1	41.5	47.4	2.25	4.98	1.654
	5	6	34	33.5	34.7	0.50	1.46	1.532
	6	6	37.6	35.5	39.5	1.44	3.82	1.575
	7	6	48.8	47	51	1.33	2.72	1.689
	8	6	42.8	41	46	1.72	4.02	1.632
	9	6	70.6	68.5	74	2.06	2.92	1.849
	10	6	57.7	54	61	2.68	4.64	1.761
	11	6	58.3	54	62	2.75	4.72	1.766
	12	6	13.3	10.3	16	2.14	16.12	1.124
	13	6	12.7	10.5	16	2.32	18.29	1.103
	14	6	34.6	33.5	35.2	0.61	1.85	1.539

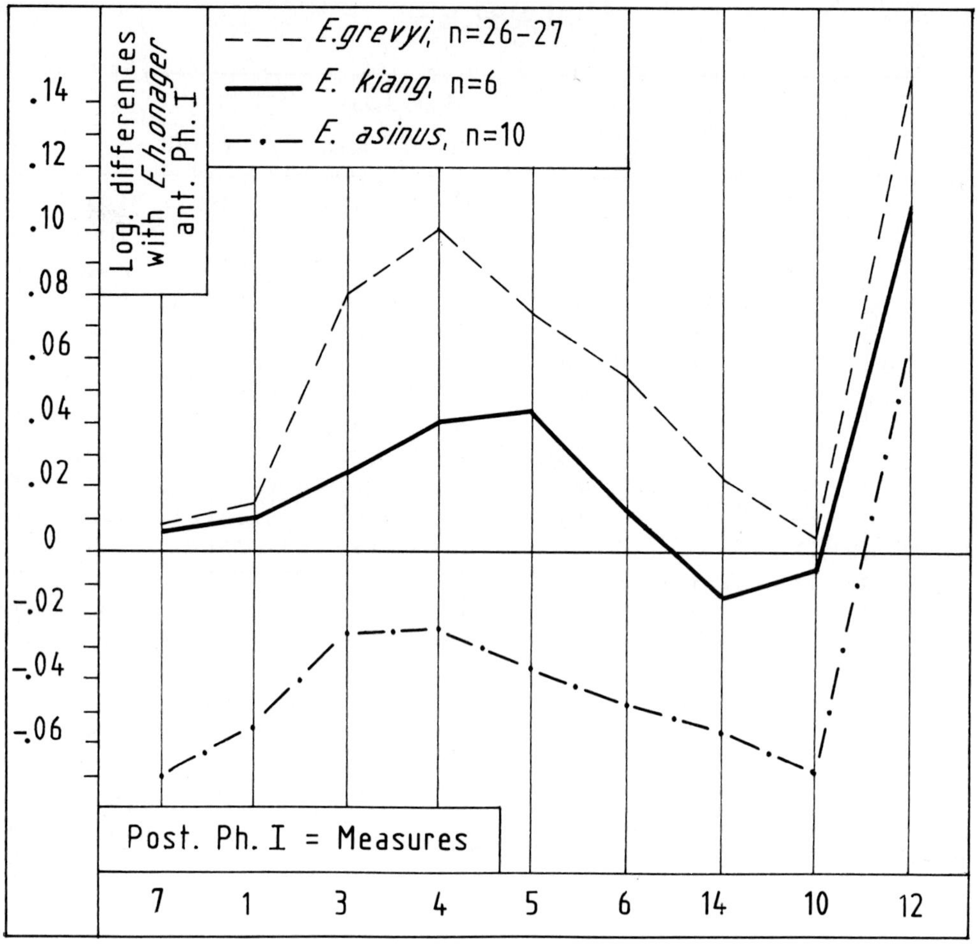

Fig. 10 Ratio diagrams for the average values of slender posterior first phalanges. Same species and abbreviations as in Figure 7.

6.3. Particular Subgroups

6.3.1. Zebras and *Equus stenonis*

Ratio diagrams as well as multivariate analyses show that the three species of zebras have similar first phalanges. They differ mostly by size as is shown by the principal components analysis (Figure 11) where posterior specimens are classed from small (Burchell's zebra) to large (Grevy's zebras) along the first axis (62 percent of inertia).

When the posterior phalanges of the three zebras are studied together with *E.stenonis* cf. *vireti* in a correspondence analysis (Figure 12), the four species are arrayed in a crescent-like scatter on the plane defined by the first two axes (57 percent and 16 percent of inertia). *E.stenonis,* characterized by long infra-tuberosital segments, is at one end of the crescent. In the middle

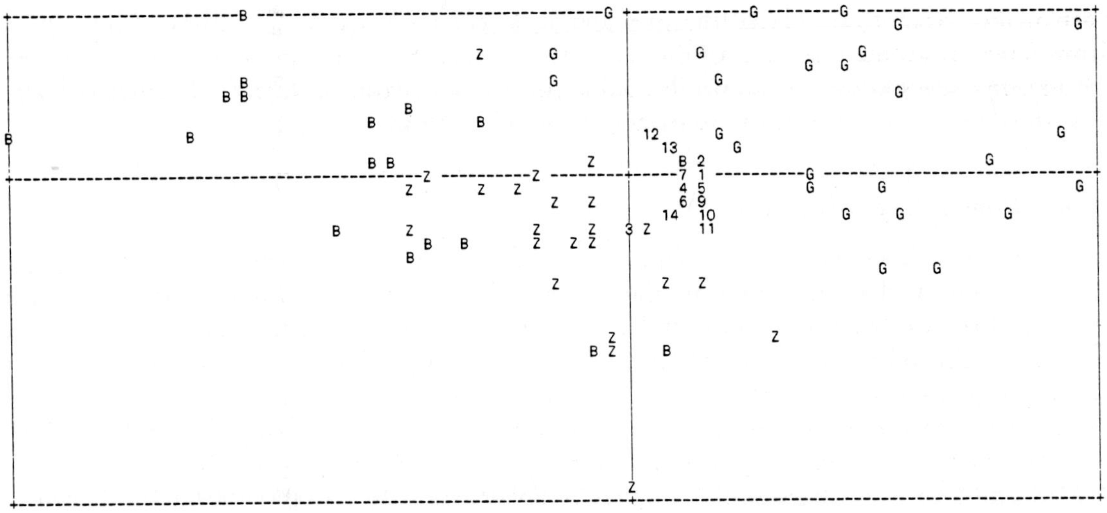

Fig. 11 Principal components analysis for 14 measures on 68 posterior first phalanges belonging to modern zebras. First axis horizontal, second axis vertical. B = Kenyan *E.burchelli*; G = *E.grevyi*; Z = *E.zebra*.

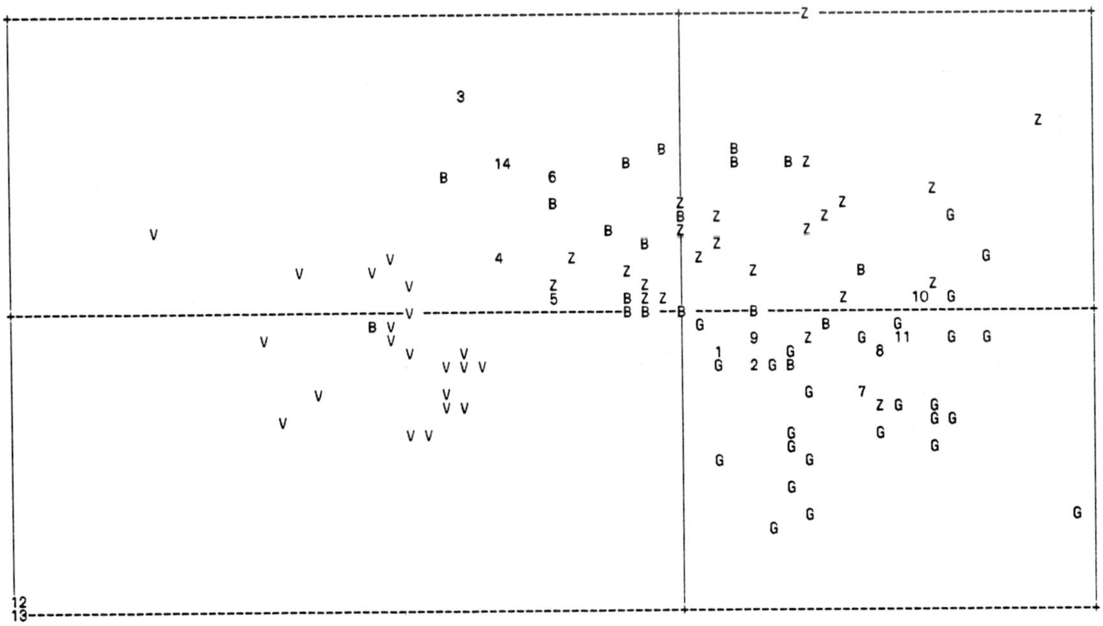

Fig. 12 Correspondence analysis for 14 measures on 89 posterior first phalanges belonging to modern and fossil *Equus*. First axis horizontal, second axis vertical. B = Kenyan *E.burchelli*; G = *E.grevyi*; V = *E.stenonis* cf. *vireti*, Z = *E.zebra*.

part of the crescent are placed the overlapping Burchell and mountain zebras. They differ from *E.grevyi* (at the other end of the crescent) by their robustness and the relative shortness of the *trigonum phalangis.* In short, the zebra most remote from *E.stenonis* cf. *vireti* is *E.grevyi.* It is also the most distinctly separate of the other zebras.

6.3.2. Domestic and wild asses

The correspondence analysis run with the 14 variables for modern ass anterior phalanges (more numerous than the posterior) shows a rather homogeneous scatter (Figure 13) where the two first axes bear 46 percent and 22 percent of the total inertia. Wild asses, however, appear usually to have more robust bones and more developed infra-tuberosital segments than do domestic asses. An interesting point concerns the differences between supposed Nubian and supposed Somali wild asses. The "Nubian" specimens (NA 10 = MU 1952-9 and NA 11 = MU 1963-133) appear very close to "Somali" specimens (SA 3 = BA 3597, SA 4 = BA 10858, and SA 9 = BL 30353). But another "Nubian" ass (NA 3 = BM 1904-6-12-1) comes very close to one domestic "White Egyptian ass" (DA 52 = AC 1893-634). Moreover, one

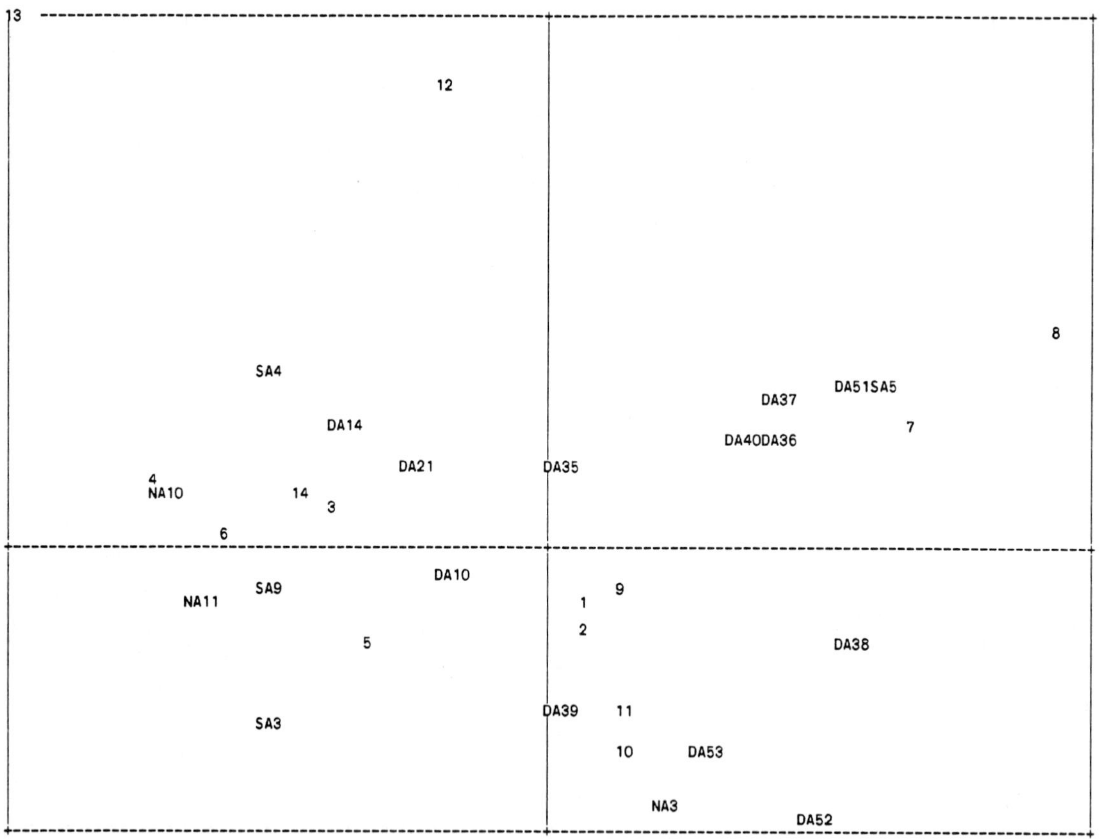

Fig. 13 Correspondence analysis for 14 measures on 19 anterior first phalanges belonging to modern wild and domestic asses. First axis horizontal, second axis vertical. DA = domestic asses; NA = Nubian wild asses; SA = Somali wild asses.

"Somali" specimen (SA 5 = BE 216) is placed close to the other "White ass" of our sample (DA 51 = AC 1875-28). It will certainly be interesting to compare these forms more thoroughly.

The principal components analysis of the same data distinguishes small and large specimens. Most of the domestic specimens are small and most of wild are large. There are, however, one small wild (MU 1952-29) and four large domestic (the two "White Egyptian," AMNH 15675, and 135017) specimens.

Another set of analyses was run with a restricted number of variables (8) which allowed us to add the fossil phalanges from Ikrit. In the correspondence analysis, the first axis expresses slenderness. Most of Ikrit phalanges, although domestic, are placed close to the wild (robust) bones. Because of their small size, however, they appear together with the domestic forms in the principal components analysis.

6.3.3. Half-asses

The correspondence analysis run with 14 variables on 52 posterior phalanges also shows a rather homogeneous cloud. The first axis (52 percent of the inertia) opposes breadths and depths to lengths, especially lengths of the *trigonum phalangis.* Mongolian hemiones and Syrian hemippes are on the "slender side" while most of the Iranian onagers are on the "broad side". Kiangs and kulans, however, fall very near the middle or on both sides. In any event, the samples are very poor: 2 hemippes, 3 kulans, and 3 khurs only. The second axis (20 percent of inertia) opposes supra- and infra-tuberosital lengths but these have no taxonomic relevance. The same facts appear using the hierarchic classification based on the chi-square metric. Hemiones and hemippes are grouped together; onagers, khurs, and kulans are in another group; kiangs are found in both groups.

6.3.4. Half-asses and domestic asses

One of the practical problems encountered by archaeozoologists working on the Middle East is distinguishing between the bones of asses and half-asses. Even as we were working on this paper, P.-Y. Gagnier asked one of us to identify a small sub-adult phalanx found in Syria at Ras-al-Bassit (30 K north of Lattaquié) in levels he believed at the time to be about 2500 years old (we learned later that they belong actually to the late Bronze age (1600-1200 B.C.: Gagnier 1986)). At "first glance" the phalanx seemed to be anterior but we were not very optimistic that we could determine if it had belonged to a domestic ass or to the extinct hemippe. Nevertheless, we drew the ratio diagrams and ran a correspondence analysis using this specimen and phalanges from half-asses and domestic asses. The ratio diagrams confirmed our supposition that the phalanx was an anterior one and showed that it bears remarkable similarities to hemippe phalanges, especially to that from the Museum of Comparative Zoology, Harvard (Figure 14). A comparison with Figure 9 shows that in hemippe, anterior phalanges are slenderer than in asses and have a longer supra-tuberosital segment. The same characters are also to be found in kiangs (Fig. 9).

In the correspondence analysis (Figure 15) done with the restricted set of variables because we lack one measure for the Ras-al-Bassit specimen, domestic asses are scattered all over the first plane. Most specimens, however, lie on the "robust side" of F1, together with most of the onagers and khurs. On the "slender side" of F1 are placed most of the kiangs, Mongolian

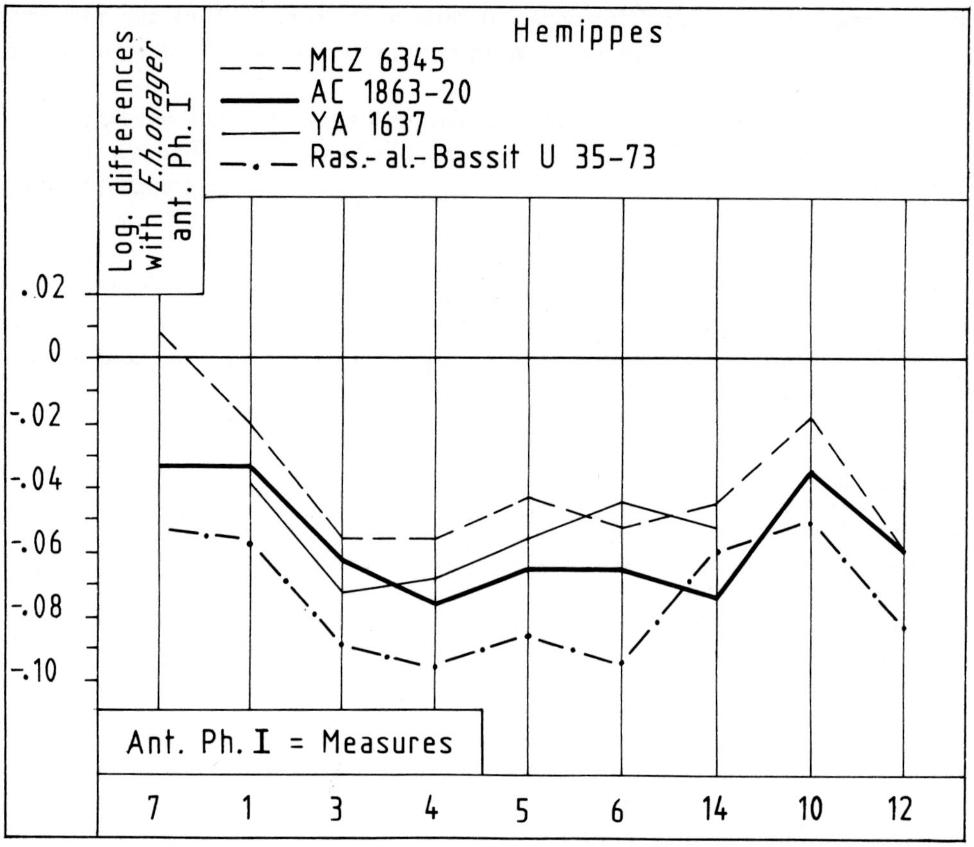

Fig. 14 Ratio diagrams of individual measurements for the anterior first phalanges of modern adult *E.hemionus hemippus* and one subadult specimen dated to 1200 to 1600 B.C.

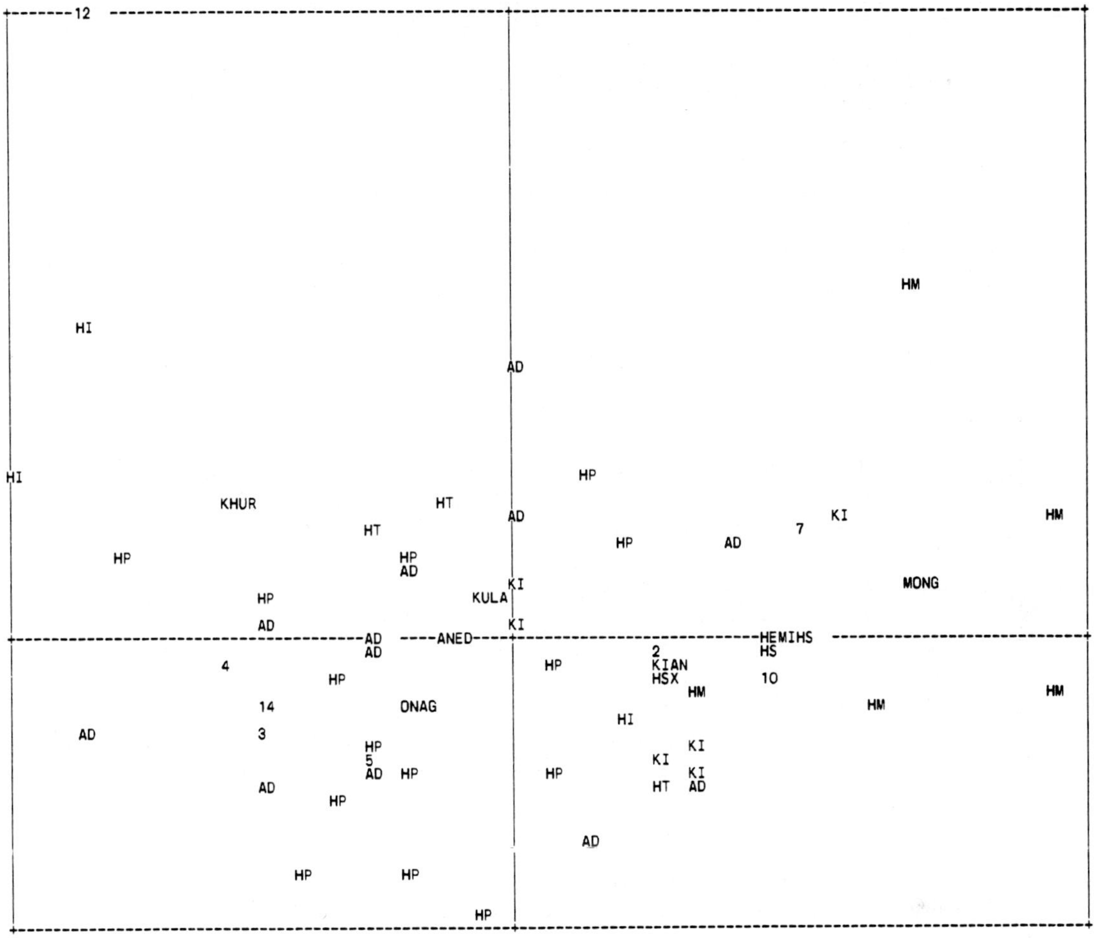

Fig. 15 Correspondence analysis for 14 measures on 43 anterior first phalanges belonging to half-asses and domestic asses. First axis horizontal, second axis vertical. DA = domestic asses; HI = *E.hemionus khur*; HM = *E.hemionus hemionus*; HP = *E.hemionus onager,* HS: *E.hemionus hemippus*; HT – *E.hcmionus kulan*; KI – *E.kiang*, IISX = possible *E.hemionus hemippus* from Ras-al-Bassit.

hemiones, and the hemippes. The Ras-al-Bassit phalanx lies in the vicinity of the hemippe phalanges. We believe, therefore, that it did belong to an hemippe although the possibility that it was a donkey cannot be eliminated by this analysis. Thus, it seems possible to distinguish Mongolian hemiones and Syrian hemippes from other taxa because of their slenderness, but all other specimens have very similar proportions. In most of the cases, only size enables one to discriminate half-ass from domestic ass phalanges. We must stress, however, that our samples are not good; it could be that better samples would bring better results.

6.3.5. Horses

On Figure 16 (first plane of a correspondence analysis with the 14 variables) appear the individual posterior specimens (represented by dashes, slashes, crosses, etc.) as well as the "aver-

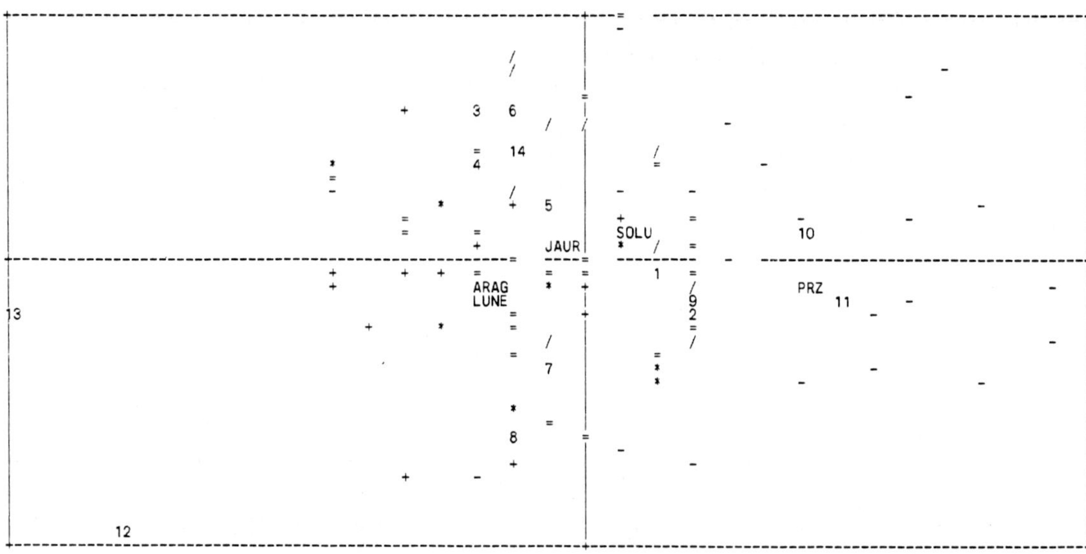

Fig. 16 Correspondence analysis for 14 measures on 81 posterior first phalanges belonging to modern and fossil horses. First axis horizontal, second axis vertical.
* = Arago; + = Lunel-Viel; = = Jaurens; / = Solutré; – = *E.przewalskii.*

age" posterior phalanges for the horses from Arago, Lunel-Viel, Jaurens, and Solutré and for *E.przewalskii.* It is remarkable that the average points lie on a parabolic line and in chronlogical order from the oldest Arago and Lunel-Viel (Mindel to Riss?) specimens to the youngest (modern) *E.przewalskii* through the Jaurens and Solutré horses (Würm). We are well aware that there is significant overlap of all these populations (conspicuous in Fig. 16), but we do not think that the chronological sequence of the "average" phalanges is purely a matter of chance. This kind of configuration, called Guttman's effect (Benzécri 1973: vol. 2: 484, 599) is characteristic of continuous modifications from any cause (climatic, spatial, temporal, etc.) within a sample. In the present case, they may well reflect horse evolution.

The clearest differences between the oldest and the youngest samples are (again!) related to those characters that express anteroposterior dimorphism. The phalanges of *E.przewalskii* are more "anterior-like" (they have longer supra-tuberosital segments) than are the phalanges coming from Arago and Lunel-Viel. If we accept the sequence of Figure 16 as being significant, there may be an evolutionary trend within the horse group with a progressive change of more "posterior-like" phalanges to more "anterior-like" ones. In this regard, it is well to note that the oldest *Equus* phalanges analyzed in this study, those of *E.stenonis,* are also the most "posterior-like" (Fig. 6). Could the same evolutionary trend exist not only within the horse group but also within the genus *Equus*?

6.4. Conclusion

The phalanges of the fossil *E.stenonis* (circa 2 mya) are clearly distinct from those of all modern species due to their overall more "posterior" morphology and their shorter *trigonum phalangis.*

The same sort of distinction, although less marked, exists when considering the phalanges of the fossil horses and those of modern *E.przewalskii.*
The modern species differ one from another mostly in degree of gracility and in size.
Two facts already noted during the study of the skulls (EISENMANN 1980) and of the metapodials (EISENMANN 1979) are confirmed with the phalanges: 1) the zebra morphologically the most similar to *E.stenonis* is not *E.grevyi* but *E.burchelli* and 2) the modern species most similar to horses is also *E.burchelli.*

7. Tentative functional interpretation and conclusion

The anteroposterior dimorphism in *Equus* first phalanges is a well established fact, but the characters involved in this dimorphism (longer supra-tuberosital segment and longer *trigonum phalangis* in the anterior phalanges) appear, at times, to be related to other factors. Indeed, we have found a more or less pronounced "posteriorization" of the phalanges in the following cases:

E.stenonis versus all other modern *Equus* species (Fig. 6);
E.burchelli versus *E.grevyi* and *E.zebra* (Fig. 7, 9, 12, and unpublished ratio diagrams);
E.zebra zebra versus *E.zebra hartmannae*;
wild asses versus modern donkeys (Fig. 13);
fossil donkeys versus modern donkeys;
fossil horses versus *E.przewalskii* (Fig. 16).

In order to try to explain this circumstance, we have looked for possible functional significance in the osteological characters involved and noted especially what sort of ligaments and/or tendons are inserted in the region of the supra-articular tuberosities and of the *trigonum phalangis.*
On the "V-scar" of the *trigonum phalangis* are inserted the oblique sesamoidean ligaments (CAMP/SMITH 1942: 74 and fig. 12) that belong to the group of distal sesamoido-phalangeal ligaments (BARONE 1968: 197).
In the region of the supra-articular tuberosities, there are various structures "inserted in or in close association with" the *scutum medium* (a fibro-cartilaginous plate situated below and behind the head of the second phalanx): the *tendo flexor digitorum sublimis,* the *ligamentum sesamoideum rectum,* and two pairs of volar ligaments (CAMP/SMITH 1942: 75). Most of these, however, are inserted on the volar surface of the bone, behind the supra-articular tuberosities. Following BARONE (1968: 196–198), the tuberosities themselves provide points of insertion only for the lower part of the lateral volar ligament and for the collateral ligament of the pastern associated with the suspensory ligament of the navicular sesamoid.
DENOIX (1984: 13) stresses the functional importance of the collateral ligament and of the volar ligaments related to the *scutum medium* as well as of the oblique sesamoidean ligaments. All of these structures are clearly important in limiting lateral movement and in sustaining the foot during the "automatic spring effect" described by CAMP/SMITH (1942: 95). The significance of the anteroposterior dimorphism of their points of insertion, however, is not clear.
Another approach to this problem is to look in a more general way for differences in the functions of the fore and hind limbs.

Classically, the anterior limb bears and receives the weight of the body whereas the posterior limb provides propulsion (Barone 1966: 41, 489, 625). Horses and their relatives belong to that large group of animals with their center of gravity nearer the forefeet (Gray 1959: 65). In the horse, about 55 percent of the body weight is supported by the front limbs and 45 percent by the hind limbs (Denoix, personal communication). We thus propose that the differences that exist between anterior and posterior phalanges are in some way related to differential division of weight between the fore and hind limbs. Anterior phalanges presenting a relatively "posterior" morphology might indicate that the center of gravity lay farther to the rear (i.e., the cranial part of the animal was relatively less heavily loaded and the caudal part was relativeley more heavily loaded). Such a circumstance would help to explain why characters related to anteroposterior dimorphism seem also, at times, to be related to very different factors (e.g., subspecific differences, differentiation between wild and domestic, modern and fossil).

One way of testing our hypothesis would be to determine the relative anterior to posterior load of the above mentioned taxa by weighing the front and hind parts of the living animal (making it step on scales successively with its front and hind feet). Naturally, this could be done (easily or not!) only with the modern species. At the moment, however, we have no data of this kind and all we can do is to suggest some indirect arguments. Thus, *E.burchelli* has a relatively small head when compared to *E.zebra* and especially when compared to *E.grevyi.* Could it be that the reduced weight of the head contributes to a relatively heavier hind part of the animal?

Domesticated asses are used to carrying additional weight (people or goods). Could the absence of such "extra-loading" account for the more "posterior" morphology of phalanges in wild asses? The equally "posterior" morphology of the fossil donkeys from Ikrit could indicate that these freshly domesticated animals were not yet fully adapted to their new weight-bearing requirements. Finally, the difference between "pure" *E.przewalskii* and animals coming from mixed breeding lines could be explained in the same way, namely that pure wild horses have never carried extra loads while in the case of mixed lines, some ancestors were domesticated and thus had done so.

In conclusion, it is probable that more than one factor is responsible for our observations and that the load-factor, if real, is not the only factor at work. It is, however, the only one that we are able to suggest at the moment.

8. Acknowledgements

The phalanges of modern *Equus* come from the museums and collections listed below. The name of each institution is preceded by the letter code that is used in the tables of measurements in the appendix to this volume.

AA : Museum of Zoology, University of Michigan, Ann Arbor, Michigan, USA
AC : Laboratoire d'Anatomie Comparée du Muséum National d'Histoire Naturelle (MNHN), Paris, France
AD : Natural History Museum, Addis Adeba, Ethiopia
AM : Zoologisch Museum, Amsterdam, The Netherlands
BA : Naturhistorisches Museum, Basel, Switzerland

BE : Naturhistorisches Museum, Bern, Switzerland
BL : Zoologisches Museum der Humboldt Universität, Berlin, DDR
BM : British Museum (Natural History), London, Great Britain
CH : Field Museum of Natural History, Chicago, Illinois, USA
EV : Ecole nationale Vétérinaire, Maisons-Alfort, France
FR : Naturmuseum und Forschungsinstitut Senckenberg, Frankfurt, BRD
GE : Muséum d'Histoire Naturelle, Genève, Switzerland
HA : Zoologisches Institut und Zoologisches Museum, Hamburg, BRD
HV : Museum of Comparative Zoology, Harvard University, Cambridge, Massachusetts, U.S.A.
KI : Institut für Haustierkunde, Kiel, BRD
LD : Rijksmuseum van Natuurlijke Historie, Leiden, The Netherlands
LG : Zoological Institute of the USSR Academy of Sciences, Leningrad, USSR
LY : Muséum d'Histoire Naturelle, Lyon, France
MA : Laboratoire des Mammifères et Oiseaux du MNHN, Paris, France
MU : Zoologische Sammlung des Bayerischen Staates, München, BRD
NA : National Museums of Kenya, Nairobi, Kenya
NY : American Museum of Natural History, New York, USA
PA : Institut de Paléontologie du MNHN, Paris, France
YA : Peabody Museum, Yale University, New Haven, Connecticut, USA
ZU : Zoologisches Museum der Universität, Zürich, Switzerland.

The phalanges of *E.stenonis* from La Puebla de Valverde belong to the collections of the Institut de Paléontologie of the MNHN, Paris. Those of the caballine forms from La Caune de l'Arago, Lunel-Viel, Jaurens, and Solutré belong to the collections of the Laboratoire d'Anthropologie of the Faculté de Médecine, Secteur Nord, Marseille; the Laboratoire de Géologie du Quaternaire of the Faculté des Sciences de Luminy, Marseille; and to the Département des Sciences de la Terre of the Université Claude Bernard, Lyon. Those of the fossil donkey from Ikrit were studied in the Department of Zoology, Hebrew University, Jerusalem.

We are deeply thankful to all curators, colleagues and friends who have given us access to their collections and who, in different ways, have helped us to study them. For the fossil collections, we thank in particular M.-F. Bonifay, C. Mourer-Chauviré, E. Crégut, S. Davis, C. de Giuli, P.Y. Gagnier, C. Guérin, E. Heintz, H. de Lumley, R. Meadow, and A.-M. Moigne. We also thank J. Combier, J.-M. Denoix, and O.A. Ryder for interesting comments and suggestions.

This study, carried out under the auspices of U.A. 12 and 184 of the CNRS, benefited in many ways from the assitance of the R.C.P. 717 and the unit 08 717 of the CNRS, the Louis Leakey Foundation, and the Wenner Gren Foundation. The graphs are the work of F. Pilard, and E. Molin helped with the statistical computation of tables and ratio diagrams. Special thanks are due to R. Meadow who instigated the work and corrected and greatly improved the original English text.

9. Literature

Barone, R., 1966, Anatomie comparée des Mammifères domestiques, Tome 1: Ostéologie; Lyon: École Nationale Vétérinaire, Laboratoire d'Anatomie.

Barone, R., 1968, Anatomie comparée des Mammifères domestiques. Tome 2: Arthrologie et Myologie; Lyon: École Nationale Vétérinaire, Laboratoire d'Anatomie.

Benzécri, J.-P./*et al.,* 1973, L'Analyse des Données: Tome 1: La taxinomie, Tome 2: L'analyse des correspondances; Paris: Dunod.

Bonifay, E./Bonifay, M.-F., 1965, Age du gisement de Mammifères fossiles de Lunel-Viel (Hérault). Comptes Rendus de l'Académie des Sciences, Paris, **260** (9), 3441–3444.

Bonifay, M.-F., 1980, Le cheval du Pléistocène moyen des grottes de Lunel-Viel (Hérault): *Equus mosbachensis palustris* n. ssp. Gallia-Préhistoire, **23**, 233–281.

Bouchez, R./Lopez Caranza, E./Amosse, J./Piboule, M./Cornu, A./Diebolt, J./Galland, D./ Rey, P./de Lumley, H./Ramirez Pareja, J. A./Guérin, C., 1984, Datation par spectométrie RPE d'émail dentaire fossile dans le domaine de 50.000 ans à plusieurs millions d'années. Revue d'Archéométrie **8,** 70–79.

Camp, C. L./Smith, N., 1942, Phylogeny and function of the digital ligaments of the horse. Memoirs of the University of California, **13** (2), 69–124.

Chaline, J., 1971, L'âge des hominiens de la Caune de l'Arago à Tautavel (Pyrénées-Orientales) d'après l'étude des rongeurs. Comptes rendus de l'Académie des Sciences, Paris, D **272,** 1743–1746.

Chaline, J., 1981, Les rongeurs de la Caune de l'Arago à Tautavel et leur place dans la biostratigraphie européenne. *In*: Datations absolues et analyses isotopiques en Préhistoire, méthodes et limites: datation du remplissage de la Caune de l'Arago à Tautavel; (Prétirage); Tautavel: Colloque International du CNRS, 22–28 juin 1981, 193–203.

Combier, J., 1955, Solutré: les fouilles de 1907 à 1925, mise au point stratigraphique et typologique. In: Thoral, M.,/Riquet, R./Combier, J., eds., Solutré; Lyon: Laboratoire de Géologie de l'Université de Lyon (Travaux: 2), 93–224.

Combier, J./Thévenot, J.-P., 1976, Solutré. In: Livret-guide de l'excursion A8, Bassin du Rhône, Paléolithique et Néolithique; Gap: IXème Congrès de l'Union Internationale des Sciences préhistoriques et protohistoriques, 111–120.

Crégut, E., 1980 a, La faune de mammifères du gisement pléistocène moyen anté-rissien de la Caune de l'Arago (Tautavel, Pyrénées-Orientales, France). Comptes rendus de l'Académie des Sciences, Paris, D **290,** 751–754.

Crégut, E., 1980 b, *Equus mosbachensis tautavelensis* nov. subsp., un nouvel *Equidae* (*Mammalia, Perissodactyla*) du gisement pléistocène moyen ante-rissien de la Caune de l'Arago (Tautavel, Pyrénées Orientales, France). Géobios, **13** (1), 121–127.

de Lumley, H. (ed.), 1979, Tautavel. Dossiers de l'Archéologie, **36,** 1–111.

de Lumley, H./de Lumley, M.-A. 1971, Découverte de restes humains anténéanderthaliens datés du début du Riss à la Caune de l'Arago (Tautavel, Pyrénées-Orientales). Comptes rendus de l'Académie des Sciences, Paris, D **272,** 1739–1742.

Denoix, J.-M., 1984, Analyse de contraintes osseuses en régions métacarpienne et phalangienne chez le Cheval: Étude sur membre isolé soumis à compression. (Mémoire dans le cadre du C.E.S. de Biomécanique de l'appareil locomoteur et Cinésiologie); Lyon: Université Claude Bernard, 1–209.

Eisenmann, V., 1979, Les métapodes d'*Equus sensu lato* (Mammalia, Perissodactyla). Géobios **12,** (6)863–886.

Eisenmann, V., 1980, Les Chevaux (*Equus sensu lato*) fossiles et actuels: crânes et dents jugales supérieures; (Cahiers de Paléontologie); Paris: Éditions du CNRS.

Eisenmann, V., 1981, Étude des dents jugales inférieures des *Equus* (Mammalia, Perissodactyla) actuels et fossiles. Palaeovertebrata **10** (3–4), 127–226.

EISENMANN, V., 1986, Comparative osteology of modern and fossil horses, half-asses, and asses. In: MEADOW, R. H./UERPMANN, H.-P., eds., Equids in the Ancient World; (Beihefte zum Tübinger Atlas des Vorderen Orients, Reihe A **19/1**); Wiesbaden: Dr. Ludwig Reichert Verlag, 67–116.

EISENMANN, V./DE GIULI, C., 1974, Caractères distinctifs des premières phalanges antérieures et postérieures chez certains Équidés actuels et fossiles. Bulletin de la Société géologique de France, 7ème série **16** (4), 352–361.

EISENMANN, V./CRÉGUT-BONNOURE, E./MOIGNE, A.-M., 1985, *Equus mosbachensis* et les grands Chevaux de la Caune de l'Arago et de Lunel-Viel: crâniologie comparée. Bulletin du Muséum National d'Histoire Naturelle, Paris, Section C **7** (4ème série, no. 2), 157–173.

GAGNIER, P.-Y., 1986, Les restes fauniques du site archéologique de Ras al Bassit (Syrie). Syria **63** (3–4), 247–255.

GAUTIER, F./HEINTZ, E., 1974, Le gisement villafranchien de La Puebla de Valverde (Province de Teruel, Espagne). Bulletin du Muséum National d'Histoire Naturelle, Sciences de la Terre, **36** (228), 113–136.

GRAY, J., 1959, How Animals Move; London: Penguin Books.

GROMOVA, V., 1949, Istorija Loshadej (roda *Equus*) v Starom Svete. Chast' 2. Evoljutsija i klasifikatsija roda. Trudy Paleontologitcheskovo Instituta Akademii Nauk S.S.S.R. **17** (2), 1–162.

GUÉRIN, C., 1980, Les Rhinocéros (Mammalia, Perissodactyla) du Miocène terminal au Pléistocène supérieur en Europe occidentale. Comparaison avec les espèces actuelles. Documents du Laboratoire de Géologie de la Faculté des Sciences de Lyon, **79** (1–3), 1–1185.

HEINTZ, E., 1978, La faune villafranchienne de la Puebla de Valverde, Teruel, Espagne. Composition qualitative et quantitative. Géologie méditerranéenne **5** (2), 277–280.

JAMBU, M./LEBEAUX, M.-O., 1983, Cluster Analysis and Data Analysis; Amsterdam: North-Holland.

KEDNEY, G., 1981, Morphological differences of the fore and hind first phalanges of *Equus.* (Manuscript).

MOIGNE, A.-M., 1983, Taphonomie des faunes quaternaires de la Caune de l'Arago, Tautavel; Université de Paris, Paris (Thèse de 3ème cycle, Géologie du Quaternaire).

MOURER-CHAUVIRÉ, C., 1980, Le gisement pléistocène supérieur de la grotte de Jaurens, à Nespouls, Corrèze, France: Les Équidés (Mammalia, Perissodactyla). Nouvelles Archives du Muséum d'Histoire Naturelle de Lyon, **18**, 17–60.

PRAT, F., 1957, Sur la discrimination des phalanges antérieures et postérieures d'Equidés. Procès-Verbaux de la Société Linnéenne de Bordeaux **97**, 22–25.

PRAT, F., 1968, Recherches sur les Équidés pléistocènes de France; Université de Bordeaux, Bordeaux (Thèse Doctorat d'État).

RENAULT-MISKOWSKY, J., 1980, Le paléoenvironnement de l'Homme de Tautavel (Caune de l'Arago, Tautavel, Pyrénées-Orientales): la faune et la flore en Roussillon au Pléistocène moyen. Comptes rendus de l'Académie des Sciences, Paris, D **290**, 747–750.

SIMPSON, G. G., 1941, Explanation of ratio diagrams. American Museum Novitates **1136**, 23–25.

SONDAAR, P., 1968, The osteology of the manus of fossil and recent Equidae with special reference to phylogeny and function. Verhandelingen der Koninklijke Nederlandse Akademie van Wetenschappen, Afd Naturkunde **25** (1), 1–76.

VOLF, J., 1960–1985, Pedigree Book of the Przewalski horse; Prague: Zoological Gardens of Prague.

JÉRÔME DIVE, Institut de Paléontologie Humaine, 1 rue René-Panhard, 75013 Paris, France
VÉRA EISENMANN, UA 12 du CNRS, Institut de Paléontologie, 8 rue Buffon, 75005 Paris, France

APPENDIX

Equus przewalskii

Phalanx 1 anterior

Coll.no. /	Measurement:	1	2	3	4	5	6	7	8	9	10	11	12	13	14
AC 1932.46	1	76.0	68.7	32.2	51.2	32.3	42.4	53.0	46.0	67.0	56.0	56.5	10.1	9.7	40.7
AC 1935.486	2	75.2	67.1	31.7	49.2	34.0	44.7	46.0	41.0	66.1	(54.)	54.4	10.2	11.4	41.3
AC 1941.322	3	74.0	65.6	31.3	49.6	32.9	44.2	48.0	43.0	64.7	55.2	52.8	10.2	11.9	39.0
AC 1962.228	4	79.9	71.5	33.1	50.1	35.5	44.5	52.0	46.0	70.9	60.1	59.5	10.5	8.1	43.7
MA 1964.107	7	82.0	74.5	33.5	52.5	35.5	46.5	59.0	52.0	73.0	64.0	63.0	9.0	10.5	44.0
AC 1973.109	9	74.0	68.0	33.5	49.0	31.5	43.5	50.0	45.0	61.2	55.0	55.0	12.0	13.0	40.0
BM 1907.5.15.1	10	79.0	72.0	32.0	50.0	33.5	43.2	53.0	45.5	70.0	57.0	57.5	12.0	12.0	38.7
BM 1945.6.11.1	11	81.0	73.5	33.0	52.5	34.5	45.5	54.0	47.0	72.0	62.0	61.0	11.0	10.0	42.5
BM 1963.1.25.1	13	82.5	76.0	33.0	50.0	36.0	45.0	54.0	48.0	74.0	61.5	62.0	11.5	10.5	41.5
LD 359	14	76.0	68.0	34.0	48.0	34.0	43.0	52.0	47.0	67.0	53.0	53.0	15.0	15.0	41.5
BL 60363	19	78.0	71.0	34.5	52.0	35.0	45.0	49.0	42.0	68.5	60.0	59.0	10.0	9.5	40.0
FR 35389	20	78.0	70.0	31.0	47.0	33.0	42.0	51.0	45.0	70.0	60.0	60.0	10.0	10.0	40.0
MU 1953.147	22	84.0	78.0	33.0	50.5	37.0	45.0	54.0	49.0	76.0	63.0	63.0	12.0	12.0	43.0
BL 60606	39	75.0	69.0	32.0	47.5	32.0	42.0	49.0	42.0	66.0	58.0	58.0	9.0	9.0	38.1
NY 32686	41	81.0	74.0	36.5	51.0	35.5	46.2	55.0	47.0	72.0	58.0	59.0	14.0	13.0	42.5
NY 32696	42	81.0	75.0	37.0	50.0	36.0	47.0	53.0	46.0	74.0	58.0	57.0	15.0	14.0	43.0
NY 21523	43	77.0	69.0	32.0	48.0	34.0	43.0	54.0	43.0	67.0	58.0	57.0	10.0	10.0	41.0
AC 1929.35	44	75.1	70.5	33.0	55.0	35.0	47.1	47.0	42.0	67.4	57.0	54.5	11.3	11.5	42.1
AC 1929.37	45	75.9	68.4	32.1	49.9	32.7	43.6	52.0	45.0	67.1	57.3	54.0	10.9	11.8	39.1
NY 204071	47	80.0	74.0	32.0	48.5	33.5	42.0	50.0	45.0	72.0	62.0	62.0	11.0	11.0	41.0
NY 80062	48	76.0	70.0	31.0	48.0	33.0	41.0	49.0	44.0	68.0	58.0	58.0	11.5	10.0	38.5
NY 90198	49	74.0	67.0	29.0	46.0	32.0	40.5	46.0	40.0	65.0	57.0	57.0	9.0	8.5	37.0
BA 10881	52	78.0	71.0	33.0	50.0	34.0	44.0	53.0	46.0	68.0	60.0	59.0	11.0	10.0	40.0
BA 10877	53	78.0	72.0	33.2	50.0	33.0	44.0	--	44.0	68.0	60.0	61.0	11.0	9.5	42.9
HV 51058	54	77.0	70.0	31.0	47.0	33.0	41.0	50.0	43.0	68.0	58.0	58.0	10.0	10.0	39.0
BE 93-75	57	82.0	75.0	30.0	48.0	34.0	42.0	54.0	49.0	73.0	63.0	63.0	9.5	10.0	40.0
BE 76-80 (2a)	58	81.0	75.0	33.0	50.0	34.0	44.0	51.0	43.0	71.0	63.0	62.0	10.0	10.0	41.0

Phalanx 1 posterior

Coll.no. / Measurement:		1	2	3	4	5	6	7	8	9	10	11	12	13	14
AC 1932.46	1	70.2	66.0	30.6	49.7	35.0	39.9	45.0	39.0	62.0	51.0	51.9	11.9	9.6	38.5
AC 1935.486	2	72.1	62.7	31.0	50.2	35.3	42.9	41.0	34.0	61.0	50.9	48.2	12.1	14.6	39.0
AC 1941.322	3	71.0	63.6	30.9	49.5	35.5	41.9	43.0	37.0	60.7	50.1	48.6	14.0	13.0	37.1
AC 1962.228	4	76.3	69.0	32.9	51.0	38.3	44.3	45.0	39.0	66.8	55.3	53.8	11.8	10.9	42.0
MA 1964.107	7	79.5	72.0	32.0	52.0	37.0	44.0	54.0	48.0	69.0	58.5	58.0	12.0	11.0	42.0
AC 1973.109	9	71.0	65.0	31.0	48.0	33.0	40.5	48.0	43.0	63.0	52.0	53.0	12.0	11.0	37.5
BM 1907.5.15.1	10	73.5	70.0	30.5	50.5	36.0	39.0	46.0	41.0	65.0	50.0	51.0	16.0	14.0	37.5
BM 1945.6.11.1	11	75.0	69.0	32.0	53.0	37.0	43.5	46.0	40.0	67.0	52.0	53.0	15.0	12.5	40.0
BM 1963.1.25.1	13	78.0	72.0	33.0	50.5	38.5	44.0	50.0	43.0	70.0	55.0	56.0	15.5	12.0	40.0
LD 359	14	70.0	63.0	32.0	48.0	36.0	42.0	44.0	36.0	(58)	47.0	46.0	15.0	16.0	39.0
BL 60363	19	74.0	68.0	34.0	51.5	36.0	43.5	45.0	38.0	64.0	54.0	54.0	11.0	11.0	39.0
FR 35389	20	74.0	67.0	31.0	47.0	34.5	40.0	47.0	40.0	64.0	54.0	56.0	11.0	10.0	39.0
MU 1953.147	22	81.5	75.0	32.0	53.0	41.0	44.0	53.0	45.0	72.0	56.0	56.0	17.0	14.5	41.0
BL 60606	39	72.0	67.0	31.5	48.0	34.0	40.2	46.0	40.0	63.5	53.0	52.5	12.5	12.0	37.0
NY 32686	41	75.0	70.0	33.0	50.0	37.0	43.0	50.0	46.0	65.0	51.0	51.0	18.0	14.0	41.0
NY 32696	42	76.0	71.0	36.0	51.0	38.0	45.0	46.0	41.0	67.0	52.0	52.0	15.0	12.0	41.0
NY 21523	43	70.0	64.0	30.0	50.0	35.0	41.0	44.0	36.0	62.0	50.0	49.0	13.0	11.0	38.5
AC 1929.35	44	72.4	66.5	32.9	51.0	36.4	42.8	43.0	38.0	64.0	51.0	49.9	12.7	12.4	39.0
AC 1929.37	45	71.2	65.0	30.6	50.2	34.8	40.1	45.0	38.0	61.3	51.0	49.7	12.3	11.0	37.0
NY 204071	47	75.0	69.0	32.0	49.0	35.5	40.5	46.0	39.0	65.0	55.0	56.0	14.0	12.0	39.5
NY 80062	48	70.0	64.0	30.0	48.2	35.0	39.5	45.0	37.0	61.0	48.0	49.0	13.0	11.5	37.0
NY 90198	49	68.0	63.0	27.5	47.0	33.0	39.0	44.0	38.0	60.0	50.0	50.0	11.0	11.0	36.0
BA 10881	52	74.0	68.0	31.5	51.0	36.0	43.0	47.0	40.0	64.0	53.0	53.0	13.0	15.0	38.0
BA 10877	53	76.0	69.0	33.0	50.5	34.0	43.2	48.0	40.0	65.0	56.0	56.5	12.0	10.5	41.0
HV 51058	54	73.0	67.0	30.0	49.0	35.0	39.0	47.0	40.0	64.0	51.0	52.0	14.0	13.0	37.0
BE 93-75	57	76.0	69.0	29.0	49.5	35.0	40.1	47.0	41.0	68.0	56.0	56.0	12.0	12.0	39.0
BE 76-80 (2a)	58	76.0	70.0	32.0	50.0	35.0	42.0	45.0	38.0	65.0	54.0	55.0	14.0	13.0	39.0

Equus kiang

Phalanx 1 anterior

Coll.no. / Measurement:		1	2	3	4	5	6	7	8	9	10	11	12	13	14
AC 1963.363	1	86.0	79.1	25.5	46.1	34.0	39.0	55.0	49.5	78.0	66.0	67.0	10.2	9.0	36.1
BM 976 e	5	86.0	79.0	26.5	42.0	33.5	37.5	52.0	48.0	78.5	67.0	67.0	10.5	9.0	36.0
AM 985	17	85.0	78.0	26.2	45.0	33.2	40.0	54.0	50.0	77.0	65.0	66.0	12.0	11.0	38.0
BL 32172	32	85.5	78.0	27.0	45.0	33.0	41.0	55.0	50.0	77.5	64.0	66.0	11.5	10.5	38.0
MU 572	39	87.5	81.0	26.0	44.0	33.0	40.0	58.0	53.0	81.0	70.0	69.0	12.0	11.0	36.5
BM 1855.1.20.1		83.0	78.0	27.0	43.0	32.2	40.2	55.0	50.0	77.0	65.0	66.0	10.0	9.0	37.0

Phalanx 1 posterior

Coll.no. / Measurement:		1	2	3	4	5	6	7	8	9	10	11	12	13	14
AC 1963.363	1	77.3	71.5	26.0	47.4	34.7	38.8	49.0	42.0	68.5	59.0	57.5	10.3	12.0	35.0
BM 976 e	5	80.0	76.5	25.5	41.5	33.5	(35.5)	48.0	41.0	72.0	61.0	60.5	11.5	10.5	33.5
AM 985	17	77.0	71.0	26.2	47.0	34.0	37.0	47.0	42.0	69.0	56.0	58.0	14.0	12.0	35.2
BL 32172	32	78.0	71.0	26.0	46.0	34.5	38.0	49.0	43.0	70.0	54.0	54.0	15.0	16.0	35.0
MU 572	39	82.0	76.0	25.5	45.3	33.5	37.0	51.0	46.0	74.0	60.0	62.0	16.0	15.0	34.2
BM 1855.1.20.1		77.0	71.0	26.5	43.5	34.0	39.5	49.0	43.0	70.0	56.5	58.0	13.0	10.5	34.5

Equus hemionus hemionus

Phalanx 1 anterior

Coll.no. /	Measurement:	1	2	3	4	5	6	7	8	9	10	11	12	13	14
BA 3529	1	82.0	76.5	24.0	41.0	31.2	34.8	58.0	52.0	75.0	68.0	68.0	13.0	13.0	35.0
LG 14741	5	88.7	--	25.8	--	--	--	59.0	--	--	--	--	--	--	--
NY 57208	13	90.0	82.0	26.5	43.0	32.5	38.5	60.0	54.0	82.0	71.0	70.0	10.0	10.0	36.5
NY 57201	14	85.0	78.0	27.0	43.5	33.5	37.5	57.0	52.0	75.0	65.0	67.0	10.5	10.0	37.5
NY 57212	16	85.0	77.0	26.0	41.0	33.0	38.0	57.0	52.0	77.0	67.0	67.0	10.0	9.5	36.5
NY 57214	17	86.0	79.0	26.0	41.0	31.3	39.0	62.0	58.0	80.0	67.0	69.0	10.5	10.0	35.2

Phalanx 1 posterior

Coll.no. /	Measurement:	1	2	3	4	5	6	7	8	9	10	11	12	13	14
BA 3529	1	74.8	69.0	24.5	42.1	31.0	33.0	44.0	39.0	67.0	59.5	62.0	14.0	12.0	33.0
LG 14741	5	80.2	--	25.1	--	--	--	50.0	--	--	--	--	--	--	--
NY 57208	13	82.0	75.0	26.2	44.0	33.0	37.0	57.0	52.0	75.0	62.0	63.0	12.5	11.5	33.0
NY 57201	14	77.0	70.0	26.0	41.0	32.0	36.5	51.0	45.0	69.0	55.0	58.0	15.0	11.5	34.0
NY 57212	16	77.0	68.0	25.5	41.0	34.0	36.5	53.0	48.0	68.0	57.0	58.0	12.0	10.0	33.2
NY 57214	17	78.0	72.0	23.0	42.0	31.0	34.0	55.0	50.0	72.0	60.0	61.0	11.0	11.0	32.0

Equus hemionus kulan

Phalanx 1 anterior

Coll.no. / Measurement:		1	2	3	4	5	6	7	8	9	10	11	12	13	14
KI 3480	2	78.5	70.5	25.0	42.0	31.0	38.5	53.0	48.0	70.0	61.0	61.0	9.0	10.0	35.0
MU 1962-203	4	81.0	74.0	25.0	43.0	33.0	38.5	51.0	45.0	74.0	62.0	61.0	12.5	10.5	35.5
Tbilissi vieux	6	80.5	74.8	25.3	42.5	32.6	37.4	53.0	--	--	--	--	--	--	--
Tbilissi jeune	7	79.5	71.2	25.7	42.2	31.0	35.5	48.0	--	--	--	--	--	--	--
LG 19046	8	79.9	--	24.9	--	--	--	53.0	--	--	--	--	--	--	--
BM 1971.2210	9	79.0	73.0	25.0	42.0	30.5	38.5	48.0	46.0	74.0	60.0	61.0	12.0	10.0	35.0

Phalanx 1 posterior

Coll.no. / Measurement:		1	2	3	4	5	6	7	8	9	10	11	12	13	14
KI 3480	2	73.0	66.0	24.0	43.0	32.0	36.0	46.0	41.0	65.0	54.0	54.0	13.0	13.0	32.5
MU 1962-203	4	72.5	65.5	23.5	42.2	32.0	36.2	45.0	40.0	66.0	54.0	53.0	12.0	11.5	33.0
Tbilissi vieux	6	74.6	68.4	23.9	43.0	33.2	35.9	45.0	--	--	--	--	--	--	--
Tbilissi jeune	7	73.2	65.2	25.5	42.7	32.5	33.8	39.0	--	--	--	--	--	--	--
LG 19046	8	73.7	--	23.4	--	--	--	45.0	--	--	--	--	--	--	--
BM 1971.2210	9	74.0	68.0	25.5	43.0	32.0	36.5	43.0	39.0	67.0	54.0	54.0	13.0	11.5	32.5

Equus hemionus khur

Phalanx 1 anterior

Coll.no. / Measurement:		1	2	3	4	5	6	7	8	9	10	11	12	13	14
AC 549	1	78.8	71.5	23.7	42.0	29.6	36.2	50.0	45.5	73.2	59.5	59.5	9.5	11.5	34.0
BM 1957.7.18.1.	10	75.0	67.0	25.0	43.5	30.1	39.0	46.0	42.0	69.0	55.0	55.0	12.5	11.0	36.0
MU 1965.207	11	78.0	72.0	27.0	44.0	32.0	37.5	51.0	44.0	69.0	57.0	59.0	14.0	13.0	37.0

Phalanx 1 posterior

Coll.no. / Measurement:		1	2	3	4	5	6	7	8	9	10	11	12	13	14
AC 549	1	72.4	66.0	23.2	42.5	30.2	34.6	45.0	40.0	67.0	54.5	55.5	10.0	9.5	33.0
BM 1957.7.18.1	10	70.5	64.0	23.5	43.5	31.0	37.5	41.5	36.5	63.5	49.0	50.0	15.0	12.0	33.5
MU 1965.207	11	72.0	66.0	26.0	44.0	31.0	36.0	42.0	35.0	63.0	52.0	52.0	15.0	15.0	34.5

Equus hemionus onager

Phalanx 1 anterior

Coll.no. /	Measurement:	1	2	3	4	5	6	7	8	9	10	11	12	13	14
AC 1893.509	1	76.0	70.0	23.9	42.1	30.9	37.0	47.5	42.5	67.9	58.5	58.0	9.8	10.0	35.2
AC 1901.9	2	73.3	66.4	23.7	39.6	30.0	34.1	45.0	42.0	67.0	56.0	56.0	8.8	8.9	33.3
AC 1902.487	3	76.5	69.0	24.9	41.3	30.0	37.2	47.0	44.5	70.1	58.0	57.5	10.0	9.0	35.0
LD 12507	7	75.0	70.0	23.0	38.5	28.5	34.0	50.0	46.0	67.5	58.0	58.0	10.5	10.0	34.5
AM 17667	8	82.0	76.0	23.0	43.0	32.0	37.0	53.0	47.0	75.0	62.0	63.0	12.0	10.5	36.0
LY 383	12	78.0	70.5	25.5	41.0	29.5	37.0	52.5	46.5	70.0	60.0	59.5	10.0	10.0	37.0
KI 1576	13	75.0	66.0	24.5	41.0	30.0	35.0	47.0	41.0	66.0	56.0	55.0	11.0	10.5	34.5
HA 5881	15	79.0	72.0	26.0	43.0	32.5	39.0	47.0	40.0	70.0	61.0	61.0	10.0	10.0	37.0
HA 7045	16	73.5	67.0	24.0	39.0	31.0	36.0	46.0	41.0	67.5	56.0	55.0	10.5	10.5	34.0
HA 7158	17	77.0	70.0	26.0	43.0	33.0	38.0	51.0	45.0	71.0	56.0	57.0	12.0	11.0	38.5
NY 35670	19	75.0	68.0	24.0	41.0	29.0	35.0	48.0	43.0	69.0	58.0	58.0	11.0	11.0	34.5
KI 1662	20	78.5	72.0	24.0	39.0	31.0	36.0	47.0	39.0	70.0	61.0	61.0	10.0	10.0	35.5
YA 5098	21	75.0	68.0	25.0	41.6	29.9	36.0	--	43.0	--	--	--	--	--	34.0
AC 1978.50	23	77.0	71.0	26.0	44.0	34.0	41.8	45.0	40.0	69.5	59.0	59.0	10.0	11.0	37.5
AC 1975.100	24	77.0	71.0	25.1	40.1	31.5	36.8	48.0	42.0	68.0	60.0	60.0	9.0	10.0	37.0

Phalanx 1 posterior

Coll.no. /	Measurement:	1	2	3	4	5	6	7	8	9	10	11	12	13	14
AC 1893.509	1	70.4	65.0	23.7	41.3	31.2	34.0	41.5	37.3	62.5	51.8	52.7	11.6	10.0	33.0
AC 1901.9	2	66.5	60.7	23.9	41.2	29.0	32.6	40.0	36.0	59.8	50.0	50.5	9.0	8.0	31.3
AC 1902.487	3	69.0	63.7	24.1	42.3	30.1	33.8	41.0	38.0	63.5	51.0	52.0	11.0	9.0	33.0
LD 12507	7	70.0	65.5	23.0	39.0	29.0	33.0	41.0	36.0	63.0	52.0	52.0	12.5	12.0	32.0
AM 17667	8	78.0	69.0	23.5	45.0	32.5	36.0	47.0	41.0	69.0	56.0	57.0	13.0	13.0	33.5
LY 383	12	71.5	65.5	25.0	42.5	29.5	35.0	42.0	36.5	64.0	51.0	53.0	14.5	12.0	34.5
KI 1576	13	71.0	63.0	23.2	42.0	31.0	35.0	42.0	36.0	63.0	50.0	50.5	13.0	12.0	32.5
HA 5881	15	72.0	65.0	26.0	45.0	32.5	38.0	39.0	33.0	63.0	54.0	52.5	11.5	12.0	34.5
HA 7045	16	69.0	63.0	23.5	40.0	30.0	34.0	41.0	36.0	61.0	50.0	50.0	11.0	11.0	31.0
HA 7158	17	71.0	64.0	26.5	44.5	32.0	37.0	43.0	37.0	63.0	51.0	50.0	13.0	13.0	36.0
HA 35670	19	72.0	66.0	24.0	44.0	31.0	33.0	45.0	40.0	65.0	52.0	53.0	14.0	12.0	32.5
KI 1662	20	72.0	66.0	24.0	41.0	32.0	34.0	37.0	32.0	65.0	53.0	54.0	12.5	11.5	33.0
YA 5098	21	70.0	63.0	25.0	42.5	31.0	34.0	--	36.0	--	--	--	--	--	32.0
AC 1978.50	23	74.0	67.0	24.7	44.3	32.0	40.0	46.0	41.0	64.0	55.0	55.0	11.0	11.0	34.7
AC 1975.100	24	72.5	67.0	25.0	41.1	31.5	35.8	42.0	34.0	63.0	53.5	54.0	11.0	11.0	35.5

Equus hemionus hemippus

Phalanx 1 anterior

Coll.no. /	Measurement:	1	2	3	4	5	6	7	8	9	10	11	12	13	14
AC 65	2	70.8	65.5	21.3	34.4	23.5	31.5	44.5	42.0	66.0	54.0	55.0	9.0	9.0	30.0
YA 1637	3	70.0	64.0	20.8	35.0	27.0	33.0	--	42.0	--	--	--	--	--	31.6
HV 6345	5	73.0	66.0	21.5	36.0	28.0	32.5	49.0	43.0	68.0	56.0	56.0	9.0	9.0	32.0
Ras al Bassit		67.0	60.0	20.0	33.0	25.0	29.5	42.5	38.0	60.0	52.0	--	8.5	7.5	31.0

Phalanx 1 posterior

Coll.no. /	Measurement:	1	2	3	4	5	6	7	8	9	10	11	12	13	14
AC 65	2	66.7	62.4	21.1	35.1	26.5	29.1	40.5	37.0	61.5	49.0	51.0	11.0	9.0	27.0
YA 1637	3	65.2	60.0	20.3	36.3	26.9	30.2	--	35.0	--	--	--	--	--	29.0
HV 6345	5	68.5	64.0	21.5	36.5	27.0	30.0	44.0	39.0	63.0	51.0	51.5	12.0	11.0	28.5

Equus africanus africanus

Phalanx 1 anterior

Coll.no. /	Measurement:	1	2	3	4	5	6	7	8	9	10	11	12	13	14
BM 1904.6.12.1	3	82.2	75.5	26.5	42.0	32.5	37.0	51.0	45.0	74.0	64.0	65.0	10.0	9.5	37.0
MU 1952.9	10	67.5	59.5	22.5	37.0	26.5	33.5	40.0	34.0	58.0	50.0	50.0	9.0	11.0	32.0
MU 1963.133	11	78.0	72.0	26.0	44.0	32.0	40.1	47.0	40.0	69.0	59.0	59.0	11.0	10.5	39.0

Phalanx 1 posterior

Coll.no. /	Measurement:	1	2	3	4	5	6	7	8	9	10	11	12	13	14
BM 1904.6.12.1	3	77.5	71.0	27.0	43.0	32.5	37.0	45.0	40.0	71.0	58.0	57.5	12.0	10.5	34.0
MU 1952.9	10	62.0	54.0	23.5	38.0	28.0	34.0	38.0	32.0	53.0	44.0	45.0	12.0	11.0	30.0
MU 1963.133	11	75.0	67.0	26.0	45.0	32.5	38.0	43.0	38.0	66.0	53.0	54.0	15.0	13.5	35.0

Equus africanus somaliensis

Phalanx 1 anterior

Coll.no. / Measurement:		1	2	3	4	5	6	7	8	9	10	11	12	13	14
BA 3597	3	75.3	68.6	26.5	43.1	30.7	37.0	45.0	40.0	68.5	63.5	64.5	11.5	11.0	37.0
BA 10858	4	74.0	68.0	26.0	42.0	30.0	37.0	45.0	41.0	68.0	54.0	56.0	12.0	11.0	35.5
BE 216	5	80.0	73.0	26.0	40.5	31.0	37.0	54.0	51.0	73.0	62.0	62.0	12.0	11.0	36.5
BL 30253	9	78.0	72.0	26.0	44.0	30.5	38.0	46.0	41.0	71.0	59.0	60.0	11.0	11.0	37.5
CH 1428	15	80.0	72.0	26.5	43.2	30.5	37.7	--	51.0	--	--	--	--	--	36.0

Phalanx 1 posterior

Coll.no. / Measurement:		1	2	3	4	5	6	7	8	9	10	11	12	13	14
BA 3597	3	70.1	64.9	26.5	43.5	31.9	36.0	40.3	35.0	63.0	56.0	58.5	14.0	12.0	35.0
BA 10858	4	70.0	65.0	25.5	44.0	31.5	36.5	41.5	37.5	63.0	49.0	49.0	15.0	13.0	33.0
BE 216	5	74.5	68.5	25.5	42.0	30.5	35.0	45.0	41.0	68.0	54.0	55.0	14.0	13.0	35.0
BL 30253	9	75.0	69.0	26.5	47.0	32.0	37.0	43.0	37.0	68.0	52.0	54.0	14.0	14.5	36.0
CH 1428	15	77.0	70.0	25.5	44.7	32.5	36.5	--	44.0	--	--	--	--	--	35.0

Equus asinus

Phalanx 1 anterior

Coll.no. /	Measurement:	1	2	3	4	5	6	7	8	9	10	11	12	13	14
NA 3952	10	66.4	62.5	24.1	37.5	28.2	34.0	43.0	38.0	61.0	51.0	52.0	9.0	8.5	32.0
LY 834	11	75.5	68.0	24.5	38.0	29.0	34.0	46.0	--	--	58.5	59.0	10.0	9.0	33.5
BE 5.11.1952	14	64.0	57.0	22.0	37.0	27.0	33.5	41.0	37.0	57.0	47.0	50.0	9.0	9.0	32.0
AC 1933.397	21	70.0	63.0	23.0	39.0	27.5	35.0	45.0	39.0	63.0	53.0	54.0	10.0	10.0	34.0
PA 1979	35	72.5	66.0	23.5	39.0	28.0	35.7	47.0	42.0	64.0	55.0	55.5	10.0	10.0	35.0
NY 15675	36	76.0	68.0	25.0	39.0	29.0	36.0	51.0	46.0	68.0	58.0	59.0	11.0	10.0	35.0
NY 135017	37	76.0	70.0	24.0	39.0	28.0	34.0	50.0	45.0	69.0	58.0	59.0	12.0	11.0	34.5
NY 204141	38	74.0	65.0	22.0	35.2	27.2	33.0	47.0	42.0	64.0	57.0	57.0	10.0	9.0	32.0
NY 100280	39	73.0	66.0	26.0	38.5	29.0	35.0	46.0	40.0	65.0	56.0	57.0	9.5	8.0	34.1
YA 1622	40	65.0	59.0	23.1	35.0	26.1	31.0	44.0	40.0	57.0	49.0	50.0	9.0	8.1	31.0
AC 1875.28	51	77.0	70.5	26.0	42.0	29.0	35.0	52.0	49.0	71.5	58.0	60.0	11.0	10.0	36.8
AC 1893.634	52	82.0	76.0	25.8	41.1	32.5	38.6	52.0	47.4	74.5	65.0	64.8	10.0	8.5	36.2
GE 797.52	53	59.0	54.0	18.0	30.2	23.0	27.3	37.0	33.0	53.0	45.0	46.5	6.7	7.0	27.3

Phalanx 1 posterior

Coll.no. /	Measurement:	1	2	3	4	5	6	7	8	9	10	11	12	13	14
NA 3952	10	62.1	57.1	23.2	37.1	27.3	32.5	34.0	29.0	56.0	44.5	46.0	12.0	10.0	30.0
LY 834	11	70.0	64.5	23.0	39.0	28.0	32.0	43.0	37.5	63.5	53.0	53.0	12.5	12.0	30.5
BE 5.11.1952	14	61.0	56.0	22.0	36.5	27.0	33.0	35.0	30.5	53.5	44.0	45.5	11.5	10.5	29.0
AC 1933.397	21	66.0	59.0	22.3	39.3	27.5	34.0	40.0	35.0	59.0	47.0	47.5	14.0	13.0	31.7
PA 1979	35	68.0	63.0	22.0	40.0	28.0	33.0	42.0	36.5	60.0	52.0	52.0	11.0	10.5	32.0
NY 15675	36	70.0	64.0	24.0	40.0	30.0	33.0	46.0	41.5	63.0	53.0	52.0	12.0	10.0	31.5
NY 135017	37	72.0	66.0	23.0	40.0	29.0	32.5	43.0	38.0	65.0	55.0	54.0	12.0	11.0	32.0
NY 100280	39	68.0	60.5	25.0	39.5	28.0	32.0	41.0	36.5	60.0	50.0	50.0	11.0	10.0	31.0
YA 1622	40	62.0	57.0	22.0	35.0	27.0	29.0	40.0	34.0	55.0	44.0	45.0	12.0	10.0	29.0
AC 1875.28	51	73.5	65.5	28.6	46.0	30.8	36.2	44.5	40.0	65.1	54.0	54.0	13.0	13.0	36.0
AC 1893.634	52	76.0	70.0	25.6	43.2	32.3	38.2	44.5	40.2	70.1	56.0	58.0	12.0	10.0	35.0
GE 797.52	53	57.0	52.0	18.0	31.0	24.0	25.2	35.0	31.0	52.0	42.0	42.0	9.0	10.0	24.2

Equus zebra zebra

Phalanx 1 anterior

Coll.no. / Measurement:		1	2	3	4	5	6	7	8	9	10	11	12	13	14
LG 111	9	75.8	68.8	27.1	44.1	30.0	37.4	51.0	--	--	--	--	--	--	--
BM 46.3.23.10	24	77.0	69.0	29.5	47.5	34.0	(42)	49.0	44.0	69.0	--	60.0	--	9.0	41.0
BL 329	48	76.5	67.5	30.0	47.0	32.0	40.0	46.0	42.0	68.5	55.0	56.0	11.5	11.5	39.5
BL 13415	49	75.0	68.0	29.0	45.0	30.5	38.1	47.0	42.0	66.0	57.0	57.0	10.5	9.0	38.0
NY 83602	63	80.0	71.0	30.0	48.0	34.0	42.0	51.0	43.0	70.0	59.0	59.0	13.0	11.0	40.5
NY 99700	65	78.0	71.0	30.0	47.0	33.0	40.0	49.0	44.0	70.0	60.0	60.0	11.5	11.0	40.0
NY 42753	66	78.0	71.0	27.0	42.5	30.5	37.0	56.0	50.0	70.0	61.0	61.0	11.0	10.0	38.1
NY 90199	68	78.5	72.0	29.5	46.0	33.0	39.0	48.0	41.0	70.0	59.0	59.0	11.5	10.0	39.5

Phalanx 1 posterior

Coll.no. / Measurement:		1	2	3	4	5	6	7	8	9	10	11	12	13	14
LG 111	9	71.2	65.3	27.8	44.6	32.0	37.8	40.0	--	--	--	--	--	--	--
BM 46.3.23.10	24	74.0	66.0	30.0	49.0	35.0	42.0	44.0	38.0	64.0	50.0	53.0	15.5	13.0	38.0
BL 329	48	72.5	65.0	30.5	48.0	34.0	39.0	44.0	40.5	66.0	51.0	51.0	13.5	14.0	38.0
BL 13415	49	72.0	66.0	30.0	47.0	33.0	38.5	43.0	39.0	64.0	53.0	54.0	12.0	11.0	37.0
NY 83602	63	74.0	67.0	31.0	49.0	34.0	41.0	48.0	41.0	66.0	52.0	53.0	15.0	13.0	38.5
NY 99700	65	74.0	67.0	30.0	49.0	35.0	39.0	45.0	40.0	66.0	55.0	54.0	12.5	11.5	38.1
NY 42753	66	75.0	69.0	28.5	45.0	33.0	36.0	46.0	39.0	68.0	55.0	56.0	14.0	12.0	36.5
NY 90199	68	73.0	67.0	29.0	46.0	34.0	38.5	41.0	36.0	66.0	53.0	54.0	11.5	12.0	36.2

Equus zebra hartmanni

Phalanx 1 anterior

Coll.no. /	Measurement:	1	2	3	4	5	6	7	8	9	10	11	12	13	14
AC 1932.364	1	81.0	74.7	30.4	51.2	33.4	43.0	50.5	44.2	73.0	61.5	61.0	11.0	11.6	41.3
AC 1933.562	2	83.4	76.1	32.3	50.7	33.2	42.8	54.0	48.0	74.5	65.0	65.8	11.0	9.6	41.5
AC 1936.325	3	79.1	72.2	31.3	44.9	31.6	39.2	51.5	47.6	72.1	61.2	61.6	9.9	9.0	39.5
AC 1948.54	6	78.5	70.5	31.2	48.6	32.5	41.3	48.5	42.4	68.3	60.3	61.0	9.8	9.0	40.3
AM 7691	17	80.0	73.0	29.0	47.0	30.5	40.0	52.0	47.0	71.0	63.0	63.0	10.5	10.0	38.0
BM 28.9.11.416	28	80.0	72.0	31.0	51.0	33.5	42.0	51.0	45.0	71.0	60.5	60.5	11.0	11.0	40.5
BM 63.6.13.1	30	84.5	75.5	32.5	49.0	34.0	41.2	54.0	46.0	74.0	65.0	66.0	11.0	10.0	40.2
AC 1919.59	31	79.0	73.0	29.5	47.9	32.2	41.8	49.5	45.3	72.5	60.0	61.1	10.3	8.6	41.5
BE 1961.25	33	79.5	70.0	30.0	48.0	32.0	41.0	51.0	46.0	70.5	61.0	60.0	10.0	10.0	40.0
BL 53065	42	--	--	32.0	49.0	34.1	--	49.0	45.0	--	--	--	--	--	--
BL 55246	47	83.0	73.5	30.2	47.0	33.0	38.5	50.0	43.0	72.0	64.0	64.0	9.0	8.0	38.5
MU 1954.179	55	80.0	70.0	33.0	50.0	32.5	41.0	54.0	48.0	69.0	59.0	60.0	12.5	11.0	41.0
MU 1956.8	56	78.0	69.0	28.5	47.0	31.0	39.5	50.0	43.0	69.0	60.0	60.5	11.0	10.5	41.0
MU 1960.237	57	81.5	74.0	31.5	49.0	33.0	42.0	52.0	45.0	72.0	61.0	62.0	13.0	12.0	40.5
BA 10918	69	81.0	73.0	31.0	47.2	32.0	40.0	54.0	48.0	73.0	62.0	62.0	12.0	12.0	40.0
NY 90240	67	81.0	74.0	31.0	47.0	31.0	40.0	52.0	45.0	72.0	61.0	62.0	14.0	12.0	41.0

Phalanx 1 posterior

Coll.no. /	Measurement:	1	2	3	4	5	6	7	8	9	10	11	12	13	14
AC 1932.364	1	75.9	69.6	30.7	51.5	36.2	41.0	43.9	39.6	69.2	54.6	55.2	13.5	12.5	40.0
AC 1933.562	2	79.6	73.0	31.9	49.7	35.7	41.8	46.0	41.0	72.0	60.0	59.8	12.4	11.3	40.0
AC 1936.325	3	75.0	70.0	32.1	48.2	35.6	39.6	46.0	42.8	69.2	57.7	57.2	10.8	9.1	38.0
AC 1948.54	6	75.1	68.5	34.1	50.2	35.3	43.9	43.5	38.5	67.0	56.2	55.6	11.1	10.7	39.5
AM 7691	17	76.0	70.0	30.0	48.5	33.5	40.0	45.0	40.0	68.0	57.0	57.0	13.5	13.5	37.0
BM 28.9.11.416	28	74.5	68.0	30.0	51.0	35.0	41.2	41.0	37.0	67.0	54.0	56.0	14.5	11.5	38.5
BM 63.6.13.1	30	79.0	71.0	32.0	50.5	36.0	41.1	44.0	37.0	70.0	56.0	59.0	15.5	12.0	39.5
AC 1919.59	31	74.0	70.5	31.9	49.8	34.4	41.0	44.0	41.0	68.7	54.0	55.4	12.9	10.9	39.5
BE 1961.25	33	74.5	66.0	30.5	49.0	34.0	40.0	44.0	40.0	66.5	54.0	52.0	14.0	14.0	38.7
BL 53065	42	76.0	69.0	30.1	50.0	34.5	41.3	43.0	38.0	68.0	55.0	54.0	13.0	13.0	38.0
BL 55246	47	77.0	69.0	30.0	48.5	34.5	38.0	44.0	37.0	67.0	56.0	57.0	13.0	12.0	36.5
MU 1954.179	55	74.5	66.5	32.0	50.0	33.0	40.3	44.0	40.0	66.0	54.0	54.0	14.0	13.0	39.0
MU 1956.8	56	73.0	65.0	30.0	49.5	33.0	38.5	43.0	36.0	64.0	51.0	52.0	14.5	12.5	37.0
MU 1960.237	57	75.0	69.0	30.5	51.0	35.0	40.5	41.0	34.0	67.0	52.0	55.0	15.0	13.0	38.0
BA 10918	69	76.0	70.0	30.5	47.5	33.6	40.0	42.0	38.0	69.0	58.0	55.0	13.0	14.0	39.0
NY 90240	67	78.5	70.0	31.0	47.5	34.0	38.5	47.0	41.0	70.0	58.0	58.0	12.0	12.5	39.5

Equus burchelli

Phalanx 1 anterior

Coll.no. /	Measurement:	1	2	3	4	5	6	7	8	9	10	11	12	13	14
MU 1954-7	2	72.0	64.0	30.0	46.5	31.5	38.0	44.0	38.0	64.0	50.5	51.0	13.5	12.0	37.0
NA 3932	6	75.0	68.0	29.0	49.5	30.5	39.0	48.0	42.0	66.0	57.0	57.0	9.0	8.0	39.0
NA 3934	7	71.5	66.0	27.0	42.0	31.0	36.5	48.0	43.0	65.0	54.0	54.0	10.0	9.5	36.0
NA 3936	8	75.0	69.0	29.5	47.5	32.0	40.0	48.0	43.0	68.0	57.0	57.0	10.5	10.0	37.2
NA 3937	9	78.5	72.0	31.0	47.5	33.0	40.0	50.0	44.0	70.0	60.0	59.5	10.0	9.0	40.0
NA 3938	10	72.0	66.5	30.0	47.0	30.0	38.5	47.0	42.0	65.0	54.0	55.0	10.0	9.0	38.2
NA 3941	11	74.0	67.0	28.5	44.5	31.0	37.5	51.0	46.0	65.5	56.0	57.0	10.0	7.0	36.0
NA 3942	12	73.0	69.0	29.5	44.5	31.5	38.5	47.0	43.0	67.0	54.5	54.5	11.0	11.5	37.0
NA 3944	13	74.5	68.0	29.0	45.5	31.0	38.0	49.0	44.0	67.0	55.0	56.0	12.0	11.0	38.0
NA 3945	14	77.0	69.0	30.0	46.0	31.5	38.5	51.0	46.0	69.0	55.0	58.0	13.0	9.0	38.0
NA 3948	15	78.0	72.0	30.0	46.0	30.5	40.5	49.0	45.0	69.0	60.0	61.0	11.0	9.0	39.0
NA 3950	16	79.0	73.5	30.0	49.5	34.5	41.5	51.0	46.0	71.0	58.5	59.0	12.0	10.0	40.0
NA 3958	17	78.5	72.5	31.0	49.5	35.0	41.5	47.0	42.0	71.0	60.0	58.5	11.0	11.5	39.1
NA 3949 2 1/2	42	70.0	61.0	29.0	44.0	30.0	38.0	45.0	40.0	62.0	51.0	52.0	10.5	9.0	36.0
NA 3946	46	69.5	64.5	29.0	46.5	29.5	38.0	42.0	38.0	64.5	52.5	53.0	10.0	9.5	35.0
NA 3956	47	75.0	69.0	29.0	47.5	32.0	39.0	51.0	47.0	68.0	55.0	55.5	11.0	10.0	38.0
NA 3957	48	73.5	67.0	29.0	47.0	30.5	38.0	53.0	48.0	66.0	56.0	55.5	11.0	10.5	37.0
NA 3943	50	77.0	69.0	30.0	48.0	33.5	41.5	49.0	43.0	68.0	58.0	57.5	9.5	10.0	37.5
NA 3955	51	82.5	75.0	30.5	50.0	33.0	41.5	54.0	49.0	76.0	63.0	63.0	12.0	9.0	37.5

Phalanx 1 posterior

Coll.no. / Measurement:		1	2	3	4	5	6	7	8	9	10	11	12	13	14
MU 1954-7	2	68.0	60.0	29.0	47.5	32.0	38.5	41.0	34.0	60.0	47.0	47.0	16.0	14.0	36.0
NA 3932	6	70.5	65.0	30.0	51.0	33.5	39.5	42.0	37.0	62.0	50.5	51.0	13.0	11.0	37.5
NA 3934	7	68.5	62.0	28.0	43.0	32.0	36.0	43.0	39.0	62.0	51.0	50.0	12.0	11.0	35.2
NA 3936	8	70.0	65.0	29.0	48.0	34.5	41.5	41.0	36.0	62.5	49.5	51.0	14.0	12.0	35.5
NA 3937	9	74.0	68.0	32.5	48.5	35.5	42.0	45.0	39.0	66.0	54.0	54.5	12.5	12.5	39.1
NA 3938	10	69.0	65.5	31.5	49.0	33.5	38.0	42.0	37.0	60.0	48.5	48.0	13.0	13.0	36.0
NA 3941	11	70.5	63.5	27.5	46.0	32.5	33.0	43.0	38.0	63.0	49.0	51.0	14.0	10.5	35.0
NA 3942	12	67.0	62.5	29.0	45.5	32.5	38.0	43.0	38.0	60.0	46.0	49.0	15.0	11.0	34.5
NA 3944	13	70.5	65.0	30.5	47.0	33.5	38.0	45.0	39.0	62.5	48.0	51.0	15.0	12.0	37.0
NA 3945	14	71.0	64.0	29.5	45.5	33.0	38.5	45.0	38.0	63.0	48.5	51.0	15.0	12.0	37.0
NA 3948	15	72.5	66.0	30.0	48.0	33.0	40.5	41.0	37.0	64.0	53.0	55.3	13.0	10.5	37.0
NA 3950	16	75.0	68.0	30.5	50.5	35.5	41.0	43.0	39.0	68.0	52.0	54.0	15.0	12.0	39.0
NA 3958	17	75.5	69.5	32.5	51.5	36.5	43.0	44.0	39.0	67.0	53.5	54.5	14.5	12.0	40.0
NA 3949 2 1/2	42	63.5	56.0	29.5	44.0	30.5	36.5	39.0	34.0	56.0	41.5	44.0	14.0	11.0	34.0
NA 3946	46	66.5	60.0	28.3	47.0	32.5	37.5	41.0	35.0	60.0	48.0	49.0	12.0	11.5	34.0
NA 3956	47	72.0	66.5	30.0	49.5	33.5	39.0	45.0	42.0	66.0	50.0	51.5	15.0	12.0	36.2
NA 3957	48	70.0	65.0	29.5	47.0	33.5	38.5	46.0	41.0	63.0	50.5	51.5	14.0	12.0	35.2
NA 3943	50	73.0	64.0	30.0	48.0	35.0	43.0	43.0	37.0	63.0	53.0	53.0	12.5	11.5	36.0
NA 3955	51	77.5	68.5	31.0	50.5	33.5	42.5	48.0	44.0	69.0	56.5	56.5	15.0	12.0	36.2

Equus grevyi

Phalanx 1 anterior

Coll.no. / Measurement:		1	2	3	4	5	6	7	8	9	10	11	12	13	14
AC 1913.58	1	92.0	85.0	31.0	50.0	37.0	44.0	61.0	55.0	83.0	72.0	72.5	10.5	10.3	40.0
AC 1931.392	2	88.0	83.0	31.0	54.0	37.0	43.0	53.5	48.0	81.5	67.0	68.0	11.5	10.5	41.0
AC 1939.75	4	88.5	83.0	33.0	51.5	35.0	44.5	54.0	48.5	81.0	68.5	69.0	11.5	10.9	41.0
LD 885	7	85.5	77.0	29.0	49.0	34.0	42.5	55.0	50.0	76.0	64.0	66.0	12.0	11.0	39.0
NA 3965	14	82.0	76.5	29.0	48.0	33.5	40.5	57.0	52.0	74.5	64.0	64.5	11.0	10.5	38.5
NA 3967	15	84.0	78.0	30.0	48.5	34.0	41.2	51.0	46.0	76.0	64.0	64.5	11.5	12.5	39.0
NA 3968	16	85.5	78.5	32.0	51.5	36.5	43.5	55.0	49.0	76.0	62.5	65.5	13.0	12.0	--
NA 5145	17	90.0	82.0	32.0	54.0	37.0	42.0	60.0	55.0	81.0	67.0	69.0	15.0	12.0	42.0
BM 1962.8.174	23	87.0	82.0	30.0	50.5	36.5	42.0	54.0	48.0	78.0	66.5	67.5	12.0	11.0	39.5
BA 10876	26	82.0	75.5	28.5	48.5	34.5	40.0	52.0	46.0	74.5	62.0	63.0	12.0	11.0	39.5
BA 10873	27	84.5	77.0	29.5	48.0	34.5	42.0	59.0	54.0	76.0	62.5	64.0	12.0	12.0	40.0
BE 1923.177	28	89.0	83.0	32.0	52.0	36.5	43.0	59.0	53.0	81.0	69.0	70.0	12.0	12.0	42.0
ZU 10918	29	85.0	79.0	32.0	53.0	36.0	45.0	52.0	47.0	76.0	66.5	68.0	10.0	9.0	41.0
HA 6879	31	88.0	81.5	33.0	53.0	37.0	43.5	53.0	48.0	79.0	68.0	68.0	12.0	11.0	41.0
HA 7111	48	85.0	79.0	30.0	50.0	36.0	41.0	53.0	50.0	77.0	66.0	67.0	12.5	12.0	40.5
MU 1953.96	50	90.0	82.0	32.0	51.0	36.0	42.0	61.0	56.0	80.5	68.0	68.0	14.0	13.0	39.0
MU 1965.113	51	87.0	81.0	30.5	50.0	36.5	40.0	55.0	49.0	78.0	65.0	65.5	15.0	13.5	40.0
NA 3966	57	83.0	77.5	30.5	50.0	33.5	42.0	54.0	49.0	75.5	62.5	65.0	12.0	10.0	40.0
NA 3970	66	85.0	79.0	32.0	51.5	36.5	42.5	53.0	47.0	78.5	65.0	65.0	12.0	12.0	--
BL 5.2.14	67	86.0	80.0	30.0	50.5	35.0	42.2	54.0	47.0	77.0	66.0	67.0	12.0	11.0	39.5
AD 1974.1	68	88.0	81.0	31.5	50.0	35.1	45.0	53.0	48.0	80.0	65.0	67.0	15.0	14.0	43.0
NY 82038	70	88.0	79.0	32.0	52.0	37.0	43.0	59.0	52.0	77.0	65.0	67.0	13.0	12.0	42.0
NY 204065	71	87.0	77.0	31.5	50.5	35.0	43.0	55.0	47.0	75.0	65.0	66.0	10.5	10.5	39.5
NY 70086	72	86.0	79.0	30.0	54.0	36.0	43.0	52.0	47.5	79.0	68.0	68.0	10.0	10.0	40.0
NY 90166	73	80.0	74.0	28.5	48.5	33.5	40.0	50.0	41.0	70.0	62.0	63.0	10.0	9.0	37.5
NY 90345	74	83.0	77.0	30.0	50.0	33.0	41.0	53.0	45.0	74.0	64.0	65.0	11.0	10.0	39.0
NY 82037	75	89.0	82.0	30.5	52.0	35.5	42.0	56.0	49.0	80.0	65.0	67.0	14.5	13.5	41.0
BA 10899	76	86.0	79.0	30.0	49.0	35.0	42.5	55.0	47.0	76.0	66.0	67.0	11.0	10.5	41.5

Phalanx 1 posterior

Coll.no. / Measurement:		1	2	3	4	5	6	7	8	9	10	11	12	13	14
AC 1913.58	1	88.0	81.0	31.0	52.0	38.5	42.0	56.0	49.0	77.0	63.0	66.0	14.5	13.0	37.0
AC 1931.392	2	83.5	77.0	29.5	54.5	36.5	42.5	47.0	40.5	74.0	62.5	63.5	13.0	12.0	38.5
AC 1939.75	4	85.5	78.5	31.0	53.5	37.5	44.0	55.0	43.0	76.5	63.0	65.0	14.7	14.0	39.5
LD 885	7	83.0	76.0	28.5	50.5	36.5	42.0	49.0	44.0	74.0	61.0	62.0	15.0	13.0	37.0
NA 3965	14	76.5	71.0	27.5	48.5	34.5	39.5	48.0	42.0	68.0	55.0	57.0	15.0	12.0	36.0
NA 3967	15	79.0	73.5	29.0	50.5	35.5	40.5	48.0	41.0	70.5	58.0	60.0	15.0	12.0	36.0
NA 3968	16	79.0	73.0	31.5	52.5	37.5	43.5	49.0	42.0	69.5	56.0	58.5	15.5	14.0	39.0
NA 5145	17	83.5	78.0	31.0	54.0	37.0	42.0	51.0	44.0	74.0	60.5	61.5	15.5	15.5	40.0
BM 1962.8.174	23	82.5	78.0	29.0	53.0	38.0	41.5	50.0	43.0	74.0	60.0	62.0	15.0	12.0	37.0
BA 10876	26	75.0	69.0	28.0	49.0	35.0	40.0	45.0	40.0	67.5	55.0	54.5	13.5	13.0	36.0
BA 10873	27	79.0	72.0	28.5	48.5	36.0	40.0	49.0	46.0	71.0	55.0	57.0	16.0	14.0	37.5
BE 1923.177	28	85.0	78.0	29.5	53.5	38.0	43.5	53.0	46.0	78.0	62.0	64.0	15.0	13.0	39.5
ZU 10918	29	83.0	76.0	31.0	54.0	37.0	44.5	50.0	43.0	72.0	63.0	64.0	13.0	11.5	38.5
HA 6879	31	84.5	77.0	31.5	54.0	39.0	44.5	48.0	42.0	75.0	61.0	62.0	15.0	15.0	39.0
HA 7111	48	81.0	75.0	29.0	50.0	36.0	42.0	47.0	42.0	72.0	61.0	61.0	12.5	13.0	38.0
MU 1953.96	50	82.0	74.0	30.0	52.5	36.5	41.0	52.0	47.0	72.0	59.0	60.0	16.0	15.5	37.0
MU 1965.113	51	82.0	75.0	29.0	52.5	37.0	39.0	48.0	40.0	71.5	58.0	60.0	15.5	14.0	38.0
NA 3966	57	78.0	72.0	29.0	50.5	35.5	41.5	49.0	42.0	69.0	56.0	58.0	15.0	11.0	38.0
NA 3970	66	79.5	73.0	31.0	52.5	38.0	42.0	48.0	40.0	70.0	59.0	60.5	14.0	12.0	--
BL 5.2.14	67	82.0	76.0	29.0	52.0	37.0	41.0	48.0	40.0	73.0	59.0	61.0	15.2	14.5	37.0
NY 82038	70	81.0	71.0	31.0	52.5	38.5	42.5	49.0	42.0	72.0	57.0	59.0	15.0	15.0	39.0
NY 204065	71	80.0	72.0	30.5	51.0	37.0	42.0	50.0	42.0	70.0	61.0	59.0	13.0	13.5	37.5
NY 70086	72	82.0	75.0	30.0	55.0	38.0	43.5	49.0	42.0	73.0	64.0	65.0	12.0	11.0	38.0
NY 90166	73	76.0	76.0	28.0	49.0	35.0	40.0	44.0	36.0	65.0	55.0	57.0	14.5	12.0	34.5
NY 90345	74	79.0	74.0	30.0	50.0	35.0	40.0	47.0	40.0	70.0	56.0	57.0	16.0	14.5	36.0
NY 82037	75	82.5	76.0	29.0	52.5	35.5	40.0	48.0	42.0	73.0	58.0	62.0	17.0	15.0	37.2
BA 10899	76	80.0	73.5	29.0	49.0	35.0	41.5	49.0	42.0	70.5	60.0	60.0	13.0	12.0	37.5